Majestic RIVER

Majestic RIVER

MUNGO PARK *and the* EXPLORATION *of the* NIGER

CHARLES W. J. WITHERS

BIRLINN

For my mother

First published in 2022 by
Birlinn Limited
West Newington House
10 Newington Road
Edinburgh
EH9 1QS

www.birlinn.co.uk

ISBN: 978 1 78027 799 8

British Library Cataloguing-in-Publication Data

A catalogue record for this book is available from the British Library

Typeset by Initial Typesetting Services, Edinburgh

Papers used by Birlinn are from well-managed forests and other responsible sources

Printed and bound by Gutenberg Press, Malta

Contents

List of illustrations

Figures

Colour plates (Section 1)

Colour plates (Section 2)

Note on names and spelling

Many of the places mentioned by the Niger's explorers and in Park's two books no longer exist and for several of those that do, the names have changed in their spelling, or changed entirely, since Park and those who followed him wrote them down. Timbuktu, for example, appears in several different spellings in the sources discussed. Park's birthplace in the Scottish Borders appears in different forms – 'Foulshiels', 'Fowlshiels', 'Fowlsheals' – the variant spellings sometimes being used even by members of Park's family. I have kept to the modern form, Foulshiels. Arabic and African personal names are given as they appear in the sources consulted. I have quoted extensively, but always with the reader in mind, so that we can get as close as possible to the conversations, politics, and personal relationships that lie behind Park's life and the Niger's exploration. In doing so, I have left intact the original spelling of words and the original form of sentences.

Note on the maps

Maps are important in exploration and to the stories we tell about it. Many of the maps reproduced in this book are large, and so in their original format are either folded, or exist only as fragile manuscripts or were printed, often poorly, in the periodicals of the day. Maps in this period were not all drawn in 'standard' ways using modern conventions of scale and orientation. The importance of the many maps included rests less in their specific detail and more because they reveal the slow 'emergence' of the Niger in the European geographical imagination – even when the mapped position of the Niger was wholly wrong.

The idea of a great river, rising in the western mountains of Africa and flowing towards the centre of that vast continent; whose course in that direction is ascertained for a considerable distance, beyond which information is silent, and speculation is left at large to indulge in the wildest conjectures – has something of the *unbounded* and *mysterious*, which powerfully attracts curiosity and takes a strong hold of the imagination.

J. Whishaw, 'Account of the Life of Mungo Park', in M. Park, *The Journal of a Mission to the Interior of Africa, in the Year 1805* (1815)

Preface and acknowledgements

Mungo Park and I have been keeping one another company for quite some while. I cannot remember when first I came across him or read his *Travels in the Interior Districts of Africa*, but I have been researching him, taking notes on the Niger's exploration, and thinking about Park, his successes, his failures, his still mysterious death, and his varied afterlife, for over twenty years. This book is the result.

Others have turned to Mungo Park before me. In a preliminary note at the start of his 1934 biography of Park, Lewis Grassic Gibbon wrote 'I think I have read almost everything by or about Mungo Park – everything which still survives in print or manuscript.' I should hesitate, I know, before making the same claim – but I think it equally true. I can at least say with confidence that I have included much new material on or about Park that has appeared in the nearly ninety years since Grassic Gibbon wrote, especially on Park's afterlife, including Gibbon's and other biographers' views of Park's achievements.

I have extended Park's engagement with the Niger to include those who followed him in exploring that river, hoping as they did so not only to solve that part of the Niger problem which Park left unanswered but also to determine how and where he had died. As they did so, geographers of one type or another – 'in-the-field' explorers or 'armchair geographers' – slowly revealed the course and termination of the river Niger and, equally slowly, reduced Europeans' ignorance of Africa.

This is, then, a book about Mungo Park and the 'unbounded and mysterious' Niger, as Park's first biographer described it, but it is also about the Niger problem, the Niger's explorers, and about exploration itself.

All books are collaborative efforts – Park's 1799 *Travels* certainly was. I owe much to a great many people for their insight, courtesies, and practical assistance over the years as rough notes and poorly expressed arguments began to coalesce into what I hope is a coherent narrative. Fraser MacDonald, Innes Keighren, Richard Fardon, David McClay, Dane Kennedy, and David Livingstone at various times listened, corresponded, and advised, and I am grateful to them. I acknowledge with thanks the staffs of the British Library, the National Archives, the National Library of Scotland, Cambridge University Library, the Centre for Research Collections in the University of Edinburgh Library, Special Collections in the University of St Andrews Library, the British Museum of Natural History Archives, the Library of Congress Maps Division, Borders Council Archives within Borders Museum and Archive Services, Fran Baker at Chatsworth House Archives, Crispin Powell, archivist to His Grace the Duke of Buccleuch, Margaret Wilkes for assistance with the archives of the Royal Scottish Geographical Society, and Kirsty Archer-Thompson and Claudia Bolling in the Abbotsford Archives. For their assistance with material in the keeping of the Royal Botanic Garden Edinburgh, I am grateful to Graham Hardy, David Long, Lynsey Wilson, and Leonie Paterson. I am grateful to Philip Dodds for drawing to my attention the evidence on Park's 1799 book in the Bell and Bradfute papers in Edinburgh City Archives. I am grateful to Karina Williamson for her permission to cite as I have from her article on Park with Mark Duffill. I have several times tried to contact Mark Duffill over this article and his biography of Park, but without success, and I hope that this acknowledgement of his fine work will suffice. I am grateful to Raymond Howgego for information on Damberger and the inclusion of elements of Park's travels within Damberger's fraudulent narrative of African travels. For assistance with the translations of Park's *Travels*, I acknowledge the help of Dan Hopkins over the Danish edition, and Ib Friis for advice on the Swedish edition.

Acknowledgement of permission to reproduce the many illustrations in the book appears alongside the images themselves. It would be a considerable discourtesy, however, not to thank several individuals who were especially helpful in this respect, often in circumstances inconvenient to themselves as they and their institutions faced restrictions upon access to the original material and to their reproduction. It is a pleasure in this respect to acknowledge Eugene Rae and Joy Wheeler in the Royal Geographical Society (with the Institute of British Geographers), Lucinda Lax and Helen Smailes of the National Galleries of Scotland, Laura de Beaté of the Borders Museum and Archives Services, Megan Barford and Beatrice Okoro of the National Maritime Museum, Paul Cox of the National Portrait Gallery London, Paul Johnson of the National Archives, and, in the National Library of Scotland, Chris Fleet in the Map Library and Hazel Stewart in the Library's Imaging Services. Chris Simmonds was immensely helpful in helping prepare the illustrations for publication.

I am honoured that Nicholas Crane and Dane Kennedy accepted the invitation to write a few words about the book and I thank them for their endorsements, Nick as a best-selling author and TV geographer, Dane as a leading figure in the histories of exploration. Thanks also go to the staff at Birlinn, especially Hugh Andrew for his encouragement over the book, and to Andrew Simmons for his patience and guidance. James Rose was an excellent copy-editor.

My final thanks go to four people without whom this book probably would not have been written in the ways it has, or perhaps at all. Andrew Grout has acted as a research assistant to me, formally and informally, for many years and has, again, proven himself peerless in tracking down unusual and hard-to-find Park material. Henry Noltie, always a patient listener and source of sound advice as well as being a walking companion and a fine writer, drew my attention to the Park material in the Royal Botanic Garden Edinburgh. Years ago, Jim Cameron of the University of Western Australia generously passed to me his notes on Barrow and the Niger's exploration: I hope I have done his own research justice in the use I have made of it. Last, but always first, I owe so much to my wife Anne. She took several of the photographs that appear in the book and, uncomplainingly, has had to live with me and Mungo Park for quite some while.

Introduction

'Still but a wide extended blank'

Mungo Park and a 2,000-year-old geographical problem

This is a book about geography, exploration, and death. It is about the life, death, and afterlife of the Scottish explorer Mungo Park who in 1796 solved the first part of a 2,000-year-old geographical puzzle and who, sometime in 1806, died failing to solve the second. It is about the achievements and the failures of those men who, following Park, sought to solve what to Europeans was the world's greatest geographical puzzle – and who died trying. And it is about those men who did solve that puzzle without leaving home. This is a story of geography, biography, authorship, exploration, and the river Niger.

By the end of the eighteenth century, the Niger was a 2,000-year-old geographical problem in two parts: which way did the river run, and where did the river end? Classical authors such as Ptolemy and Herodotus had written about the river but come to no firm conclusion on either point. Medieval scholars fared little better, even when they could draw on African sources. Did the river flow east to west as some surmised, or west to east as others believed? Where, if it did, did the Niger enter the sea or did it, as some argued, flow into an inland lake in north Africa and from there into the Nile? Simply, no one in Europe seemed to agree where this major river ran and in which direction. The fact that indigenous peoples knew the river by different names only confused matters further. For those who thought the river flowed eastwards to join with the Nile, the Niger seemed to hold the key to the possibility of trans-African commerce, from the Gambia and Senegal in west Africa to Alexandria in Egypt and, from there, to the Mediterranean and to Europe.

As tales of African gold and ivory reached Europe and as geographers and map-makers failed to agree over how to describe and depict the river, knowing where the Niger ran and where it ended mattered. Its precise expression varied over time, but for interested contemporaries in the late eighteenth and early nineteenth centuries the river Niger was the world's single greatest geographical problem.

The route to its solution began in a London pub. The Association for Promoting the Discovery of the Interior Parts of Africa, known usually as the

African Association, was established on 9 June 1788 in the St Alban's Tavern, off London's Pall Mall. The dozen members of the Association, a mix of nobility, clergy, and parliamentarians led by the prominent naturalist Joseph Banks, had previously met as the Saturday's Club for the convivial discussion of current and world affairs. As Henry Beaufoy MP, first secretary to the Association, made clear, the conversation that day centred upon the world's geography, ignorance of it in general and, specifically, ignorance of Africa's interior. In their view, the recently completed Pacific voyages of James Cook had been so successful 'that nothing worthy of research by Sea, the Poles themselves excepted, remains to be examined'. Banks would have agreed no doubt: his standing as Britain's pre-eminent natural historian and someone who had the ear of politicians derived from his having sailed with Cook to the southern oceans.

By contrast, knowledge of the world's continents was poor as Beaufoy reminded those present: 'by Land, the objects of discovery are still so vast, as to include at least a third of the habitable surface of the earth: for much of Asia, a still larger proportion of America, and almost the whole of Africa, are unvisited and unknown'. Such general geographical ignorance was reprehensible but that of Africa in particular 'must be considered as a degree of reproach upon the present age'.

Recognising what they did not know and determined to do something about it, the nine like-minded men present that June day in 1788 agreed upon a resolution: 'That as no species of information is more ardently desired, or more generally useful, than that which improves the science of Geography; and as the vast continent of Africa, notwithstanding the efforts of the ancients, and the wishes of the moderns, is still in a great measure unexplored, the members of this Club do form themselves into an Association for Promoting the discovery of the inland parts of that quarter of the world.' Banks and Beaufoy, together with Andrew Stuart, a lawyer who served on the Board of Trade, and Richard Watson, the Bishop of Llandaff and Archdeacon at Ely Cathedral, formed the Association's first Committee of Management.[1]

This resolution and the attitudes of mind that lay behind it – about improved geographical understanding, about ancient authorities, and about Africa – spoke to a peculiarly European and even a distinctively British perspective on the geography of the world at the end of the eighteenth century. It was certainly one widely held. As he looked back in his *Modern Geography* of 1802 at the late eighteenth-century voyages of Cook, Lapérouse, and Bougainville and to the effects of such European maritime exploration, geographer and essayist John Pinkerton noted how the recent geographical discoveries of Australasia and Polynesia together constituted a new 'fifth part of the world'. The ancients knew the world in terms of three parts: Asia,

Europe, and Africa. From the end of the fifteenth century, Columbus and others had added a fourth, the Americas. By the eighteenth century, "Old world" though it was, Africa was the poor relation among the Enlightenment world's emergent geographies.

Parts of Africa were known to Europeans. North Africa along the Mediterranean coast and Egypt down the Nile was well known to European traders. Important information on southern Africa had recently come to light through the work of the Swede Anders Sparrman, parts of whose 1785 publication documented his travels with James Cook on the latter's second voyage. Abyssinia would become better known following the travels of the Scot James Bruce whose published account was even then 'preparing for the press' as Beaufoy told his pub companions. Africa south of the Sahara remained largely hidden from European view, however, excepting what little was known around coastal trading centres run by the British, Danish, Dutch, French, and Portuguese. As Beaufoy further noted, 'notwithstanding the progress of discovery on the coasts and borders of that vast continent, the map of its interior is still but a wide extended blank . . . The course of the Niger, the places of its rise and termination, and even its existence as a separate stream, are still undetermined.'[2]

Africa's sub-Saharan interior was unknown in part because its physical geography discouraged ease of access: an extensive desert to the north, dense forests beyond the coastal margins and, in its equatorial and tropical regions, a climate that for Europeans could occasion disease and heavy mortality. The sheer territorial extent of the continent meant locations, distances and travel times were uncertain, for outsiders at least. To the geographical authority James Rennell, whose work with Mungo Park would do much to lift the veil of European ignorance on Africa, the fact that the geography of Africa was little known 'is to be attributed more to natural causes, than to any absolute want of attention on the part of Geographers'. For him, Africa's little-known geography, especially its interior, seemed to reflect Divine disposition and the limited location of human affairs: 'Formed by the Creator, with a contour and surface totally unlike the other Continents, its interior parts elude all nautic[al] research; whilst the wars and commerce in which Europeans have taken part, have been confined to very circumscribed parts of its borders.' Because they knew so little, Europeans often got things about Africa wrong.[3]

For the English traveller William Browne who travelled extensively in northern Africa in the early 1790s, 'errors in African geography are numerous and proceed from various causes': provinces had more than one name, orthography varied, and there seemed to be no agreed or standard measurements for the passage of time or linear distance. His was a view based, like that of Banks, Beaufoy, and Pinkerton, upon European ignorance and from a concern to read Africa's geography only in European terms. Africa's size and the diversity of its physical geography was a puzzle to Europeans. What was known seemed strange and, because strange, was often regarded as outside and below what Europeans took for granted as normal and acceptable. In part too, what seemed to be myriad kingdoms, states, and political identities presented a complex social geography, poorly understood by Europeans. And because they were poorly understood by Europeans, Africa's human populations were as commonly misrepresented by them, usually in terms of racial hierarchies which positioned white Europeans at the top, black Africans at the bottom.[4]

Beaufoy's 'reproach upon the present age' was not simply about European geographical ignorance concerning Africa and the course of the Niger. The Association was founded at a time of rising concern over slavery and the slave trade – several of its members were involved in the abolitionist movement. Abolitionism was rooted in moral outrage against the horrors inflicted on Africans, especially along the 'Slave Coast' – today's Togo, Benin, and the western coast of Nigeria. It stemmed also from concerns about the economic impact upon European capitalism since, were slavery and the slave trade to be abolished, the economies of western Africa would be hard pressed to pay for British goods. As Mungo Park would discover, public interest in his 1799 *Travels in the Interior Districts of Africa* following his first Niger trip centred almost as much upon what he wrote (and did not write) on slavery as it did upon his account of the river's exploration.

Interest in the Niger was fundamentally and enduringly a question of geography. While geographical ignorance and a concern to promote natural philosophy and humanitarianism lay at the heart of late eighteenth-century British interests in the Niger, commercial interests were also evident. They would become more so over time, from the 1830s especially. The exploration of the Niger was always centrally about the geography of the river, but it was never only so. At one time or another and to different degrees, slavery and its abolition, ethnography, the politics of west-central Africa, correcting Classical authorities, delimiting the physical geography of the interior, promoting missionary activities, commercial opportunism, and imperial advantage were each and all motives for knowing where the Niger ran and where it ended.

The British search for the course of the Niger that began with the African Association – in the decades that followed Park's first travels, one might almost term it an obsession – was never a simple single narrative from geographical problem to colonial possession.[5] It must be understood in terms of the motives for exploration as they were differently voiced and made real at different times, in relation to what different explorers did and did not achieve, and in terms of how the river itself was given shape and meaning, in books, in maps, and in the public's reception of explorers' printed words.

Majestic River is a study in biography, in exploration history, of the authorship of exploration narratives, and the commemoration of exploration. It is, centrally, a biography of Mungo Park, his geographical achievements and his later failure and death. It considers his reception and continuing afterlife as, long after the geographical discovery which brought him fame, Park became the subject of biographical assessment and public commemoration. It is an account of the men and expeditions who explored the river Niger and west Africa in the decades following Park's death, and of the different ideas concerning the geography of the river and why they mattered so much to those who held them.

It is also, if less strongly, a biography of the Niger itself as this 2,600-mile-long river took shape in contemporaries' geographical imagination, in texts, and on maps. There is good reason to see why the Niger interested and puzzled Europe's geographers and politicians. Over its 2,600-mile course from its origin in the mountains of west Africa, it flows north-eastwards through the semi-desert sahel and the southern margins of the Sahara before dramatically changing direction to head south-east and then to flow almost directly south. This distinctive and sharp change of direction is the result of river capture: what is now one river was in the past two, the waters of a former river having been 'captured' by what is today the Niger. The Niger's volume of flow and dimensions vary seasonally, a fact which Europeans were slow to comprehend, and which bears upon the circumstances of Park's death. Flooding begins in September and peaks in November with the waters finally receding by May in each year. Travel was severely limited by high waters. Trade and slave caravans were often delayed for months in waiting for the waters to recede.

In different parts of its reach, the Niger had different names, a fact which Europeans were also slow to understand. The river even smelt different in different places, the pungent and 'pestiferous air' of its lower delta in marked contrast to the dry heat of the summer and the torrential downpours and flooded banks of the rainy season in its upper reaches. Most puzzling of all, no one in Europe knew in which direction the Niger flowed or where it ended.

Successful geographical exploration does not require that the explorer should leave home and die in the undertaking although, in the case of the Niger, many did. Several people sought the solution to the Niger problem – and got the answer right – without leaving home. It requires, above all, that one should be believed, and that both the resultant narrative and the author should be thought credible and trustworthy in the eyes of others. The significance of the Niger problem and of that river's exploration rests not alone in what was done 'in the field', in Africa, and in the act and process of exploration itself. It lies also in what happened 'back home', in the pre-planning of exploration, in later writing about it, and in how explorers, exploration, and exploration narratives were made, and made sense of, in life and after death through the work of intermediaries such as editors, reviewers, and publishers. It lies, too, in the work of those who traced the Niger's course without ever seeing the river or visiting Africa.

Mungo Park and the Niger's exploration have been the subject of some attention before now. I have drawn from this work in what follows. Among modern biographies, Lupton's *Mungo Park African Traveler* (1979) offers the most detailed account of Park's life and death and his two African journeys. Rich on the living Park, it has little on Park's treatment by earlier biographers, and nothing on Park's afterlife. More recent research by Duffill and others has illuminated Park's literary abilities and ambitions before he left for Africa. Duffill's short biography of Park concentrates upon Park's two African journeys and his 1799 book. The workings of the African Association have been the subject of detailed analysis and editorial commentary by Robin Hallett in 1964.[6]

The Niger's early exploration has been most fully discussed by British historian Edward Bovill. Bovill's *The Niger Explored* (1968) begins with Park's death. From its opening sentence, the reader is left in no doubt over Park's importance: 'In the stirring history of African exploration there was no more dramatic event than the discovery of the Niger by Mungo Park in 1796.' Bovill's *Missions to the Niger* (1964–6) details the Niger's exploration by Friedrich Hornemann, Alexander Gordon Laing, and the 1822–5 Bornu Mission of Dixon Denham, Hugh Clapperton, and Walter Oudney. Excellent and thorough as his work is, Bovill is silent on the explorers' afterlives, the Niger's exploration after 1830, and, notably, on the views of those who solved the Niger problem without leaving home.[7]

Anthony Sattin's *The Gates of Africa* (2003) is concerned more with explorers' attempts to reach Timbuktu than with the Niger. Sanche de Gramont's history of the exploration of the Niger in his 1975 *The Strong Brown God* is enlivened by his own journey down the river's 2,600-mile course 'by car, boat, dug-out, train, truck, and camel'. From de Gramont's own experiences and through use of contemporaries' words from the historic present, the Niger emerges through its exploration, the flow of the river and the structure of the narrative aligned one with another. But de Gramont makes little mention of explorers' narratives or their public reception, none of Park's afterlife.[8]

Similar studies of rivers emerging through their exploration and representation have been told: of mid-nineteenth-century American explorer John Wesley Powell and the Colorado river, of Robert Schomburgk, the

British explorer who claimed to have discovered El Dorado in Guyana in the 1830s.[9] In the main, these consider individual endeavour. In contrast, the Niger problem was not only 2,000 years old, its solution in the work of Mungo Park and others was the result of collective, even collaborative, effort, in Africa and elsewhere.

Acknowledging this and others' work as I do is to recognise that most studies of Park and of the Niger's exploration are now some forty to fifty years old. They pre-date the recent revitalisation of exploration history and geography. None pays detailed attention to Park's afterlife. Very few consider the publishing history and the public reception of the principal Niger narratives, Bovill's work on the Bornu Mission excepted. Maps and other visual evidence produced at the time have not been subject to critical appraisal. Yet the mapping of the Niger illustrates not only the emergence of the river in the European geographical imagination but also the different ways in which the river was explored: by some through arduous exploration in the field, by others through careful and sedentary speculation. In existing accounts of Park and the Niger's exploration, questions of trust, truth, epistemology, and indigenous testimony hardly figure at all. In several ways then, what Bovill described as Park's dramatic 'discovery' of the Niger and all that followed in its wake is ripe for reinterpretation and, in the case of Park's afterlife, for its first detailed assessment.

Mungo Park solved the first part of the Niger problem. He died attempting to solve the second. In the decades that followed Park's death, others followed in his wake. Many of them also failed and died for geography. This book is a study of the lives, death, and books of Mungo Park and of those men who, like him, died trying to solve a 2,000-year-old geographical problem. It is also a study of those who stayed at home to investigate the Niger problem. And because explorers did not become explorers until they became authors, this is a book about explorers' books, how they were written 'in the field' and why it matters that, in many cases, explorers' narratives were amended and altered upon their return – if they returned at all, that is.

1

'Its final destination is still unknown'

The Niger problem and the nature of exploration

European interest in the Niger by the end of the eighteenth century was not simply a reflection of how little was known about the river – even whether it was a separate river at all. It was also a matter of trust. Because what little was known was either contradictory or founded upon sources whose testimony could not always be relied upon, and because the river had not yet been seen by a reliable witness one had to place one's trust in what was written and in what one was told. European ignorance and uncertainty lay not only in the paucity of facts concerning the river, but also in how those facts were arrived at and by whom they were reported. Exploration depends upon trust in the explorer and the trustworthiness of any resultant written or verbal account of it. To Banks, Beaufoy, and the others in the African Association, however, as they turned their attention to the Niger, the written sources available were few and inconsistent.

Early textual and map evidence

The Niger was recognised as a geographical mystery in antiquity. In the fifth century BCE, the Greek historian and geographer Herodotus claimed that the Niger flowed eastwards to a lake in north central Africa and, from there, joined the Nile. In the second century AD, the geographer Ptolemy similarly wrote that the river flowed eastwards from the mountains of what today are Senegal and Gambia, but he argued that rather than join the Nile, it emptied into an inland lake in the African interior.

Classical authors were not the only ones to turn to the Niger. Given the Muslim presence in north Africa, several medieval Islamic geographers and cosmographers had studied the sub-Saharan African interior. Early Islamic map-makers depicted *al-Wāq-Wāq*, the African interior south of the Sahara, only as an unknown space, or traversed by lines of the caravan trade. Arab knowledge concerning the Niger would become better known to explorers during the early nineteenth century as would that of other indigenous groups. But before Mungo Park set out only three Islamic texts were known to European scholars.

The first was that of Abu Abdullah Muhammed al-Idrisi al-Qurtubi al-Hasani al-Sabti, known to Europeans as Xeriff Edrisis, Edrisi or, more usually, as al-Idrisi or Idrisi. His *Kitab Rujar*, or the *Book of Roger* after Roger II of Sicily who commissioned it in 1138, was largely based upon the spoken evidence of travellers and merchants to north Africa, and was completed in 1154. For centuries, the work was effectively lost to Europeans before it was re-discovered in the Renaissance. An abridged version was printed in Rome in 1592. As interest in Africa heightened, a French version was published in Paris in 1619, under the title *Geographia Nubiensis*. On the evidence of what he was told, al-Idrisi accepted Ptolemy's argument for a large lake in the African interior south of the Sahara but differed from him in asserting that what al-Idrisi termed *Nil as-Sudan* (Nile of the Sudan) flowed out of that lake towards west Africa. Sudan is here being used in its earlier geographical sense, that of the southern desert of the Sahara as a whole, rather than the modern nation state. In effect, al-Idrisi reversed the Niger's direction of flow from that proposed by Herodotus and Ptolemy.

The second source known is the *Taqwīm al-Buldān* (usually translated as 'The Sketch of the Countries'). This geography of the world as it was then understood was written in about 1310 by the Damascus-born Kurd Abu'l-Fida, known usually in English as Abulfeda. Abulfeda's description of Africa, only elements of which were available in translation by 1650, is limited to the north and north-east of that continent. On the Niger, he echoed al-Idrisi's view that the river flowed from east to west.

The third source came from al-Hasan ibn Muhammad al-Wazzan al-Zayyati, Granada-born but of Berber origin and known usually to Europeans as Leo Africanus. His *Description of Africa* was written in 1526 and first published, in Venice in 1550, as part of Giovanni Battista Ramusio's *Della Navigationi e Viaggi*. An English translation by John Cory was published in 1600 under the title *A Geographical Historie of Africa*. As a diplomat, Leo Africanus had good reason to travel. While he probably never visited all the places he mentioned, parts of his work were based upon first-hand encounter. It is known that he visited Timbuktu, for example, but that he did not venture to that great river that runs ten miles to its south.

On the Niger, he simply reported the different views held: 'This lande of Negros hath a mightie river, which taking his name of the region, is called Niger: this river taketh his originall from the east out of a certaine desert called by the foresaide Negros Seu' ['Seu' is here an abbreviation of Sudan]. 'Others will have this river to spring out of a certaine lake, and so to run westward till it exonerateth it selfe into the Ocean sea.' He further reported that 'the said river of Niger is derived out of Nilus' – that, in effect, the Niger flowed out of the Nile before disappearing underground and bursting forth 'into such a lake as is before mentioned'. But as he also recounted, 'Some others are of opinion, that this river begineth westward to spring out of a certaine mountaine, and so running east, to make at length a huge lake.' This lake was probably what today is Lake Chad. In the 1820s, Lake Chad would become of great importance to those aiming to solve the second part of the Niger problem – where did the river end? – and, in doing so, to confirm whether the Niger ran into the lake, out from it, or flowed somewhere else entirely, even 'into the Ocean Sea'.[1]

In the early 1620s, the English adventurer Richard Jobson explored the Gambia river, lured by tales of gold. Jobson hoped to interest King James and his courtiers in African trade but was disappointed by the lack of interest shown. His *The Golden Trade, or, A Discovery of the River Gambra* (1623) is interesting on the myths of west African riches – he failed to find any gold – but it contains among the first observations on the extent and nature of slavery: black merchants selling enslaved people to Portuguese and Spanish slave traders for onward shipping to the Americas. Informative though he is about the course of the Gambia river, Jobson adds nothing of significance on the course of the Niger.

The Niger appears as an object of interest in European maps of Africa during the eighteenth century. In 1700 the French geographer and map-maker Guillaume Delisle showed the Niger on his *L'Afrique dressée sur les observations de Ms. De l'Académie Royale des Sciences*. This map is a landmark in the history of Africa's mapping because of its accurate longitudinal dimensions: little was known of the interior, but the continent was at least beginning to assume its correct shape. Improved versions of this map followed in 1707, 1722, and in 1727. On the 1707 map, Delisle portrays the Nile as separate from the Niger, which suggests that he did not hold to the view that the two rivers were connected. If he thought by his depiction to scotch that rumour, Delisle failed: reports of connections between the two rivers continued even into the early nineteenth century.

In his 1722 *Carte d'Afrique*, Delisle distinguished the Niger from the Senegal and Gambia rivers: his map showed the Niger's source to lie in mountainous regions south of Timbuktu and had the Niger flowing in a west to east direction to end in an inland lake in the Borno or Bornou region of the interior. Delisle derived his evidence not from first-hand encounter but from André Brué, the French Commandant General of the Senegal

Company, who had explored coastal west Africa between 1697 and 1700, and, for his later maps, from French clergyman-polymath Jean-Baptiste Labat who, in turn, took his knowledge from the reports of Jesuit missionaries to west Africa.

Like Brué, Francis Moore drew elements of his *Travels into the Inland Parts of Africa* (1738) from personal experience in situ, as a manager in Gambia for the Royal African Company of England. Moore made several excursions into the west African interior as part of his responsibilities, but there is no evidence to suggest that he encountered the Niger. The great part of his *Travels* was a collection of what earlier authors such as al-Idrisi and Leo Africanus had written. Moore's inexpert compilation of others' accounts only added to confusion over the Niger and the geography of west Africa: '. . . the Reader . . . may from these collections . . . form his judgement of what is true, by comparing one Account with the other, and see whether there is a Probability that the Niger and the Nile flow from the same Fountains, or that the Niger and the Gambia are the same'.[2] Quite how readers were supposed to judge where Africa's rivers ran and in which direction he did not say. That the Niger and Africa's interior were objects of European interest more widely is apparent from the German map-maker Johann Hase's 1737 map, his *Africa Secundum legitimas Projectionis Stereographicæ regulas*. This was based upon Leo Africanus's work and so depicts the Niger flowing west to east, including the great bend to the south and east that the river takes, but Hase erroneously placed the river to the north of Timbuktu.

The leading French geographer and map-maker Jean-Baptiste Bourguignon d'Anville based his 1749 map *Afrique publiée sous les auspices de Monsigneur le Duc d'Orléans* in part upon Delisle's work but showed the Niger flowing north-eastwards between a source in 'Marais Nigrite' and its termination in an inland lake. Unlike other European contemporaries, d'Anville would return to the subject of the Niger in an essay ten years later. To English geographer James Rennell, later to play a significant role in Park's book and in discussions over the Niger's course, the fact that d'Anville had to rely on Classical and Islamic sources for his 1749 map was a source of some embarrassment: 'Nothing can evidence the low state of the African Geography, more than M. D'Anville's having had recourse to the Works of Ptolomy [sic] and Edrisi, to compose the Interior Part of his Map of Africa (1749).'

D'Anville's 1749 map (Figure 1.1) is nevertheless important for his insistence upon showing only what was reliably known. The consequence of this enlightened shift in the nature of mapping, to emphasise by omission what little was known, was to leave large areas of Africa's interior blank: maps with gaps. This was also a feature of maps of Africa by the English map-makers Samuel Boulton and Thomas Kitchin in 1794 and by Aaron Arrowsmith

Figure 1.1 D'Anville's 1749 map illustrates Europeans' poor understanding of the continent by the mid-eighteenth century: more complete knowledge of the north and west coasts, some of the Nile region, almost nothing of the interior. The Niger is shown running through 'Nigritie' to the south of the Sahara. Source: J.-B. d'Anville, 'Afrique: Publiée sous les auspices de Monseigneur le Duc d'Orleans premier prince dusang' (Paris, 1749). © The British Library Board, King's Topographical Collections, K. Top. 117. 8. K. 2. TAB.

AFRIQUE
PUBLIÉE SOUS LES AUSPICES
DE MONSEIGNEUR LE DUC D'ORLEANS
PREMIER PRINCE DU SANG
PAR LE Sr D'ANVILLE
MDCCXLIX
Avec Privilege
A PARIS chés l'Auteur aux Galeries du Louvre
CARAMANIE
CANDIE
CYPRE
ÉDITERRANÉE
SYRIE
Barca
Levata
ERDOA
EGYPTE
MER ROUGE
TEHAMA
YÉMEN
SOCOTORA
ABISSINIE
DUNGALA
Désert de Béhiouda
KAUGHA
GORHAM
Désert de Leu
Rme D'ADEL
Rme DE ZENDERO
Omnno-Zaidi
MACHIDAS
MARACATOS
Montagnes de la Lune
LIGNE EQUINOCTIALE
117
18
Africa

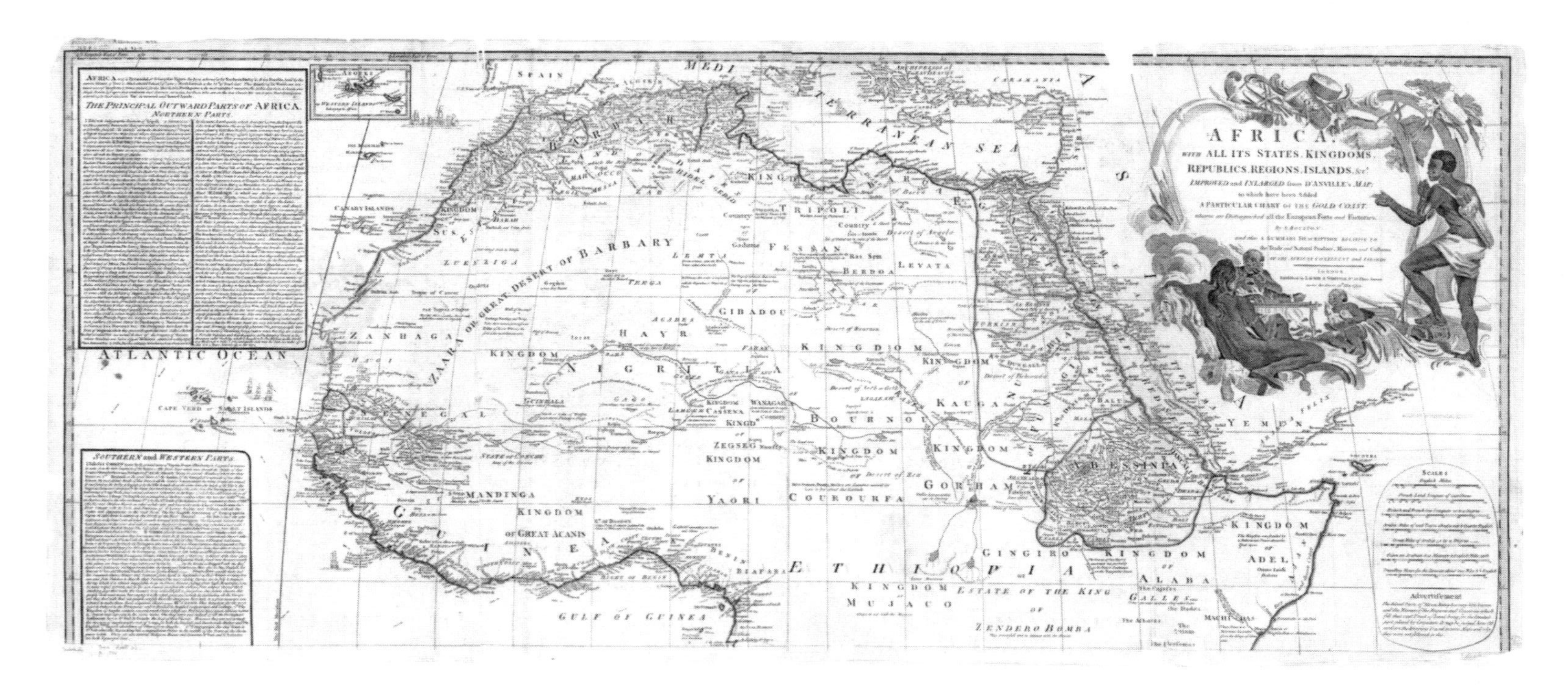

Figure 1.2 Boulton's 1794 map of Africa adds little to that of d'Anville in 1749 (*cf.* Figure 1.1) despite his claims to it being an 'Improved and Inlarged' version. His text panels offer only general descriptions of the continent. Source: S. Boulton, 'Africa with all its States, Kingdoms, Republics, Regions, Islands, &c' (London, 1794). Library of Congress, Geography and Map Division.

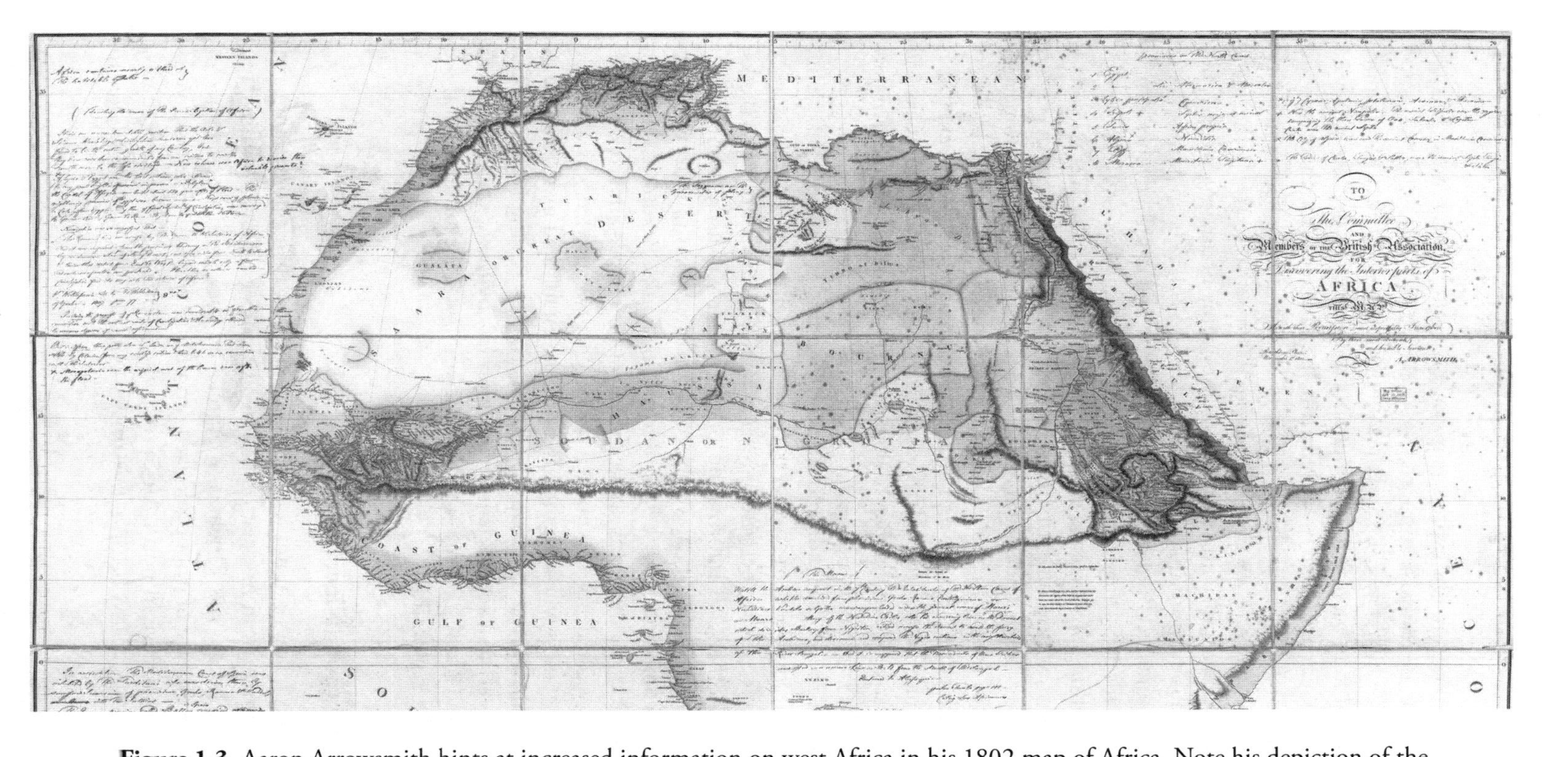

Figure 1.3 Aaron Arrowsmith hints at increased information on west Africa in his 1802 map of Africa. Note his depiction of the mythical mountain chain thought to bisect Africa: the Mountains of Kong to the west, the Mountains of the Moon to the east. Source: A. Arrowsmith, 'Africa: To the Committee and Members of the British Association discovering the interior parts of Africa this map is with their permission gratefully inscribed' (London, 1802). Library of Congress, Geography and Map Division.

in 1802. Blankness might express map-makers' honest uncertainty. It could also, of course, be read as a reproach.[3]

All this is to say that by 1788 and the foundation of the African Association, the Niger had long been written about, and within recent decades had been mapped by German, French, and British map-makers. Yet there was no shared understanding as to the direction of its flow, its possible connection to other African rivers, or even if it existed as a separate river. The maps were based on second-hand reportage or the words of ancient writers, not from direct observation or survey. D'Anville's 1749 map is an admission of European ignorance. Boulton added text panels and a rather more ornate cartouche to his 1794 map which was based on d'Anville's of 1749, but he could add nothing of substance regarding Africa itself (Figure 1.2). No one – no one, at least, whose authority in the present could be relied upon – knew where the river Niger ended. Such knowledge as there was in Europe was an amalgam of merchants' tales, contradictory accounts from Arab traders and sixteenth-century Islamic diplomats, maps which showed different directions of flow and no agreed outlet for the river, and a reliance upon Classical authorities who believed the river to flow west to east but who differed in how and where the river ended.

This is not to say that indigenous knowledge was not regarded as useful, or that the testimonies of medieval Arab merchants and diplomats were regarded with scepticism because of what was said. To the contrary: to Banks, Beaufoy, and others, all information on the Niger was useful. But not all such information was of equal weight. What mattered was not just what was said but by whom and how they derived it. For Banks, Beaufoy, and others such information as there was about the Niger problem 'came from sources whose veracity could not be vouched for'.[4] What was needed was a reliable geographical witness whose credibility would be established by several things in combination: by the use of appropriate methods – see the Niger for oneself, return safely and report definitively upon it – by the observer-reporter's own moral and social standing, and by association with the authority of those proposing the exploration.

Early exploration: Niger men before Park

Banks and his fellows did not have to wait long after their pub meeting and resolution. The African Association had two volunteer explorers within days. One, Simon Lucas, was Oriental Interpreter to the Court of St James in London. On the face of it, Lucas looked promising. He had spent three years in captivity as a slave in north Africa following capture by Barbary pirates. Upon his release, Lucas became British vice-consul and chargé d'affaires to the empire of Morocco before his return to London. He had the required linguistic abilities and first-hand familiarity with north African politics and culture.

The other was an American, John Ledyard. Cut from different cloth than Lucas, Ledyard seemed the very embodiment of the adventurous explorer. To Beaufoy, Ledyard 'seemed from his youth to have felt an invincible desire to make himself acquainted with the unknown, or imperfectly discovered regions of the globe'. Born in Connecticut, Ledyard had sailed as a British marine with Cook on his last and fatal voyage: his published account of this made him something of a celebrity as an American Oceanic explorer. Prompted by Thomas Jefferson and by Joseph Banks, Ledyard then attempted an exploration of the west coast of North America, chiefly on foot, by way of Siberia. He began this in December 1786 with an unsuccessful crossing over the ice of a not-entirely-frozen Gulf of Bothnia between Sweden and Finland. Ledyard's northern adventures ended in March 1788 with his arrest in Siberia as a French spy and deportation from Russia. Penniless and destitute in Königsberg, Ledyard survived thanks to his borrowing five guineas against Banks's account. Using this borrowed cash, he was able to return to England. Meeting him in London, Beaufoy did not pass remarks upon Ledyard's rather cavalier approach to others' finances but did comment upon Ledyard's physique and bearing: 'I was struck with the manliness of his person, the breadth of his chest, the openness of his countenance, and the inquietude of his eye.'[5] The Association had its first Niger men.

Lucas was instructed to head south from Tripoli to the Fezzan. The Fezzan is today an administrative region of south-west Libya; then, with its capital Murzuq, it was a long-established sultanate on the trans-Saharan trade route from Tripoli to Timbuktu. Ledyard was sent to Cairo, his brief the task of 'traversing from east to west, in the latitude attributed to the Niger, the widest part of the Continent of Africa'. Lucas never travelled south across the Sahara, reporting that a revolt among Bedouin tribes had made travel hazardous and that the army to quash the rebellion promised by the Bashaw, the ruling Arab official in Tripoli, never materialised. Lucas nevertheless learned about the kingdoms and polities south of the Sahara from meetings with two merchants from the Fezzan, Shereef Mohammed Bensein Hassen Fouwad and Shereef Imhammed. They assured him that, in normal circumstances, regular trade was possible across the Sahara. In adding to what was known from Leo Africanus, Lucas established – but not by virtue of his own travels – that the west African interior was accessible from Tripoli in a desert crossing but only with the assistance of others, whether merchants or under military protection and, crucially, with the support of the Bashaw. By late July 1789, Lucas was back in London.

Lucas at least returned. Ledyard never did; he never even left Cairo. He accidentally poisoned himself in January 1789, the effect of 'too powerful a dose of the acid of vitriol' which he had taken to ward off a bilious complaint caused, it was later conjectured, by boredom and depression at his 'unexpected detention, week after week, and month after month'.[6] He would not be the only one who hoped to explore northern Africa but who at first could not and then never did.

At the same time as Lucas and Ledyard were in Africa, members of the Association gleaned further information from two Moors then resident in London. One, Ben Ali, offered to escort a traveller from the Gambia to Timbuktu. His plan was taken up on behalf of the Association by a Mr Hollen Vergen and by Francois Xavier Swediaur, an Austrian-Swedish physician then resident in the coastal village of Port Seton in Scotland. Both men were 'animated by an earnest desire of promoting the great object of the Association', but their scheme never materialised. Rather like Ledyard, but not fatally, Swediaur became unwell in the autumn of 1789 with a severe colic caused, apparently, by drinking adulterated wine.

The other London-based Moor was Asseed El Hage Ab Salaam Shabeeny, known as Shabeni, and like Ben Ali a Moroccan-born merchant. He found himself in London in early 1790 following his capture on a ship bound for Hamburg. Both he and Ali told Banks, Beaufoy, and others of the commercial opportunities possible in trade with the kingdoms of the African interior, in and around Timbuktu, Bornu, and the hitherto unknown kingdom of 'Housa'. Early in 1790, a proposal by two English merchants, a Dr Cramond and a Mr Walwyn, who spoke of visiting Ghadames, the Berber market town in the west of the Fezzan, may have been based on what was told to them by these two London-based Moors. But the plan came to nothing.

In July 1790, the Association was approached by an Irish major, Daniel Houghton. Houghton was known to Banks, had experience of Morocco where in 1772 he had been British consul to the emperor, and he had served at the British outpost at Gorée in west Africa following that island's capture from the French. What Beaufoy saw in Ledyard, he similarly recognised in Houghton: 'a natural intrepidity of character, that seems inaccessible to fear, and an easy flow of constitutional good humour.' In September 1790, Houghton was issued with detailed instructions from Beaufoy in the form of fifty-four questions under eleven headings. Beaufoy's queries included half a dozen 'Questions respecting the Niger'. Two questions in particular encapsulated contemporaries' long-standing ignorance regarding the Niger: 'Does the Neel il Abeed (the Niger) flow from, or towards the Setting Sun?'; 'Does it empty itself into the Sea? or does it end in a great Lake? or is it lost in the sands of a Desert?'

Queries in this form, to be answered by reliable informants, were then becoming an established method of natural and social enquiry for those who could not visit the places they wished to know about. They were, as Beaufoy made clear, written instructions for Houghton's benefit, designed to direct him in soliciting further information regarding that 'considerable Empire, distinguished by the name of *Houssa*' and of the Niger, 'the Committee being also desirous to be informed of the Rise, the Course and the Termination of the Niger as well as of the various Nations that inhabit its borders'. Beaufoy's 'Queries' represent the first in a series of written orders and instructions designed to direct the Niger's explorers in what to see and how to proceed – attempts 'at home' to regulate the explorer over what was to be done 'out there'.[7]

In contrast to the intended trans-Saharan routes of Lucas and Ledyard – north to south for the former, westwards from Egypt for the latter – Houghton proposed a more direct route via west Africa and the Gambia river. Taking with him a variety of goods for trade, he made his way to Medina, capital of the kingdom of Woolli, and today part of Dakar in modern-day Senegal. He was by all accounts well-received there. Houghton then suffered a series of misfortunes. His interpreter made off with his horse and three of the asses used to transport the trade goods. He was injured in the face and arm after a gun he had purchased locally exploded in his hands. He lost most of his remaining goods and possessions in a fire that destroyed much of Medina. The Niger's exploration commonly involved such mishaps: Houghton's were the first of many.

Undaunted, he continued into the neighbouring kingdom of Bondou, only to be hampered by the fallout from conflict between the kingdoms of Bondou and Bambouk. In Bambouk, Houghton exchanged his few remaining goods for gold dust, was presented with a purse of gold from the king and he agreed a plan with a local trader who would accompany Houghton to Timbuktu and return with him to the Gambia, there to be paid £125 as a premium for ensuring Houghton's safe conduct. All this we know from letters which reached Dr Laidley, Britain's consul in the Gambia, the letters then being despatched to London. In a letter of 6 May 1791 written from Medina, Houghton remarked upon masted vessels on the river he was going to explore, heading 'eastward, to the centre of Africa'. His last letter is dated 1 September 1791. Houghton spoke of having been robbed of all his remaining goods but was 'on his way to Tombuctoo'. Whether he got there and, in doing so, encountered the Niger, is unknown.[8]

The circumstances and location of Houghton's death remain unclear. Different reports suggest Houghton died of starvation, from dysentery, or that he was murdered. His few letters nevertheless provide important information on the Niger, both on its likely source and its easterly direction of flow. As the *Proceedings of the African Association* note, 'we now have an assurance that the Niger has its rise in a chain of mountains which bound the eastern side of the kingdom of Bambouk', that it was a separate river and that it ran 'in a contrary direction from that of the Senegal and the Gambia' – that is, the Niger seemed to flow eastwards. But Houghton had not survived to return to confirm this. And the Niger's termination remained a mystery: 'the place of its final destination is still unknown: for whether it reaches the ocean; or is lost . . . in the immensity of the Desert; or whether . . . it terminates in a vast inland sea, are questions on which there still hangs an unpenetrated cloud.'[9]

Houghton's evidence, and his death (which was not confirmed for some years), left the African Association with problems at once political and geographical. At a meeting on 26 May 1792, those committee members present formally noted how Houghton's discoveries 'have furnished a valuable addition to Geographical Science; and afford a prospect of important advantages to the Commercial Interests of the Kingdom'. This phrasing is noteworthy. So is the resolution that followed: 'That the Committee be empowered to make, in the name of the Association, whatever application to Government they may think advisable for rendering the late discoveries of Major Houghton effectually serviceable to the Commercial Interests of the Empire.'[10]

Here, unequivocally expressed, is the connection between new geographical information, politics, and commercial advantage that underlies the Niger's exploration. There is, even, a hint that geography was the necessary handmaiden to empire – as was more evidently the case in Africa's exploration from the second half of the nineteenth century. As the Association soon understood, however, and as later Niger men and expeditions would experience, there was often a gap between stated resolutions and the instructions issued to explorers prior to departure and what was accomplished by them in the field.

Houghton had the misfortune to die before he could return to endorse what his correspondence reported. To confirm that information – specifically, to confirm the direction of the Niger's flow at first hand and to dispel the 'unpenetrated cloud' that still hung over the Niger's termination – and to advance the Association's commercial intentions, Beaufoy approached leading figures in the government with what was, essentially, a political proposition. This was to appoint a consul to the kingdom of Bambouk, or Senegambia as it became known, a resident British official who would oversee the hoped-for 'extensive and lucrative Trade' with the peoples between the Gambia and the Niger. The man appointed as Britain's consul general in Senegambia was James Willis.

James Willis never saw Africa. He never even left England. But his failure was another's opportunity. Willis's plans included having as a travelling companion the latest figure to come forward to serve the interests of the Association. Banks had appointed him on 23 July 1794. As Willis dithered, Banks's and the Association's latest 'Geographical Missionary' acted and, on 22 May 1795, he sailed for west Africa. His name was Mungo Park.

My purpose in summarising the Niger problem and the workings of the African Association between its foundation in 1788 and Park's departure for Africa in 1795 is to make clear the nature and longevity of the Niger problem and the complexity of its projected solution. Before Mungo Park set sail for Africa, the Niger problem had been the subject of close but intermittent attention by the African Association for seven years, by others for two millennia. Not all Niger men were explorers in the sense of that term as travellers 'in the field'. Information about the Niger came from talking to Arab merchants in London, in Tripoli, in Murzuq, and in the kingdom of Bambouk. It came from reading the works of Classical authors. It came from consulting French, English, and German maps and geographical articles. For those men who did travel to Africa, linguistic ability was helpful, whether Arabic as for Lucas or as Houghton found to his advantage a smattering of Mandingo, the language commonly spoken in west Africa. Houghton was provided with scientific instruments so that he could give longitudinal precision to the location of places and measure topographical features: after all, to function optimally, market economies depend upon knowing where the markets and the people who attend them are, how long travel between places takes, and in which direction.

Knowledge about the Niger was slowly and differently accumulating but it was not of equal value, nor was it treated equally. Primary importance was attached to the idea – and to the ideal – of the explorer as an 'in-the-field' figure possessed, like Ledyard, of the qualities of 'manliness' and 'natural intrepidity', someone whose credibility and authority came from personal experience derived from empirical encounter.

By the time Park left for Africa, knowledge for certain about the Niger's source, its course, and its termination had not been secured by anyone who had survived the undertaking. Interpreting Mungo Park's achievements in 1796 in solving the first part of the Niger problem as a success – which it was – depends in part upon our acknowledging and explaining his predecessors' failure. In turn, Park's death in 1806 and the fact that he left the second part unsolved motivated others. His failure then – which it was – initiated changes in the character of the Niger's exploration, and in who funded and directed it, even in what exploration was held to be.

The nature of exploration

By the late eighteenth century, exploration in the primary sense in which it was then understood – geographical fieldwork, travel, direct encounter – was considered reliable because it was a form of rational empirical enquiry. It presumed trust in explorers' visual acuity: for seeing the object in question for themselves. It presumed trust in their moral integrity: in reporting faithfully upon what was seen. There was, in consequence, from those persons proposing and directing exploration, a preparedness to trust not just in the character of the explorer but in the content of his words, spoken at first or in letters but, in time, written and printed. Exploration narratives were held to straightforwardly reflect the nature of the exploration undertaken. There was, in short, an assumed and direct correspondence between social status, morality, epistemology, and textual authority in what exploration was held to be. Travel, it was commonly assumed, made truth.

In recent years, these assumptions have been subject to critical review. The nature of exploration and the authorship of exploration narratives have been subject to reinterpretation. Even the term 'explorer' has been scrutinised. In its connotations of someone sent out by others to determine the extent and content of a given territory and to report back faithfully upon that information in one way or another, the term 'explorer' was never in common parlance during the eighteenth century: the French, for example, employed the term *voyageur naturaliste*. The term 'explorer' became common only from the early nineteenth century and did so in association with what very largely were European or Western-based notions of territorial enquiry into the rest of the world. One modern author has argued that its use was as a 'back-formation', that is, it was used after and not before the events themselves, as the author of the work in question described himself or was so described by others: one only became an explorer in becoming an author.[11]

This is an important point. Exploration counts for little unless one's words make a final journey – into print. Questions about exploration, authorship, truth and trust, and the construction of a printed exploration narrative find particularly clear expression in the story of Mungo Park, in his 1799 *Travels*, and in the Niger's slow unveiling in words and on maps. For Park and for others, writing in the field, on the move so to say, could be equally complex. Note-taking during exploration was vital as an immediate record of what was undertaken and as a guide to memory, but we should not assume it to have been either straightforward or easily accomplished. What is an archive source for modern scholars – the travel narrative – was, often, a combination of diary, a register of locational information, and a record of distances travelled.

It was seldom the simple and direct result of writing 'on the spot'. Exploration is often recoverable only from its written traces and these vary. As he sought clues to Park's death during his first Niger travels on the Bornu Mission in the early 1820s, for example, Hugh Clapperton undertook what others have described as his 'in the raw' writing in two formats, a log and a journal. Clapperton altered the second, on a more-or-less daily basis, as circumstances permitted, from records contained in the hurriedly prepared first. The fair copy text was itself later amended elsewhere, partly by Clapperton, partly by others, before it was published.[12]

The recent emphasis in studies of exploration upon the production and the reception of its printed outcomes has had the effect of making book history a central feature of exploration history. It is, strictly, more a larger matter of textual and print history than it is of book history alone since periodicals and newspapers carried reviews of expedition reports. To understand the Niger's exploration, we need to understand how and where the facts of its exploration were recorded, how they travelled into print (or did not), through which intermediaries, and how they were received by others. Where we can, we must undertake, as it were, a biography of the book and the other texts involved – their making and reception and in some cases their multiple or 'aggregative' authorship – as well as study the individuals involved.[13]

The nature of exploration as a process involving human encounter and cultural difference has similarly been the subject of renewed focus. Importantly, recognition is now more commonly given to indigenous agency and to indigenous knowledge. Even so, many narratives of Africa's exploration evidence what we can think of as a kind of epistemological effacement, a 'double movement'. That is, there is at best tacit recognition by the explorer-author of the assistance received from indigenous peoples in the form of porters and interpreters and, crucially, as guides, but their knowledge is seldom accorded equivalent status to that of the explorer, even if it is acknowledged at all. The agency of such persons is clear, but their names have not survived in the historical record: the cast list of exploration is intrinsically uneven in this respect and perhaps never fully recoverable. Some things and names we cannot now know. Because exploration involved indigenous agency, albeit unequally, it is an appropriate term to employ: 'discovery' is altogether more problematic.

If indigenous people are now commonly recognised as auxiliaries or intermediaries, by which terms we may understand how they acted to assist the explorer and so, perhaps, were complicit in empire's empirical advances, they were still autonomous agents and custodians of indigenous knowledge. On occasion, this could present explorers with moments of incommensurability and misunderstanding. Most Niger explorers benefitted from the assistance of those peoples whose territories they passed through. Park certainly did:

he only survived thanks to the kindness, amongst others, of several African women and an Arab slaver. But not all the Niger's explorers recognised and valued the indigenous knowledge that came their way, in several instances as drawn maps, more commonly by word of mouth.[14]

The scientific instruments that explorers used in the field were no less 'intermediaries' or mediators between explorers and the geographies they encountered, simply of a different kind. Many explorers carried a watch, compass, and other precision devices such as a quadrant or sextant with which to calculate longitude and latitude and so determine their position. Thermometers were used to take temperature readings, barometers to calculate topographic elevation. A series of recorded longitudinal positions and topographical notes might be sufficient to produce a sketch map: detailed map work did not take place in the field but was done elsewhere upon the explorer's return. The use of instruments lent precision to narrative claims, made 'blank spaces' measurable and so knowable. In the form of maps, instrumentation made those spaces reproducible, portable, as accurate as possible and, in the fullness of time perhaps, governable.[15]

Just as explorers could and did falter, however, so instruments broke or were lost, watches ran down, and devices had to be repaired or discarded. Explorers' accounts commonly mention the value and use made of the instruments they took with them. Park was despondent when, following an attack by thieves, his compass was returned to him, shattered and broken. The credibility of author-explorers could be enhanced by making authoritative claims to instrumental competence, from which (it was commonly assumed) their measurements and the resultant narrative and map was similarly accurate and precise. In many instances, however, the narratives that followed exploration often incorporated claims to accurate and 'up-to-date' measurements which were arrived at despite the wholesale failure of the instrument used or from incorrect readings as devices failed to function properly given high temperatures or clumsy handling.

Here, too, the historian of exploration is presented with important implications to do with trust, failure, and the authoritativeness of explorer-authors and of their narrative accounts. If the narrative was written and its accompanying map drawn up despite instruments being 'crack'd' and breaking down – as, perhaps, the author-explorer did too – can we fully trust the claims of the published account? How should we understand failure, whether of the device, the explorer, or the expedition?[16]

Often, failure was attributed to the facts of geography – to the heat, the climate, the distances to be travelled – as well as to the loss of scientific instruments, limited provisions, or human frailty. Dixon Denham, who travelled with Walter Oudney and Hugh Clapperton in search of the Niger in the mid-1820s, apportioned blame for his exploratory failures upon everyone but himself including, knowingly and falsely, accusing Clapperton of inappropriate sexual behaviour with an Arab servant: one later historian considers Denham perhaps the most odious man in the history of geographical discovery.[17] Exploration, always politically and scientifically motivated, was equally always mediated by the relationships between people and the things they carried with them and by the efforts of individual human beings whose bodies sometimes weakened or broke down under the strain. Exploration was always physically arduous and sometimes morally taxing. Not all the Niger's explorers in the field got on one with another.

Not all those who tried to solve the Niger problem were fieldworkers. Geographical solutions could also be arrived at by sitting still and thinking hard, drawing one's conclusions from synthesising the work of others. In Britain from the late 1840s, differences between field exploration and sedentary speculation were especially clear in debates over the source of the Nile. The source of the Nile had been discussed earlier by Classical authors, notably by Ptolemy. Oral testimony was gleaned from merchants, Portuguese travellers, and map-minded German missionaries. In addition to those 'critical geographers' or 'armchair geographers' (they were also known as 'carpet geographers'), whose sources were others' texts and testimonies, the source of the Nile was also the focus of expeditionary science as explorers such as David Livingstone, John Hanning Speke, and Richard Burton sought to locate the river's source from direct observation and so fill in the blanks on the map of east central Africa. The geographical exploration of the Nile in the mid-nineteenth century was a 'conflict of methods', between explorers in the field, and sedentary geographers and map-makers who worked with others' words. Yet it is important not to see these two routes to geographical knowledge as necessarily in conflict. They might be and they commonly were, but they also could, and did, complement each other.[18]

As for the Nile between the late 1840s and the early 1870s, so for the Niger between the late 1780s and the mid-1830s: the river's exploration was a matter of work done in Africa and work done elsewhere, a matter of direct first-hand observation and of sedentary enquiry citing different evidence; of trust in one sort of information or device more than another; of compilation and mental calculation in one place, and of observation and physical hardship by others elsewhere. The exploration of the Niger was never either a simple or a single thing but was, rather, a matter of different people at different times using different methods in different places.

The Niger's exploration was always shaped by matters pertaining to its political and institutional context. This is perhaps the defining feature of recent work in exploration history: 'If there is a single thread that runs through the recently renewed scholarly interest in the subject, it is the conviction that explorers and exploration cannot fully be understood without identifying and explaining the multiple contexts within which they operated.'[19] The geographical and intellectual context to the river's exploration is clear – the 2,000-year-old Niger problem. From the early nineteenth century, the Niger's exploration was a political problem and an increasingly commercial and humanitarian one. The question of slavery and its abolition runs throughout the Niger's exploration, most evidently in the expeditions of 1832–4 and 1841. And as the context to the Niger's exploration became more complex, so different institutions became involved.

By the second decade of the nineteenth century, branches of the British government, chiefly the Admiralty, and the War and Colonial Office (the two merged in 1801), replaced the African Association as the leading bodies directing the Niger's exploration. From 1830, their role was supplemented by the Royal Geographical Society with which the African Association merged and so ceased to exist as a formal body. As institutions came and went, so did influential individuals. Henry Beaufoy died in May 1795 and so never learned of Park's success, or even of his departure for Africa. Joseph Banks died in June 1820, and so never witnessed the decade of Niger expeditions that preceded the Landers' achievements in 1830. Their later equivalents were John Barrow, Second Secretary to the Admiralty, Hanmer Warrington, Britain's consul general in Tripoli for over thirty years from 1814 and, also in Tripoli, the Bashaw, Yusuf Karamanli.

Although the Niger was the source of consistent interest between Park's first travels and the mid-1850s, there was a periodicity to the expeditions sent out. The decade-long hiatus in the Niger's exploration between Park's death in 1806 and the expedition of John Peddie in 1816 is easily explained: Britain was at war with Napoleonic France. So, too, is the revival in interest from 1816 easily explained. Following the cessation of hostilities in 1815, large numbers of military officers, many trained in the use of scientific instruments, were available and eager for exploration work. Several of 'Barrow's Boys' – Joseph Ritchie, Alexander Gordon Laing, and Hugh Clapperton among them – headed for the Niger and died in its exploration.[20]

The recent reinterpretation of exploration has also deepened and extended the central place of biography in exploration history. This is evident in the examination of the essentially autobiographical nature of many exploration narratives. It is evident, if less so, in the extension of biographical analyses to non-human agents such as scientific instruments, thus revealing their working lives during exploration and how their working or not hindered the explorer. And it is apparent in the examination of biographies themselves as historical sources and in the contexts shaping different biographical accounts.

Explorers' words commonly outlive their deeds. Biographers' accounts can make them live anew – to be, almost, a different person, their actions differently explained according to biographers' interpretative intentions. To take a metabiographical approach – a biography of biographies – is not to search for the real figure in question but to demonstrate how and why that person was refashioned and reinterpreted after death. Metabiography as a method has illuminated the lives of several explorers and scientists in this way: Henry the Navigator in the early fifteenth century, Isaac Newton in the late seventeenth century, and in the nineteenth century, Alexander von Humboldt, the celebrated Prussian traveller in South and Central America, and the African explorers Henry Morton Stanley and David Livingstone.[21]

Mungo Park died in Africa in 1806 but he has enjoyed a varied geographical afterlife over the 200 and more years since. His afterlife is evident in a statue of him in Selkirk and a monument to him and to Richard Lander in west Africa. It is evident in biographies, poems, novels, plays, and scholarly works, even in a film. Park is today the subject of Nigerian rap and dance music, his 'discovery' of the Niger there reviled as culturally and historically insensitive. In *Stunning Scotland*, a 2021 children's book in the 'Horrible Geography' series, Park is the first of several Scottish African explorers noted.[22]

Park's afterlife is remarkable, varied, and virtually unexamined before now. The remembered and commemorated Park seems, almost, to parallel other events in the Niger's exploration: the statue unveiled in March 1859 was the result of a campaign for funds that began in 1841, the year in which yet another Niger expedition ended in disaster as British imperial hubris in west Africa again ran aground in the face of technological failure, disease, and human ineptitude. The Niger was not explored easily. Books about its exploration were not easily written. Mungo Park has had a lively cultural presence long after his death on the river.

2

'The bustle of life'

Park's early years, 1771–1795

Biographers have an advantage over their historical subjects (two, if you count the fact that they are not themselves dead). That advantage is also a responsibility. Biographers can fashion the subject's life without their approval or knowledge, presenting it not only as ordered but also as somehow defined by their later achievements as if everything beforehand inevitably led, as if predestined, to the accomplishments for which they are remembered. Yet lives are not lived as a sequence of events to a known end. Biographers aim to recount the individual's life story fully and with clarity. Doing so requires an admission on their part and acknowledgement of it from the reader: namely, source evidence survives only variably; lives are shaped by contingencies and chance; the defining events for which people are remembered were often shaped by circumstances beyond themselves.

So it is with Mungo Park. While, relatively, little is known of his formative years, we know enough to see how Park came to be in Africa at all, to know that while he was something of an innocent abroad, he had experience of foreign travel and residence in the tropical world before he set off in search of the Niger. But he was not the only man that the African Association had in mind to follow in the footsteps of Lucas, Ledyard, and Houghton. In London and before Park set out, others were trying to make sense of the Niger, and to guide Park accordingly, without leaving home.

Home and education: Selkirk and Edinburgh, 1771–1791

Mungo Park was born on 11 September 1771 at Foulshiels in the parish of Selkirk, in Teviotdale in the Scottish Borders.[1] In his 1791 description of the parish as part of Sir John Sinclair's *Statistical Account of Scotland*, the year in which Park completed his medical studies at the University of Edinburgh, the Reverend Thomas Robertson, Selkirk's parish minister, depicts a rural society largely unchanged over the previous two decades. The economy of the Borders – then as today – was dominated by sheep farming with arable agriculture limited to the warmer and drier slopes. In Selkirk and in other small towns, the chief manufacture then was woollen cloth. The parish population in 1791 of about 1,700 persons, of which total 700 lived 'in the country districts', was only slightly below that recorded in Alexander Webster's mid-century census.

In the Borders as elsewhere in late eighteenth-century Scotland, the continuities of rural life could be easily disrupted by natural causes and human agency alike. A decade earlier, in 1782, a cold winter, a late and wet spring, and an even later and wetter summer had caused widespread harvest failure throughout Scotland. The resultant rise in the price of grain prompted considerable hardship in Selkirk parish – 'the crop was very deficient, and the poor were reduced to very great distress', wrote Robertson. The young Mungo Park may have encountered the distressed poor and perhaps observed too what Robertson termed 'the extinction of small farms' which had led to a rise in the number of paupers in the parish. But Foulshiels was a middling-size farm, and the Park family was relatively well-to-do. It is unlikely he and his family experienced shortage themselves.[2]

Mungo Park, who took his name from his father, was the seventh of thirteen children (at least, thirteen is the number given by early biographers on the basis of family knowledge: as Lupton points out, the parochial registers are inconsistent on the matter: 'if there was a thirteenth child he [sic] must have come later'). The younger Mungo had two older sisters, Margaret and Jean, and an older brother Archibald. He was followed by Alexander, John, Adam, Isobel, and James (who died, aged three, in 1784). Three other children born before Mungo, John and twin girls Jean and Helen, died in infancy. The Park family were members of the Burgher Secessionist Church in Selkirk – their number, while not accurately known, 'cannot be considerable' Robertson noted. The Secessionists split from the Church of Scotland in 1733 over the right of the congregation, and not the patron, to elect the minister, and would split again over internal politics in 1747 and in 1761.

How far Mungo senior, and from him the Park family, was involved in Church affairs is not known although being Seceders speaks to a certain depth of devotional commitment regarding Calvin's teachings. Park senior certainly intended that the younger Mungo should aim for a career in the Church. Park later admitted to a friend that the course of his life was a matter of providential disposition and he later had cause on more than one occasion during his Niger travels to either pray to or offer his thanks to God. But there is no evidence to suggest he was particularly devout beyond what then, for him and his peers, was a belief in the Almighty and a commonly held familiarity with the scriptures.[3]

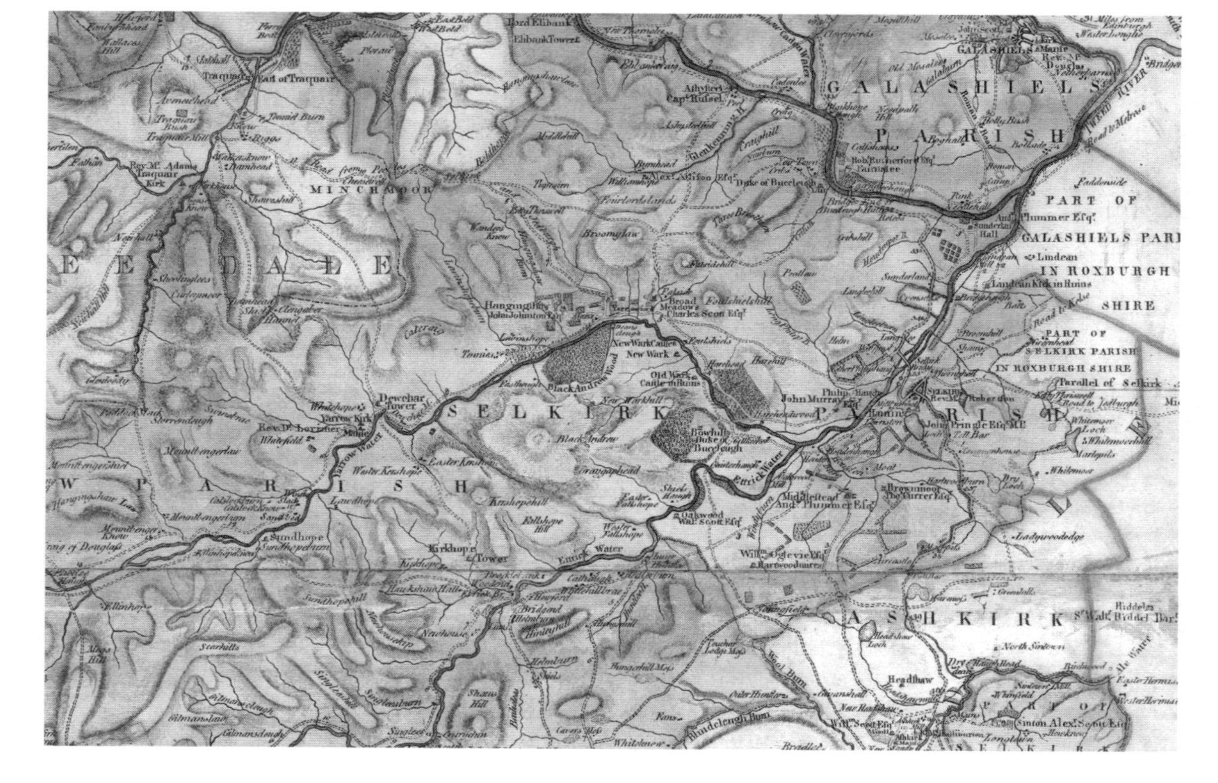

Figure 2.1 This detail from John Ainslie's 1773 map of Selkirkshire shows Foulshiels, Park's birthplace (near centre) and, on the opposite bank of the Yarrow Water, Newark Tower, marked here as 'New Wark Castle' (see Plate 1). The town of Selkirk is to the east. Source: J. Ainslie, 'Map of Selkirkshire or Ettrick Forest' (Edinburgh, 1773). Reproduced by kind permission of the National Library of Scotland.

Park was taught at home by a private tutor before moving on to Selkirk Grammar School. There, in addition to Bible studies, he learnt English, Latin, mathematics, and geography. We know that Park took delight in walks in his local countryside, at one time carving his initials on a wall of the neighbouring and long-ruined Newark Castle or Tower, its only inhabitants 'the mopping owl and the chattering daw', as Robertson had it (Plate 1). Although a map of Scotland survives with Park's writing on the reverse – 'My map of Scotland – M Park' – which suggests a familiarity with the outline of his own nation and, perhaps, an understanding of maps as a form of geographical representation, it is impossible to know which books the young Mungo used in the classroom or may have read at home.

The three geography books most widely used in late eighteenth-century Scottish school education were Thomas Salmon's *A New Geographical and Historical Grammar*, published in London in 1749 and with editions in Scotland in 1767, 1771, 1778, 1780, and 1782; Salmon's *The Modern Universal Gazeteer*, first published in Scotland in 1777 and again in 1781 and 1785; and the Brechin-born William Guthrie's *A New Geographical, Historical, and Commercial Grammar*, published first in London in 1770, with Edinburgh editions in 1790 and 1799.

Park's near contemporary Robert Burns read these books, the poet observing in a letter to a friend in 1787 how 'My knowledge of ancient story was gathered from Salmon's and Guthrie's geographical grammars.' Neither Burns nor Park is likely to have been intellectually stimulated by any of them given their rather formulaic structure, country by country and theme by theme, the characteristic of eighteenth-century geographical grammars. If he paid attention to the content on Africa, Park would have learned, from Guthrie for example, that the English presence in Guinea had opened new sources of trade, that the French had done likewise in Senegal, and that the Niger 'falls into the western ocean at Senegal, after a course of two thousand eight hundred miles'. This suggests that Guthrie may have taken his view, wrong-headed though it was, from reading al-Idrisi rather than from Classical authors. Salmon was no better: 'The great river Niger, which runs from East to West through Negroland . . . falls into the Atlantic Ocean.'[4]

Park left school in 1785 to take up a position as an apprentice to Dr Thomas Anderson, a physician and surgeon in Selkirk. To serve an apprenticeship – familiarising oneself with medicinal compounds, visiting the ill, observing treatments and perhaps even assisting in them – was common practice before studying medicine at university. Park already knew the Anderson family. He was close friends at school with Alexander Anderson, the oldest child and, on entering the Anderson household as a fourteen-year-old, would have encountered five-year-old Allison among the Anderson's eleven children. Their house in Selkirk still stands, facing Park's statue.

Park and the Andersons would be intimately connected, in life and in death. At the height of his fame, following his first Niger travels and the success of his book about it, Park would marry Allison Anderson on 2 August 1799. On his second and unsuccessful Niger journey, early on the morning of 28 October 1805, Park would record the death of 'my dear friend' Alexander Anderson, after a sickness of four months. 'I shall only observe', Park continued, 'that no event which took place over the journey, ever threw the smallest gloom over my mind, till I laid Mr. Anderson in the grave.'[5]

Park and Anderson both studied medicine at the University of Edinburgh. Park travelled the forty or so miles north to matriculate in the autumn of 1788, a year ahead of Anderson. Park's brother Alexander was at that time studying law in Edinburgh. It was an auspicious time to be a student, in Edinburgh especially. Although Glasgow, Aberdeen and even Perth were centres of enlightened thought, Edinburgh was Scotland and Europe's 'hotbed of genius' as Scottish novelist Tobias Smollett phrased it (Figure 2.2). The university was distinguished by its many leading figures in the arts, in medicine, and in natural and moral philosophy. The town's civic culture had numerous literary clubs and debating societies. Together with other learned bodies, the Medical Society and the Philosophical Society of Edinburgh (from 1783 the Royal Society of Edinburgh), actively promoted the common good: everywhere, and virtually everyone in polite and literate circles, had regard for moderate religion and an interest in critical reason.[6]

Park's biographers vary in their treatment of Park's university years. Early commentators tend to pass over them. Lupton simply records that 'Park applied himself quietly and diligently to his studies' – a view expressed as early as 1835 by one biographer – while noting, and then only briefly, that as a student of medicine, Park was taught amongst others by Alexander Monro Secundus (the second medical Monro in a dynasty of three), that he was interested in botany, took classes in chemistry, and was a regular but not heavy borrower of books from the university library. Recent biographical work has revealed more about Park's university studies – and what he did outside the lecture hall and anatomy theatre – and so has afforded a fuller picture of Park in his formative years.

Only a small proportion of the medical students who matriculated at the University of Edinburgh at the end of the eighteenth century took, or could afford to take, the full diet of seven courses that regulations required for a degree in medicine: anatomy and surgery, chemistry, botany, *materia medica* (the study of the origins and properties of substances used in medical practice) and pharmacy, medical theory, medical practice, and clinical lectures.

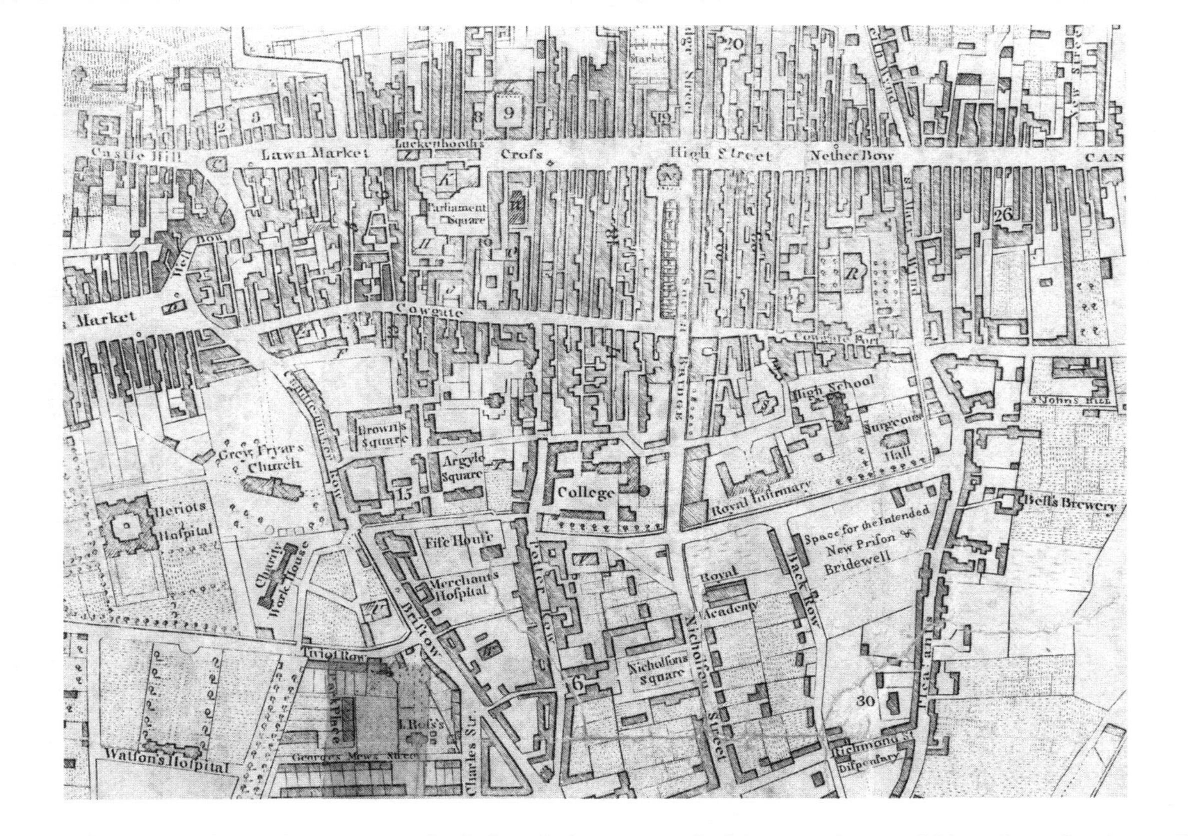

Figure 2.2 This detail from Daniel Lizars' 1787 map of Edinburgh shows several of the sites that would have been familiar to Park as a student at the 'Toun's College' as the University of Edinburgh was then known: in the centre the 'College', to the east the Royal Infirmary, Surgeon's Hall, and Bell's Brewery. Source: D. Lizars, 'Map of Edinburgh' (Edinburgh, 1787). With permission of the WS Society.

We know that between 1788 and 1791 Park took Joseph Black's course in chemistry, that of Alexander Monro Secondus in anatomy and surgery, and those of Andrew Duncan on medical theory and medical jurisprudence, James Gregory on medical practice, and Daniel Rutherford on botany.

Park also took John Walker's natural history class, being one of twenty-two students listed on 18 April 1791 for the forthcoming session. Walker normally had more students: class lists for 1786 and 1789 list thirty-eight and thirty-five students respectively, a fact which was important to Walker at least since professors had to supplement their basic salary from students' class fees. The reduced natural history class of 1791 may reflect the fact that Walker was a churchman (he was minister of Colinton parish near Edinburgh as well as the university's professor of natural history) and in 1790 had been Moderator of the General Assembly of the Church of Scotland. One year on, Walker's involvement in ecclesiastical politics may have made him something of an unknown quantity to intending natural history students. Fewer than usual perhaps, Park's natural history classmates were nevertheless international in origin, his fellow medical students and 'students of physic' coming from Geneva and South Carolina as well as from England and Scotland.[7]

The subjects which Park was taught were not then as discrete as they are today. Walker was fascinated by mineralogy, lectured in geology and oceanography as part of his natural history, and was a first-rate botanist. From Walker's class in natural history, Park would also have understood chemistry as a useful and applied science. From Rutherford as well as from Walker he would have taken a deep interest in botany, understood as the variety and utility of plants and their geographical distribution as well as the binomial system of classification recently proposed by the Swede Carl Linnaeus.[8] Walker also led field days for his natural history students, to the nearby Pentland Hills and to the seashore at Granton and Cramond. Instruction and learning took place in Black's laboratory, in the anatomy theatre, in the open air, and through reading in one's lodgings.

In these various settings, the student Park was exposed to up-to-date thinking in the form of comparative and critical empiricism, advances in theoretical reasoning, the power of acute practical observation, surgical intervention, and medical etiquette (such as it was), and the importance of systematisation and classification in natural knowledge. It is likely that he would have understood, perhaps from exposure to James Hutton's work in geology, that explanation of the age of the earth no longer rested on scriptural foundations. Walker's teaching by contrast laid great emphasis upon natural diversity as divinely ordained.

Park did not graduate in medicine – there was no obligation to do so, and it is possible he had no intention of doing so. But as the beneficiary of specific medical training, in anatomy and in surgery especially, he also received a good grounding, at once philosophical and practical, in what today would be understood as the natural and the earth sciences.[9]

Park extended what he was taught formally by participating in Edinburgh's Enlightenment culture in gown and in town. He was a member of the Chirurgo-Physical Society, a student-led medical society founded in 1788. This met at 7.00 p.m. every Friday evening between November and April to discuss dissertations on medical topics – papers orally delivered and later transcribed by the secretary. This was no drink-fuelled and rowdy student gathering. Attendance was strictly monitored, and fines were issued for non-attendance or other misdemeanour: one regulation stated that 'No excuse shall be admitted for non-attendance, except sickness, attendance at a medical lecture, or delivery, or absence more than two miles from town.' Park was an 'Extraordinary Member', admitted to that position on 12 February 1790 (Alexander Anderson was admitted an ordinary member on 26 November 1790). Park delivered three papers to the society in the period 1789–91, on jaundice, ringworm, and scurvy (the last two in April 1790).[10]

Outside this world of learning, we catch glimpses of Park's social life and his efforts in public speaking and at authorship, from letters – nineteen in all – written to a William Laidlaw between 16 November 1790 and 14 June 1791, and from several poems written by Park. Like Park, Laidlaw was from a farming background in the Borders. For the most part, the Park–Laidlaw correspondence involves Selkirk gossip and the exchange of news: Park's concern to manage his reputation following a ball near Selkirk in the autumn of 1790; Park's hangover following a good lunch on Christmas Day 1790 – and managing despite it to attend Joseph Black's class the next day; his concern at witnessing a serious accident on a building site near the university; and, in January 1791, getting himself 'pretty Canty' with friends [the worse for drink], and escorting a party from Selkirk to a dinner in Leith, accompanying them to the circus in Edinburgh and entertaining them in his lodgings at the end of it all. He visited the theatre in Edinburgh twice in February 1791, once to a comedy, once to *The Fall of Captain James Cook*, a serious pantomine which was, reported Park, 'received with unbounded applause . . . the part of Captain Cook grandly done'. Park would no doubt have been incredulous had anyone then told him that fifty years later he would himself be the subject of theatrical performances and, later still, of biographies that placed him alongside Cook in the pantheon of Great British Heroic Explorers.[11]

In mid-February 1791, Park did perform in the Pantheon, the Edinburgh debating society of that name, as part of a public debate on the subject 'Whether is the Theatre or the Circus most worthy of public notice'. Although we do not know his views on the matter, we do know that Park wrote in verse and spoke – as he reported it to Laidlaw – 'with universal applause so much indeed that I had to stop a considerable time after each verse in consequence of the ruff' [the drumming of heels as an expression of approbation]. He spoke again in the Pantheon in early March 1791 on the subject 'Is poverty, profiligacy [sic] or deformity the greater bar to marriage' and probably did so again in mid-April that year when the subject concerned the respective merits of the poets Allan Ramsay and Robert Fergusson.[12]

Six specimens of Park's own verse survive, three of which were published in his lifetime, another being printed posthumously: it is likely there were more. The outlet for Park's published verse was *The Bee; Or, Literary Weekly Intelligencer* which was published in Edinburgh between 22 December 1790 and 21 January 1794. Its founder editor and principal contributor was Dr James Anderson. Anderson embodied the intellectual range of the Enlightenment in Scotland. He was a member of the Philosophical Society of Edinburgh and of other learned bodies, an agriculturalist (and inventor of a new plough type), a correspondent of American politician George Washington, and a political economist. His writings on rent, timber production, and animal domestication engaged contemporaries including Jeremy Bentham and Thomas Malthus and, later, Karl Marx and Charles Darwin.

Under Anderson's direction, *The Bee* provided space for publications of different type and varying subject. Early issues carried extracts from John Ledyard's letters to the African Association and, later, a memorial verse in his honour. In December 1791 in an essay on 'Hints respecting the study of Geography', Anderson advanced a scheme for the teaching of 'philosophical geography' as an aid to 'accurate physical knowledge of the globe' and to 'the improvement of mankind'. Park would almost certainly have read it and the tributes to Ledyard.

Considerable space in *The Bee* was given over to literary works, including poetry. Park included poetry in his correspondence with Laidlaw, both as finished verse in the form of an untitled epitaph to Tibby Thomson, a Borders woman who had committed suicide, and as unfinished fragments in Latin. His poem 'On the Death of William Cullen, M.D.', which was published in *The Bee* on 15 June 1791, is a more sophisticated and ambitious piece. In elegiac terms, Park lauds William Cullen who was, from 1766 to 1789, professor of medicine at Edinburgh University, and James Anderson's lifelong friend. Cullen had been an outstanding teacher and a distinguished physician. Park's poem in praise of someone he never knew may have been written to curry favour with Anderson by producing in verse a tribute to match that of Anderson's encomium in prose.[13]

Park's poem in honour of Cullen was dated 4 June 1791. His letter of that date to Anderson is headed 'Braes of Yarrow' (itself the title of a poem by Scottish poet John Logan). Park had by then returned to the Borders, having finished his university studies with Walker's class on natural history. Surviving letters give no indication of Park's plans, and we do not know what Park intended to do. In all likelihood, neither did he.

If his thoughts were to use his medical training, we are given no clues as to the direction – he perhaps again assisted Thomas Anderson in Selkirk. While it is probable that he did not see himself as a Borders farmer in the longer term, his return to Foulshiels may have been to help with the farm: it was the summertime and his father was then in his mid-seventies. A subscriber as well as a contributor to *The Bee*, Park would have read or perhaps learned from other sources of Ledyard's fate – no one then knew anything for certain of Houghton's circumstances – but it is unlikely that he entertained thoughts about exploring the Niger. Then as now, however, the course of one's life can take an unexpected direction – not from what one knows, but from whom.

Botany, natural history, patronage: Scotland, London, Sumatra, 1791–1795

In addition to his own abilities, his presidency of the Royal Society and a leading role in the African Association, and the credibility that came from having explored the Pacific world with James Cook, Joseph Banks's status as the pre-eminent scientific statesman of late eighteenth-century Britain owed much to a wide network of like-minded associates. Among those who shared his botanical interests was James Dickson.

Dickson was born, like Park, in the Scottish Borders, in 1738 in Traquair, a small settlement about ten miles west of Foulshiels on the Minchmoor drove road between the valleys of the Yarrow and the Tweed (see Figure 2.1). As a young man Dickson moved to London, initially as a nurseryman gardener and by 1772 had established himself as a seedsman in Covent Garden. With Banks and others, Dickson co-founded the Linnean Society in 1788. Given access to Banks's library and with connections to the Royal Society, Dickson became a leading authority on British cryptograms (plants which reproduce by spores, such as mosses and ferns). He was, like most botanists in his day, an avid field collector and taxonomist. He knew John Walker, Park's natural history professor, and passed specimens of saxifrage to him in 1792.[14] Dickson undertook several trips to Scotland, large tracts of which, as he noted in addressing the Linnean Society in February 1793, 'are still unexplored by any naturalist'.

This statement is only partly true. John Lightfoot's *Flora Scotica* (1777) had already done much to identify Scotland's native flora and to classify it on Linnean principles, and a decade or so earlier, before he took up his professorship at Edinburgh, John Walker had conducted a survey of the Highlands, including its economic botany. Dickson acknowledged as much. But as he also stated, further 'botanical riches' remained to be discovered, in the Highlands in particular. Dickson made several plant-hunting trips to the Highlands, in 1786, 1789, and 1792. In his talk to the Linnean Society on 5 February 1793, Dickson enumerated the species found on his trip of 1792. He did not, however, name his travelling companion and fellow collector although several months earlier, on 20 November 1792, he had successfully nominated him for associate membership of the Linnean Society. By then, Mungo Park was more than Dickson's botanical associate and nominee. He was his brother-in-law, Dickson having married Margaret Park in 1786.[15]

By the late autumn of 1792, one year or so after completing his studies, Park had common interests with a leading botanist, had derived further field experience through plant collecting and travel in the Scottish Highlands, and, perhaps most importantly and thanks to Dickson, had direct acquaintance in London with a potential patron, Joseph Banks. London, as Park wrote to Laidlaw on 31 October 1792, was all 'bustle and confusion'.

In that letter, Park reflected upon his circumstances in ways which suggest both a certain wistfulness at being distant from Scotland and friends, and a willingness to accept whatever might lie ahead: 'I look forward to the bustle of life with indifference sensible that all things are guided by a superior hand and tho absent from friends, by his blessing we can converse and by his gracious presence be happy tho all the powers of this world should combine against us. I cannot tell in what link of the chain providence has placed me but wherever it is I hope I shall say thy will be done.'[16]

The hand that next guided him was strongly secular. Banks's extensive connections included leading personnel in the East India Company, and, through them, Banks secured a position for Park as assistant surgeon on one of the company's ships, the Sumatra-bound *Worcester*, early in 1793. This was probably a direct result of Banks's links with William Marsden, the orientalist, linguist, and member of the African Association, whose *History of Sumatra* (1783) was based on his residence at Fort Marlborough, the company's trading station at Benkulen on Sumatra, between 1770 and 1778.

Before he could depart, Park was required to demonstrate his medical competence in an oral examination at Surgeon's Hall in London. Writing to William Laidlaw on 2 February 1793 as he was completing his preparations for the voyage, Park asked his friend to tell Alexander Anderson that 'he need not be afraid of Surgeons hall' and that he too could have secured a place on an Indiaman given the 'scarcity of mates'.

Marsden had been much taken by Sumatra: it was, he wrote, 'surpassed by few in the bountiful indulgences of nature'. He saw his book about the island as an opportunity 'to throw some glimmering light on the path of the naturalist' and to provide philosophers 'with facts to serve as *data* in their reasonings'. Marsden's focus was the island's political history and ethnic diversity to the neglect of its natural history, the zoology particularly. This left room for others. Both Banks and Dickson probably prompted Park on what and how to collect natural specimens there to supplement Marsden's work. Accompanied for part of the way by two 74-gun naval vessels – Britain being then at war with France – the *Worcester* and its sister ship *Minerva* left Gravesend on 18 February 1793, and Portsmouth on 5 April. The *Worcester* was a relatively new vessel, launched in 1785, and under the command of Captain John Hall. The ships anchored off Benkulen on 21 August, but this was a commercial voyage, not oceanic exploration, and by mid-September, preparations were underway for the return voyage, with a cargo of pepper, arrack, and sugar together with cargo for the private trade of the captain and the ship's officers. They were underway by late October. Calling at St Helena on the return voyage, the *Worcester* returned to Gravesend on 2 May 1794.[17]

If Park kept a journal of his voyage on the *Worcester* and experiences in Sumatra, it has not survived. We do not know Park's exact responsibilities as assistant surgeon on ship and shore, nor how he spent his days in Benkulen. Park probably encountered instances of ringworm and scurvy amongst the crew and so would have been able to put to practical use what he had said about them as the subject of two of his dissertations to the Chirurgo-Physical Society. We do know that on the outward voyage he read the first volume of Dugald Stewart's *Elements of the Philosophy of the Human Mind* (1792). Dull reading for an ocean voyage perhaps, but Park had taken it, as he informed his brother, 'to amuse me at sea'. Stewart's attention to the importance of perception and the role of memory and imagination may have struck a chord, as may Marsden's attention to facts as the basis to reasoning.

Both the region and the East India Company's trading station in Benkulen were unhealthy, so Park would almost certainly have encountered dysentery, fever, malaria and other tropical diseases. We know that he had time to botanise and to zoologise. Upon his return, Park presented several specimens of plants and drawings of fish to Banks (Plate 2). Addressing the Linnean Society on the Sumatran fish on 4 November 1794, six months after returning from his voyage, Park admitted that his paper was 'the fruit of my leisure hours during my nine weeks stay' and that, this being his first attempt at ichthyology, 'the descriptions may in many places be inaccurate'.[18] If they were inaccurate, no one was in a position to correct him.[19] The more important point was that Park had proven, to himself and to others, that he could overcome the rigours of long-distance travel and residence in tropical climes to produce new knowledge about places not well understood by Europeans.

Park's Sumatra trip was a precursor to his Niger journey, but it was not a rehearsal for it. Park's world had changed, his horizons broadened, but his future was in no sense determined. The death of his father, in May 1793 at the age of seventy-nine, which Park only found out about upon his return to London and to the Dicksons in May 1794, was a blow, albeit not entirely unexpected. But upon his return, Park felt dislocated. As he wrote to his brother Alexander on 20 May 1794, 'The melancholy news, and the sudden change, from the Bustle of a sea life to the stillness of the fire side, wrought so powerfully on my mind that I found no desire to do any thing. I remained a mere inactive lump, and stared about me as if I had come newely [sic] into the world.'

Park returned to Foulshiels in late May 1794 but stayed only a few weeks. By then it is likely that his mind was more firmly set upon his future. He wrote with the news to his brother:

> I have, however, got Sir Joseph's word that if I wish to travel he will apply to the African Association and I shall go out with thier [sic] Consul to Fort St. Joseph on the river Gambia where I am to hire a trader to go with me to Tombuctoo & back again. The Association pays for everything and will allow 200 Guineas to the trader, who goes as my guide and interpreter, when we return to Fort St. Joseph. I shall write more particularly in my next but as it is a short expedition & will give me an opportunity of coming to Scotland before I set off besides the hopes of Distinguishing myself, for if I succeed I shall aquire [sic] a greater name than any ever did! & at present I see no reasonable expectation against accepting it.[20]

Park's excitement is palpable. His excitement also exceeded his grasp of Africa's geography: the consul for Senegambia, James Willis, was a government appointment, not the Association's; and Fort St Joseph lay on the Senegal river.

To judge from the records of the African Association for 23 and 31 May 1794, and this letter to his brother, Park offered his services to Banks and to the African Association sometime between 2 and 20 May 1794. At their meeting on 23 July 1794, Banks and Beaufoy having again interviewed Park earlier that day, the Association resolved to accept his 'Services to the Association as a Geographical Missionary to the interior countries of Africa'. Park was to be paid 7s 6d per day from 1 August 1794 until he reached Africa, 15s a day thereafter from the start of his African travels until his return, 'provided the day of his said return shall not exceed Two years from the day of his departure from the Gambia'. A further £200 was allowed for expenses. He was to proceed with Consul Willis to the Gambia and, from there, 'on to the River Niger . . . that he shall entertain the course, and, if possible, the rise and termination of that River'. Park was an excellent choice being, as Banks and Beaufoy noted, 'sufficiently instructed in the use of Hadley's quadrant to make the necessary observations; geographer enough to trace out his path through the Wilderness, and not unacquainted with natural history'.[21]

Getting to Africa: delays and instructions

If Park was ready, keen for the 'short expedition' as he wrote to his brother, others were not. The Association's plan was that Willis and Park should leave in September or October 1794; James Willis having been appointed Britain's consul general to Senegambia on 10 April 1794. His plan, as he laid it out to Beaufoy and to members of the government, was to send a military detachment of fifty Europeans and to build a fort for their security and as a base from which to begin trade negotiations. This resident garrison was to be assisted by two sloops whose purpose was to protect the fort and maintain communications. Delays in 1794 in hiring and provisioning the sloops and in official circles meant that he and Park missed the optimum time to reach west Africa in the dry season. Willis overspent massively – £5,816 against an allocated £3,000. Members of the Association and of the government railed against the expense, their concern over expenditure heightened by the cost of the war with France. The whole Willis affair was a fiasco and it delayed Park's intended departure for almost ten months.[22]

Park however was not idle during 1794 as Willis's plans unravelled. He prepared and delivered his paper on Sumatran fish to the Linnean Society that November. In early October, he received written instructions in the use of the sextant, and perhaps too in Hadley's quadrant, from Nevil Maskelyne, the Astronomer Royal. Maskelyne was the author-compiler of the annual *Nautical Almanac and Astronomical Ephemeris* which, in giving tables of lunar distances from the centre of the moon to selected stars or to the centre of the sun, could be employed with a sextant and by use of trigonometry to calculate latitude and longitude and so determine the observer's geographical position relative to a known prime meridian. Whether Maskelyne, resident in Greenwich, gave Park hands-on training, Park then staying in Covent Garden with his brother-in-law and wife at their seed shop, is not known.

Maskelyne certainly offered detailed guidance – 'You should have a watch with a second or moment hand. Perhaps you would not chuse [sic] to carry out so valuable a time keeper as Mr Earnshaw makes' (referring to the high-quality watches produced by London instrument maker Thomas Earnshaw). Maskelyne expected nevertheless that Park 'will doubtless take out the nautical almanac and requisite tables' and he ended his instructions in support of Park's venture: 'I wish you success in your laudible [sic] pursuit after knowledge, and shall be glad to hear of it. Send home your observations by every opportunity.'[23]

Park read what he could on west Africa and the Niger while he waited – Jobson's *Golden Trade*, Moore's *Travels into the Inland Parts of Africa*, possibly also al-Idrisi in translation in Cory's *Geographical Historie of Africa*, perhaps even re-reading Guthrie's and Salmon's geographical gazetteers. He consulted such maps as were available. The result, of course, would only have been to sow confusion given the contradictory claims made about the Niger and the general ignorance about west Africa. The Association issued him with the same detailed instructions they had given to Houghton.

Park's most important guidance came from Britain's leading geographer, James Rennell. Rennell had a life and geographical career in two parts. Between joining the Royal Navy in 1756 aged fourteen and until 1777, he was, successively, naval officer, explorer, and surveyor for the East India Company in Bengal. His *Bengal Atlas* (1780) and later Hindoostan maps set new standards in their exactness and precision – remarkable achievements, in fact, given that in 1766, Rennell had been attacked by bandits while surveying and never regained the full use of his right arm. From 1778, in London and in Brighton, and in contrast to his fieldwork in India, he lived the life of an armchair geographer, studying ancient geography, oceanography and, as an Honorary Member, acting as a geographical consultant on Africa for the African Association. Rennell would play a major part in Park's book about his first African journey. But that lay in the future. While Park was still a student, collecting plants in Scotland and, later, patrons in London and fish in Sumatra, the sedentary Rennell was at work in trying, with others, to make sense of the Niger.[24]

From his London home at 23 Suffolk Street, Rennell maintained a wide correspondence and a network of like-minded associates: Joseph Banks, William Marsden, and the hydrographer Alexander Dalrymple were among his close friends. In addition to his oceanographic and mapping interests, Rennell had long been working on the ancient and modern geography of western Asia. This involved the critical assessment of what was known by Classical authorities compared with the work of Rennell's modern contemporaries. Unsurprisingly perhaps, given the scale of what he intended and other demands upon him, this was never completed.

Only one volume of Rennell's major undertaking was published, his *Geographical System of Herodotus*, in 1800. Rennell was able to correct

Herodotus on the Niger given what Park would tell him after 1797 following his African travels: as Rennell put it, 'For, it is certain, that Herodotus had a very positive, and in some degree, circumstantial, knowledge, of the course of the river Niger; now, by the discoveries made by Mr. Park, shewn to be the same with the *Joliba*, or great inland river of Africa.'[25] As part of this work, Rennell included a map of Africa 'according to Herodotus' in which the Niger was shown running into a lake in central north Africa (Figure 2.3).

At the same time, he was preparing maps for the African Association to illustrate their slowly emerging understanding of west Africa which, in London, derived from Ben Ali and Shabeni, in Tunis from the consul and, in Tripoli, from Shereef Imhammed by way of Simon Lucas. Rennell's map appears in the first volume of the Association's *Proceedings* in 1790. It showed the routes as described by these Moorish sources, other reported routes across the Sahara, and, as an indistinct line to the south of Timbuctoo, the course of the Niger (Figure 2.4).

Rennell used his 1790 map as a base map on which to record further information, particularly that received from Houghton. Following the receipt from Banks of Houghton's news, Rennell added, in red ink in his own neat hand, Houghton's intended route compared with his actual route of travel. Rennell noted too the course of the Niger, named the 'Joliba' as it flowed north-east toward Timbuctoo, and 'Neel or Niger R' as the river turned south and east in what Rennell took as its 'supposed course'. Rennell also added in details of trans-Sahara routes and the oases and other places on those routes (Plate 4). This manuscript version was compiled by Rennell by May 1793, from Houghton's evidence of 1791 and from these other sources, in London and in Africa. In a related manuscript, we are told that the committee of the African Association had received, 'thro the medium of distinct and unconnected channels, new and interesting intelligence'. All this was used by Rennell, acting with Henry Beaufoy, to make the additions to his 1790 map. The resultant map was then sent by Beaufoy to Frederick Augusta Barnard, the King's Librarian, with an explanatory note:

> I have at last obtained from the Draughtsman, & have now the pleasure to send you a Map by Major Rennell of the late discoveries of the African Association. Those discoveries are distinguished by Red Lines from the content of their knowledge in 1790, when they published, or rather printed for the use of their Members a narrative of their Proceedings. The Memoir which accompanies the Maps [sic] is intended as an explanation of the means by which their late Intelligence was obtained, & of the important commercial advantage which that Intelligence is likely to afford.[26]

Figure 2.3 Africa according to Herodotus. James Rennell has marked the Niger as 'Western River of Herodotus' with an arrow indicating the presumed west–east direction of flow. Source: J. Rennell, *The Geographical System of Herodotus* (London, 1800). Reproduced by kind permission of the National Library of Scotland.

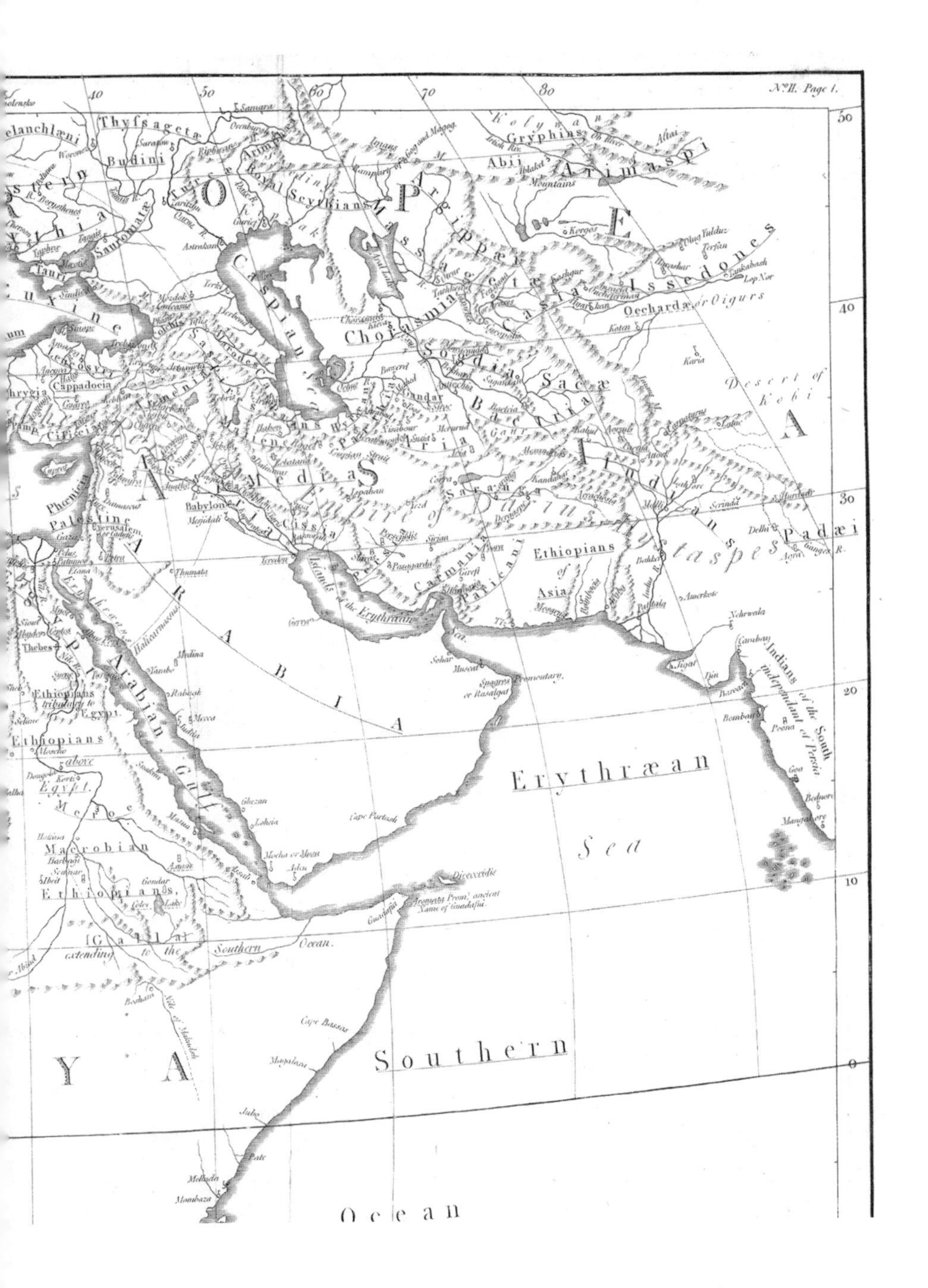

N°II. Page 1.
40
50
60
70
80
Thyssagetæ
Budini
Sauromatæ
Caspian Sea
Royal Scythians
Massagetæ
Gryphins
Arimaspi
Abii
Issedones
Oechardæ or Oigurs
Casia
Chorasmia
Sogdia
Sacæ
Bactria
Aria
Cappadocia
Cilicia
Armenia
Media
Phœnicia
Palestine
Babylon
Cissia
Persepolis
Carmania
Paricani
Ethiopians of Asia
Indians
Empire of Darius Hystaspes
Padæi
Desert of Kobi
Arabia
Arabian Gulf
Egypt
Thebes
Ethiopians
Meroe
Macrobian Ethiopians
Mecca
Medina
Mocha or Moca
Aden
Muscat
Erythræan Sea
Southern Ocean
Bombay
Goa
Delhi
Cambay
Indians independant of the South of Persia
Cape Bassas
Mombaza
Melinda
Pate
Jerusalem
Damascus
Ispahan
Kandahar
Kabul
Attock
Lahore
Multan
30
20
10
0

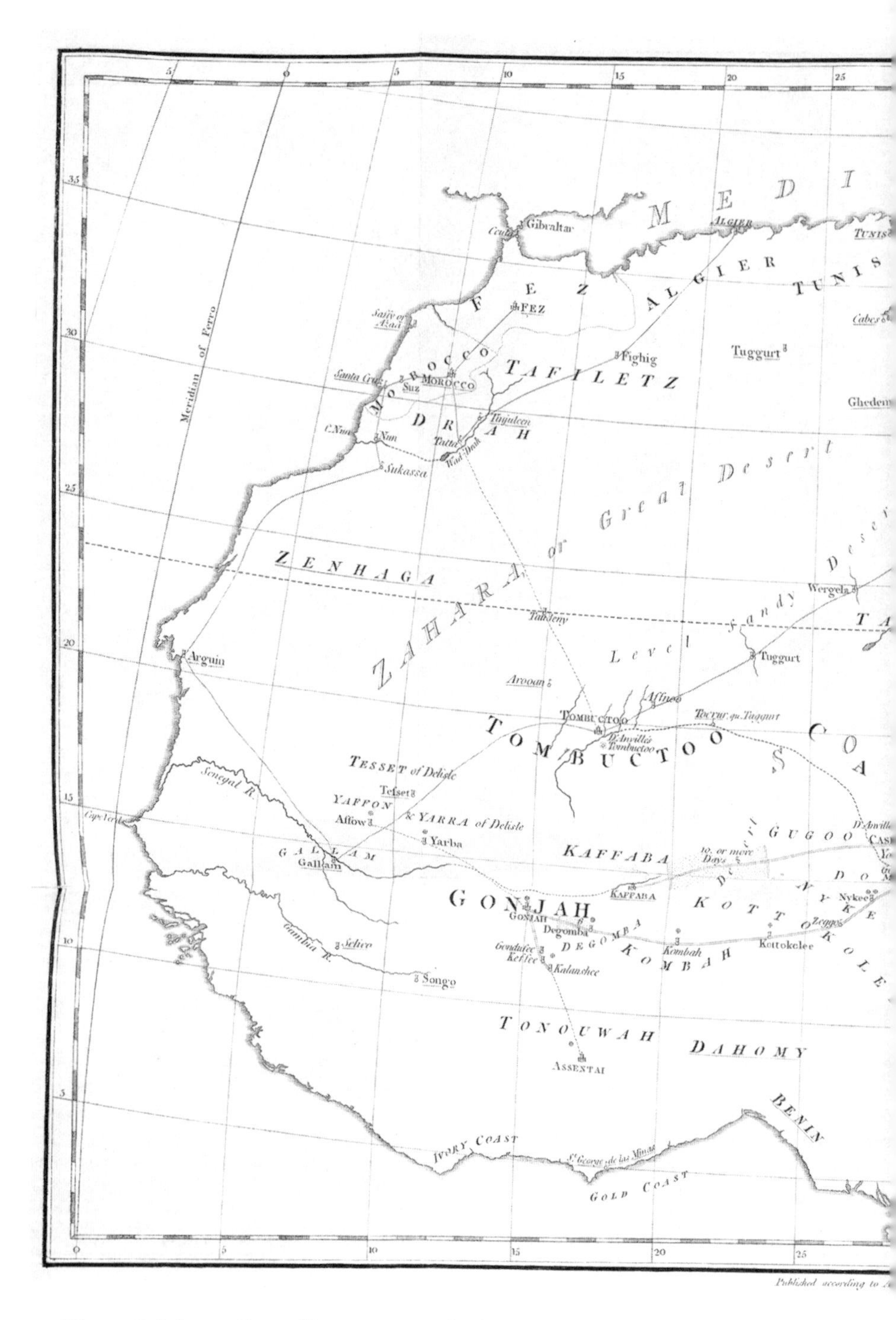

Figure 2.4 James Rennell's 1790 map of Africa shows the information gathered by then by the work of the African Association – rather little (*cf.* Figure 1.2). Source: J. Rennell, 'Sketch of the Northern Part of Africa', in *Proceedings of the Association for Promoting the Discovery of the Interior Parts of Africa* (London, 1790). Reproduced by kind permission of the National Library of Scotland.

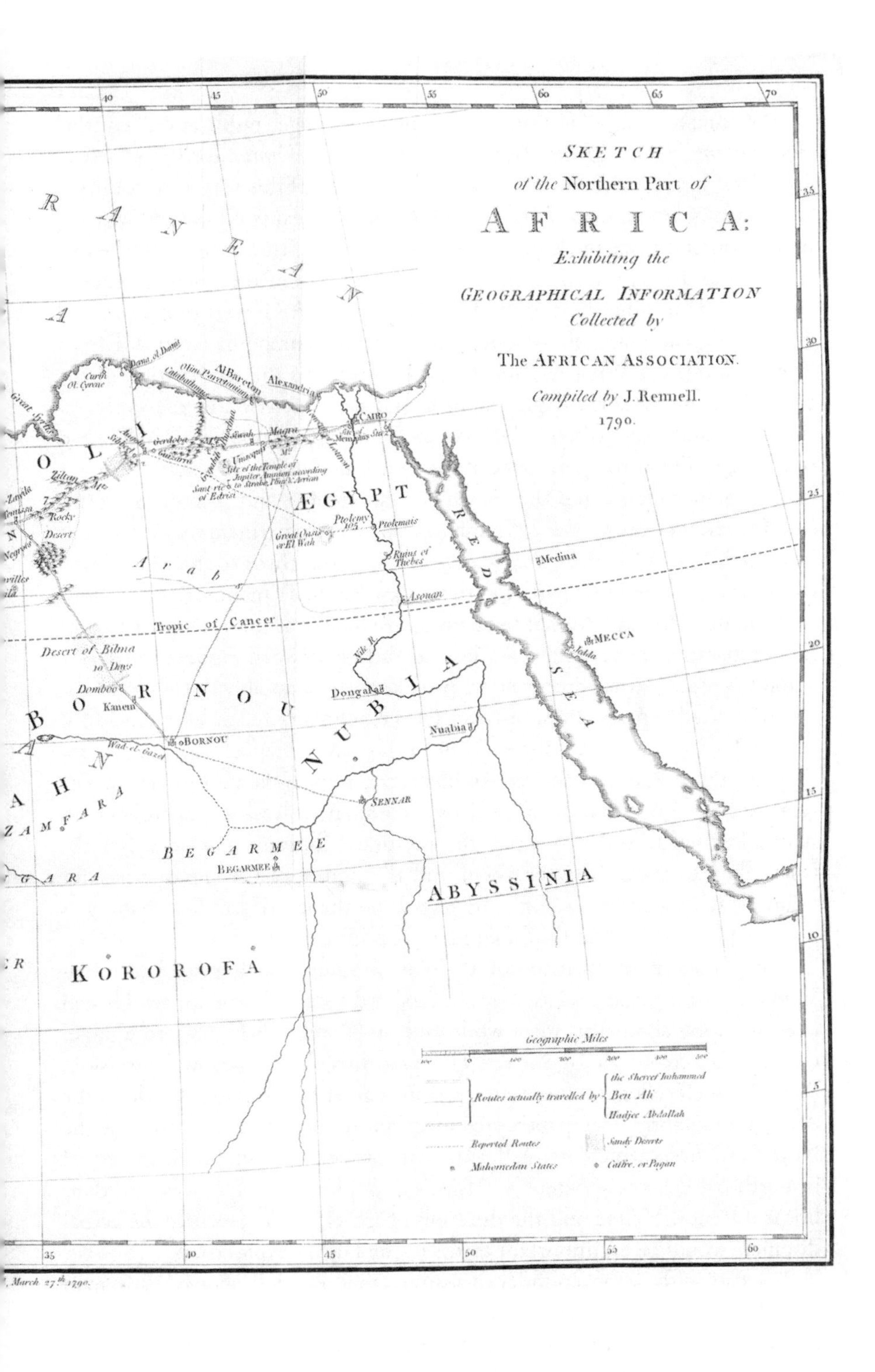
SKETCH
of the Northern Part of
AFRICA:
Exhibiting the
GEOGRAPHICAL INFORMATION
Collected by
The AFRICAN ASSOCIATION.
Compiled by J. Rennell.
1790.
Alexandria
CAIRO
ÆGYPT
RED SEA
Medina
MECCA
Jedda
Ruins of Thebes
Asouan
Ptolemais
Great Oasis or El Wah
Tropic of Cancer
Desert of Bilma
10 Days
Dombo
Kanem
BORNOU
Wad-el-Gazel
NUBIA
Dongala
Nuabia
SENNAR
BEGARMEE
ABYSSINIA
KOROROFA
ZAMFARA
Arabs
Rocky Desert
Geographic Miles
Routes actually travelled by
the Shereef Imhammed
Ben Ali
Hadjee Abdallah
Reported Routes
Sandy Deserts
Mahomedan States
Caffre, or Pagan
March 27th 1790.

The memoir in question is a thirty-one-page publication entitled *Elucidations of the African Geography; from the Communications of Major Houghton, and Mr Magra; 1791.* Textual memoirs of this sort were common accompaniments to maps in the Enlightenment, to explain how and why the map in question was put together. We are told that 'These accounts, the one collected at Tunis, the other in the neighbourhood of the river of Senegal, agree in determining the course of the Niger to be from west to east; as well as in fixing the place of its source at a very distant point westward from Tombuctoo [sic].' What remained to be known was the termination of the Niger: 'Having in some degree, approximated the position of the source of the Niger; and also ascertained the course be from west to east; our next head of inquiry will naturally be, concerning its course, and final exit.'[27]

This map evidence and the related correspondence and memoir is significant for several reasons, the 1793 manuscript version particularly. Apart from brief mention of Rennell's *Elucidations* by Lupton, these sources have been overlooked in studies of Park and the Niger. Rennell's manuscript additions to his map of Africa have not been noted. Rennell was using maps as part of his comparative methodology – Classical authorities with modern testimony – and to correct earlier European map-makers, most notably d'Anville, about whom he had expressed doubts over the Frenchman's reliance in the 1740s upon Classical authorities for want of up-to-date evidence. The manuscript map illustrates Rennell's concern to incorporate up-to-date information. The map is best understood as a kind of accumulative working document on which Rennell recorded not just the presumed location of places, but the time taken to travel between them and the nature of the terrain (see, for example, the statement '10 or more days', and the use of stippled shading as a symbol for 'sandy desert': *cf.* Figures 2.4 and Plate 4).

Independently of his map but as a contribution to it, Rennell had calculated these time distances based on the reported speed of camel travel. He had been thinking about this for a while and, in March 1791, devoted a paper on the topic to the Royal Society. His concern was accuracy and precision. What Rennell called 'the hourly rate of the camel . . . as a useful scale to the African geography' (his paper presented the results of camel travel in the Near East, from which Rennell hoped to provide insight for desert travel in north Africa) was apparent as a time geography between places, not their linear distance.[28] Time and the duration of travel, not necessarily the actual distance, would be an important factor in the Niger's exploration.

We can with some confidence assume that Rennell showed this map material and the memoir to Park as the most up-to-date evidence – almost certainly this manuscript version with his hand-written remarks on Houghton's route and the statement that the Niger flowed eastwards. Beaufoy had earlier used maps in this way to inform Ledyard about his projected route: Beaufoy recounted on meeting the American how he had 'spread the map of Africa before him, and tracing a line from Cairo to Sennar, and from thence Westward in the latitude and supposed direction of the Niger, I told him that was the route, by which I was anxious that Africa might, if possible, be explored'.[29] The map in question was the one published in 1790 (Figure 2.4). There is every reason to suppose, as Park waited for Willis and confirmation of his departure, that Rennell, probably with Beaufoy in attendance, showed Park his 1793 manuscript map and memoir in much the same way and to the same purpose.

It is reasonable to suggest then that for some months before he left England, Park not only had a sense of the likely times and distances involved in Africa's exploration, but also something of what he might expect of the physical environment to be faced there. It is also likely that he already 'knew' that the Niger flowed west to east because he had been told it did by Rennell. In one sense, his task was less exploration, more one of confirmation of what Houghton had reported. In another, it was millenia old: where did the Niger end?

3

'Geographer enough'

Park's first journey to the Niger, 1795–1797

What we know of Park's significant achievements in relation to his first African travels we know largely from his own words after the events themselves. To judge from what survives in the written record, Park was a faithful, even a diligent, journal keeper: his journey and experiences are narrated in chronological order more or less – though not for every day given the circumstances in which he found himself.

Park's exploration of the Niger was successful: he determined the course of that river from first-hand observation. His narrative about his exploration also makes clear that his travels were intermittent and interrupted. For different reasons and at different times, they involved immobility, an ailing body, and the actions of others, kindly and otherwise. His African journey was never unhindered, a safe and easy passage over foreign terrain with, at journey's end, a text guaranteed – and easily produced – as a record of one's travels. Quite the reverse, in fact.

In Africa

Park landed in west Africa, at Jillifree on the north bank of the Gambia river, on 21 June 1795, having left Portsmouth on the *Endeavour* on 22 May. Two days later, he crossed the river to Vintain – 'a place much resorted to by Europeans, on account of the great quantities of bees-wax which are brought hither for sale' – and, from there, made his way to Jonkakonda by early July and thence to Pisania and the home of Dr John Laidley. Both Jillifree and Pisania have now disappeared, but in Park's day they existed to serve the slave trade. At Pisania, Laidley and two brothers by the name of Ainsley enjoyed 'perfect security under the king's protection' and, Park noted, 'the greatest part of the trade in slaves, ivory, and gold, was in their hands'.[1]

Park's Niger exploration began with his staying put for nearly six months. One reason for this was because he was waiting for James Willis – who would never come: his appointment was cancelled in February 1796 and eventually annulled. As Park waited, he took pains to instruct himself as to what was necessary for the task ahead: further in-situ training to complement what Maskelyne had taught him in London and what he had been told by Rennell and Beaufoy. He learnt Mandingo, the language in general use throughout that part of west Africa and was helped in this by Laidley who was fluent. Park sought information on the regions he intended to visit, initially from the *slatees*, free black merchants who traded with the interior, but he came to distrust what they told him 'for they contradicted each other in the most important particulars, and all of them seemed extremely unwilling that I should prosecute my journey'. Both things, noted Park, 'increased my anxiety to ascertain the truth from my own personal observations'.

Park was never what one modern anthropologist has termed the 'chameleon fieldworker', able to blend in with the indigenous peoples he lived with and travelled among, not least because of the colour of his skin and his decision always to wear European clothes, even the heavy nankeen breeches he had travelled in from Britain. But his decision to learn one African language, to make enquiries of those familiar with the regions through which he intended to travel and, in effect, to recognise his own deficiencies and to remedy them before setting out all stood him in good stead.[2]

A further reason for Park's immobility was illness. On the evening of 31 July, while taking longitude readings, Park caught a chill. He was laid low for much of August with what he called 'a smart fever and delirium'. In his convalescence, Park took local walks to acquaint himself with the region and its peoples. These excursions may have acquainted him further with local peoples and the environment, but they afforded only brief respite from his fever. On 10 September, he was laid low again, for a further three weeks. The weather conditioned what was possible: 'I was able, when the weather would permit, to renew my botanical excursions; and when it rained, I amused myself with drawing plants, &c. in my chamber.' Many of these sketches survive (Plate 3), and reveal Park's botanical knowledge, a precise eye, and skilled draughtsmanship – techniques which he may have been taught by Rutherford and Walker in Edinburgh, perhaps honed with Dickson in Highland Scotland, and developed in Sumatra in the drawings later presented to Banks.

Laidley's companionship helped Park's recovery in what, in its heat and its sounds, was an alien environment:

> his company and conversation beguiled the tedious hours during that gloomy season, when the rain falls in torrents; when suffocating

> heats oppress by day, and when the night is spent by the terrified traveller in listening to the croaking of frogs, (of which the numbers are beyond imagination,) the shrill cry of the jackal, and the deep howling of the hyaena; a dismal concert, interrupted only by the roar of such tremendous thunder as no person can form a conception of but those who have heard it.

West Africa may have sounded strange to his ears, but Park – at first anyway – felt at home in the countryside around Pisania. His description of it captures both a sense of place in European terms and something of his feelings on its latent possibilities: 'The country itself being an immense level, and very generally covered with woods, presents a tiresome and gloomy uniformity to the eye; but although nature has denied to the inhabitants the beauties of romantic landscapes, she has bestowed on them, with a liberal hand, the more important blessings of fertility and abundance.'[3] His views on the different landscapes he encountered were often tempered by the heat, his mood, and, often, his health.

In going down with fever, recovering, only to suffer a further bout and to recover from that, Park was experiencing both the rhythms of tropical illness, almost certainly malaria whose insect-derived cause was not then understood, and 'seasoning' himself as contemporaries understood and used the term, his body and mind becoming accustomed to the climate and seasons of the tropics. There is no record of Park suffering thus during his nine weeks in Sumatra: had he, it is possible that he was inured against the worst effects of fever in west Africa.

Illness, of body and of mind, was widespread among Africa's explorers, the Niger's exploration included, and what was described as 'fever', most commonly malaria, was a common cause of mortality. We do not know if Park was weakened by his weeks of illness in Pisania or, to the contrary, better acclimatised – there is no evidence either way, and Park is silent on the matter. Park's experiences during his first Niger journey broadly divide between the well Park getting more and more frail on his outward trip, and the sick Park slowly recuperating on the return journey. His experience would be repeated, often with fatal consequences, by many of those who followed him.[4]

As he convalesced, Park listened and observed. Park may have distrusted what the *slatees* told him, but like other intending travellers, he was at least in principle dependent upon them or other merchants since their coffle, camel caravans, afforded the safest access to the African interior. These trade caravans moved in the dry season, between late November and April, and only then if sufficient goods had been accumulated. Travel was not easy at the best of times, the landscape a mix of scrub vegetation, dense jungle in which the path was hard to discern, and clearer stretches in which the full heat of the tropical sun beat down.

He watched the waters of the Gambia recede throughout October 1795 and into November – 'at first slowly, but afterwards very rapidly'. At one point, the river's waters were fifteen feet above the high-water mark of the tide. As the rainy season dissipated, Park felt restored: 'I recovered apace, and began to think of my departure; for this is reckoned the most proper season for travelling.' But, being suspicious of his travelling companions – 'the characters and dispositions of slatees, and the people that composed the caravan, were entirely unknown to me' – and because no date had been indicated as to when the caravan would depart, Park resolved to set off without them. Laidley, Park tells us, 'approved my determination'. Whether or not Laidley also cautioned him of the dangers and, if he did, whether a clearly determined Park listened, is not known. His decision speaks to a certain determination, even a foolhardiness: this was not an expedition but a personal journey. Park set off in search of the Niger, and to address the African Association's other queries, on 2 December 1795.[5]

Park did not travel alone. Upon departure he was as he put it 'provided with a Negro servant' by Laidley, by name Johnson, who spoke English and Mandingo following several years spent in England upon his release from slavery. In fact, Johnson was Park's interpreter, formally hired on a month-by-month contract, payment in his absence going to his wife. Park's servant was Demba, also provided by Laidley, who spoke the language of the Serawoollies as well as Mandingo. Others in the party included Madiboo, a free man, two unnamed *slatees* of the Serawoollie people, and Tami, a native of Kasson, who was returning there having been a blacksmith in Laidley's employ. Laidley and the Ainsley brothers initially accompanied the party, but they left Park in mid-afternoon on 3 December. Their departure gave him pause for thought: 'I had now before me', Park later reflected, 'a boundless forest, and a country, the inhabitants of which were strangers to civilized life, and to most of whom a white man was the object of curiosity or plunder.'[6]

Park encountered civilised life in many respects, just that it was different from that with which he was familiar. On 11 December 1795, he spent a pleasant evening at a *neobering*, or wrestling match, a form of sport and entertainment very common in the Mandingo countries. Presented with a liquor while watching the wrestling, Park thought it tasted 'so much like the strong-beer of my native country (and very good beer too)' and was surprised to learn from his hosts that it was made from malted corn, much as barley was malted at home. His good mood was somewhat spoilt the next morning on discovering that one of the men he had hired to secure water for their travels had made off with the money Park had given him, without filling the calabashes with water.

His musings about becoming an object both of curiosity and of plunder were prescient. At first, local kings and other rulers welcomed him. All expected tribute. The passage of merchants and traders through their territories required a tax or other payment, usually in goods. Some kings provided guides to assist Park, within their territory and to its boundaries. On 15 December, the two Serawoollies left the party.

Travel was intermittent rather than to a regular daily pace. Moments of rest were restorative, even educational. In the town of Koorkarany, Park discussed comparative literature with the local *maraboo* or priest: he was shown Arabic manuscripts, Park in turn 'shewed [sic] him Richardson's Arabic grammar, which he very much admired'.[7] As he journeyed, Park observed local customs, agricultural products, and the geography of the countries he passed through. Whatever else it is as a pioneering work of African travel and of the Niger's exploration, Park's *Travels* is the product of a keen observer for whom most of what he saw was foreign, and so had to be reported on fully and faithfully.

Park, we should understand, was an unknown quantity. Local rulers seemed perplexed by the reasons given for Park's journey. For one, the monarch of Fatteconda, capital of Bondou, the notion of travelling for curiosity 'was quite new to him'. Another, Tiggity Sego, brother to the King of Kasson, seemed to doubt the truth of what Park told him, 'thinking, I believe, that I secretly mediated some project which I was afraid to avow'. Park stood out, not because his intentions over exploring a river seemed bizarre, even unbelievable, to indigenous people or because they had concerns – accurate as it turned out – that Park was the forerunner of later and greater European expansionism. He stood out as a lone white man with only two African companions, far from home, professing a strange reason to be there, and was an oddly dressed one at that.

Park suffered his first setback on Christmas Day 1795, in Joag, a frontier town of the kingdom of Kajaaga. Park and his party were surrounded by armed horsemen who claimed that the necessary duties due to the king on entering the town had not been paid, and as a result, their goods were all forfeit. Reasonably enough, Park explained that he had not known to do so and would be happy to do so now. Even after offering them some gold, given to him by the King of Bondou, the horsemen ransacked Park's baggage, robbing him of half his goods. Park received assistance from a nephew of Demba Sego Jalla, the Mandingo King of Kasson, who offered Park protection on the journey to Kasson.

As he gave out tribute or unwillingly lost more of his trade goods and personal possessions, Park gained travelling companions. On leaving Joag, his party comprised thirty persons. His later experiences of solitude and destitution have coloured our view of Park's undertaking. But his outward travels in December 1795 and in the first months of 1796 were often as part of a group which changed in numbers and in character as people came and went.

Park's party several times found themselves caught up in local disputes, even war between neighbouring kingdoms. Park was detained for several days at the village of Teesee in Kasson as an embassy from a neighbouring kingdom demanded the inhabitants embrace Islam or risk military intervention. Park stayed over a week in Soolo in late January 1796 awaiting news of the impending war involving the kingdoms of Kasson, Kajaaga, Kaarta, and Bambarra. Delay could not be avoided when to stay was demanded and even enforced by the local king. But as Park knew, delays also held the threat of travelling in the wrong season: 'and I dreaded the thought of spending the rainy season in the interior of Africa'.[8]

On 16 February, Park and his companions found themselves in the town of Funingkedy, whose population was in a state of nervous concern about an impending attack by Moorish bandits. This duly came, the Moors driving off their best cattle. In the course of the raid, a local lad who had tried to defend the herd was shot. Park was invited to inspect the wounded boy: 'I found that the ball had passed quite through his leg, having fractured both bones a little below the knee: the poor boy was faint from the loss of blood, and his situation withal so very precarious, that I could not console his relations with any great hopes of his recovery.' Park no doubt drew upon his medical training and advised the boy's weeping mother and the others gathered there on the best course of action: 'to give him a possible chance, I observed to them that it was necessary to cut off his leg above the knee'. This was met with horror and disbelief – 'they evidently considered me as a sort of cannibal for proposing so cruel and unheard of an operation' – and the boy was left to the care of some village elders whose sole treatment was to insist that he repeat several Islamic prayers. He died that evening.[9]

Park several times encountered rumours of Houghton or people who had met him. On the morning of 18 February, he and his companions passed the village of Simbing, on the frontier of the kingdom of Ludamar, from which village Houghton had sent his last letter to Laidley. Park by now was about 560 miles from Pisania. Deserted here by his servants, who refused to follow him 'into the Moorish country', Houghton had engaged some Arab merchants to guide him towards the town of Jarra and, from there, to Timbuktu. At some point on that journey, Houghton was robbed of his remaining possessions, but the exact circumstances of his death were not known: 'Whether he actually perished of hunger, or was murdered outright by the savage Mahomedans, is not certainly known; his body was dragged into the woods, and I was shewn [sic] at a distance, the spot where his remains were left to perish.'[10]

In Jarra in mid-February 1796, Park lodged with Daman Jumma, a Gambian *slatee* who traded with Laidley. While there, he suffered a further setback with the news that Johnson and Demba resolved to go no further: 'the danger they incurred of being seized by the Moors, and sold into slavery, became ever more apparent; and I could not condemn their apprehensions.'[11] As it happens, Demba was prepared to continue: it was Johnson who had tried to induce him to return to the Gambia. If there was any hesitancy in Park's mind about proceeding further, there is no trace of it in his *Travels*.

In order that he might proceed, Park asked Daman Jumma to secure permission from Ali, king of the Moorish kingdom of Ludamar, that he might pass unmolested through Ludamar into Bambarra. Negotiations to this end took fourteen days. During the wait, Park encountered locusts for the first time. The trees, black with them, were quickly and completely stripped of their leaves: the noise of their excrement falling upon the leaves and withered grass, remarked Park, 'very much resembles a shower of rain'.[12] As he waited, Park gathered most of his papers together and trusted them to Johnson to take back to Gambia. Prudently, Park also kept a copy for myself, 'in case of accidents'. By early March, Park resolved to proceed alone and headed east, but both Demba and Johnson, who had changed his mind about leaving, caught up with Park as he departed the town of Deena en route to Benowm, Ali's place of residence.

Park and his companions reached Benowm late on the afternoon of 12 March 1796. As a settlement, it was in the semi-desert and unlike the clay and stone-built towns they had passed through in the more thickly wooded areas: 'It presented to the eye a great number of dirty looking tents, scattered, without order, over a large space of ground; and among the tents appeared large herds of camels, cattle, and goats' (Figure 3.1). Whether Park's comments on the disorder of the settlement reflect his views on the morality of its inhabitants we shall never know, but in Benowm his travels ceased for several weeks – and they came close to ending.

In Benowm, Park was the subject of public amusement, intimate inspection, and prolonged derision. Upon arrival, he was jostled by crowds such that he could hardly move. His toes and fingers were counted 'as if they doubted whether I was in truth a human being'. His clothes fascinated onlookers, the buttons especially. Each group of visitors wanted to see how they worked, and how he dressed: 'in this manner,' Park recalled, 'I was employed, dressing and undressing, buttoning and unbuttoning, from noon to night.' A hog was placed in his tent, Ali making signs that Park should kill it and dress it for supper. Seeing it as a ruse to expose his Christianity, Park graciously declined. The animal was then let loose in the tent in the hope that it would attack Park, 'for they believe that a great enmity subsists between hogs and Christians'. In this they were both wrong and disappointed: the pig charged at everyone alike before hiding beneath Ali's chair. It was later moved to Park's hut. 'I found it', noted Park, 'a very disagreeable inmate.'[13]

Figure 3.1 Ali's encampment at Benowm, where Mungo Park was held captive during his first African journey. Source: M. Park, *Travels in the Interior Districts of Africa* (London, 1799). Reproduced by kind permission of the National Library of Scotland.

In ways which reverse that trope characteristic of much Orientalist and later African exploration – the white explorer's masculine gaze upon the black African or Arab woman – Park several times found himself an object of curiosity, even wonder, and not just for his fingers and toes. In one seraglio at an early point in his journey, the whiteness of his skin and his prominent nose were the subject of attention: both, the local women insisted, 'were artificial', the first from being bathed in milk as a child, the second from pronounced and prolonged pinching, 'till it had acquired its present unsightly and unnatural conformation'. In Benowm, one party of women who had been troublesome since his arrival sought 'to ascertain, by actual inspection, whether the rite of circumcision extended to the Nazarenes (Christians), as well as to the followers of Mahomet.' Park made light of this – 'I thought it best to treat the matter jocularly' – by observing that it was not customary in his country to afford visual proof in such matters but that if one young lady would care to stay behind – at which point he gestured toward the youngest and prettiest among them – 'I would satisfy her curiosity.' The ladies dispersed, laughing, the young woman in question 'though she did not avail herself of the privilege of inspection . . . afterwards sent me some bread and milk for my supper'. The men of Benowm were affronted by Park's nankeen breeches which they thought not only inelegant, 'but, on account of their tightness, very indecent'.[14]

With each passing day, Park was subject to 'the same round of insult and irritation' as he put it: 'the boys assembled to beat the hog, and the men and women to plague the Christian'. Park railed against his captors:

> It is impossible for me to describe the behaviour of a people who study mischief as a science, and exult in the miseries and misfortunes of their fellow-creatures. It is sufficient to observe, that the rudeness, ferocity, and fanaticism, which distinguish the Moors from the rest of mankind, found here a proper subject whereon to exercise their propensities. I was a *stranger*, I was *unprotected*, and I was a *Christian*; each of these circumstances is sufficient to drive every spark of humanity from the heart of a Moor; but when all of them, as in my case, were combined in the same person, and a suspicion prevailed withal, that I had come as a *spy* into the country, the reader will easily imagine that, in such a situation, I had every thing to fear. Anxious, however, to conciliate favour, and, if possible, to afford the Moors no pretence for ill treating me, I readily complied with every command, and patiently bore every insult; but never did any period of my life pass away so heavily: from sunrise to sunset, was I obliged to suffer, with an unsullied countenance, the insults of the rudest savages on earth.[15]

'Each returning day brought fresh distresses', as Park put it. Demba and Johnson were treated equally poorly. In late March, Park's fever returned, his condition worsened by lack of food: a bowl of couscous with a little salt and water was their daily fare. In early April, a sandstorm and whirlwind blew in a wall of his already inadequate hut, which was left unrepaired. The air in the camp at Benowm was 'insufferably hot', the wind coming through the gaps in Park's hut so hot as to cause 'sensible pain'.[16]

Park began to learn to write Arabic, partly to pass the time, partly as a stratagem to lessen the ill treatment by his captors. If Park was at a loss to know how best to occupy himself, so too were his captors. At one point, he was appointed to 'the respectable office of *barber*' to the King of Ludamar and, three-inch straight razor in hand, was entrusted with shaving the head of the young prince of Ludamar. Park was understandably apprehensive. Things did not go well: 'whether from my own want of skill, or the improper shape of the instrument, I unfortunately made a slight incision in the boy's head, at the very commencement of the operation.' The king instructed Park to stop at once and ordered him from his tent. Park reckoned this 'a very fortunate circumstance' – as may the young prince – 'for I had laid it down as a rule, to make myself as useless and insignificant as possible, as the only means of recovering my liberty'.

On occasion his spirits would lift following conversations with passing merchants. Park quizzed one of them about travel times across the Sahara from Morocco to Benowm (fifty days in total). If he had in mind thoughts of Rennell's 1793 map (Plate 4) as he did so, we are not told. Hunger, at first 'a very painful sensation', led to languor and debility: 'I felt', Park recorded, 'no inclination to sleep, but was affected with a deep convulsive respiration, like constant sighing; and, what alarmed me still more, a dimness of sight, and a tendency to faint when I attempted to sit up. These symptoms did not go off for some time after I had received nourishment.'[17]

Respite but not release and the freedom to continue came from two sources – Fatima, Ali's wife, who took pity at Park's circumstances, and encroaching war which, by late May 1796, forced Ali to strike camp. Ali, with Park in train, temporarily removed to Jarra. By 1 June, Park was back 'at the house of my old acquaintance, Daman Jumma', whence he had left two months earlier. En route, Ali claimed Demba for his own household retinue but allowed Johnson to proceed with Park. Repeated pleas from Park that Demba should remain with him were to no avail – 'it was in vain to expect any thing favourable to humanity, from people who are strangers to its dictates' – and the two parted tearfully.

Park stayed at Jarra for most of June 1796. He was effectively under house arrest as the Moorish troops under Ali's command despoiled the town. As the war approached, he and his captors were forced to move into the countryside into smaller settlements. On the afternoon of 1 July, as Park was tending his horse near the village of Queira, Ali's chief slave and four Moorish horsemen arrived and were overheard planning Park's renewed imprisonment. Park had contemplated escape in late June during his second residence in Jarra: his landlord, Daman Jumma, saw no prospect of being paid and so was increasingly inhospitable; Johnson was again 'refusing to proceed'. But the circumstances were not favourable: 'If I continued where I was,' recalled Park,

'I foresaw that I must soon fall a victim to the barbarity of the Moors; and yet if I went forward singly, it was evident that I must sustain great difficulties, both from the want of means to purchase the necessaries of life, and of an interpreter to make myself understood.' What determined him to stay was the thought of failure: 'On the other hand, to return to England without accomplishing the object of my mission, was worse than either.'[18]

If some modern commentators cast Park in 'heroic mould', it is also true that, in several senses, Park was something of an anti-hero, at least in regard to how, later, he tells the story of his own travels and travails. He was an object of almost erotic enquiry to some local women: to others an object of jest and denigration. He several times admitted to illness and indecision and yet, on each occasion, he resolved to continue. At one point during his captivity as we learnt from the unfortunate incident as a would-be barber, Park feigned ignorance of anything mechanical lest he be thought of as accomplished and asked to assist his captors. In early July 1796, Park again encountered barbering – this time of himself. As he came to leave the town of Dingyee, his landlord pressed Park for a lock of his hair, having been told that white man's hair made a powerful *saphie* (token or talisman of good luck) 'that would give to the possessor all the knowledge of the white men'. Park had never heard of such a belief but consented to his landlord's request: 'my landlord's thirst for learning was such, that, with cutting and pulling, he cropped one side of my head pretty closely'. He would have done the same for the other side had Park not 'signified my disapprobation, by putting on my hat, and assuring him, that I wished to reserve some of this precious merchandise for a future occasion'.[19]

Park's book, we should remember, was put together after his return: it is, in important ways, an exercise in memory and recall. These instances reflect a sense that while Park's thoughts remained his own, his motives for learning Arabic and feigning ignorance of things mechanical likewise, his body – including his own hair – was not always his to command. It was on display, as white skin, toes, and whatever else his captors chose to gaze upon.[20]

Park was in many ways a distinct anomaly to Africans, and not just because of his nose, clothes, talismanic hair, and white skin. His account is a singularly personal narrative in which things happen to him, not one in which he is the dominant agent in an expedition designed for the formal process of exploration. The relationships he develops with others as he travels, and, particularly, when he is immobile, either held captive or recovering from illness, are unequal ones, and Park is nearly always the lesser party.[21]

The news overheard on 1 July 'was like a stroke of thunder' to Park. He determined to set off immediately in the direction of Bambarra. Johnson refused to accompany him, having by then been engaged by Daman Jumma to help conduct a coffle of slaves to Gambia. Park resolved to push on alone and packed all that he had: two shirts, two pairs of trousers, two pocket-handkerchiefs, an upper and under waistcoat, a hat, and a pair of half boots: 'these with a cloak, constituted my whole wardrobe. And I had not one single bead, nor any other article of value in my possession, to purchase victuals for myself, or corn for my horse.' At dawn on 2 July, as his would-be-captors slept, Park left Johnson – 'desiring him to take particular care of the papers I had entrusted him with' – and rode out alone.

Two miles east, just as he had 'begun to indulge the pleasing hopes of escaping', Park was overtaken by three Moors on horseback. Park was careless about what seemed his inevitable end: 'An indifference about life, and all its enjoyments, had completely benumbed my faculties, and I rode back with the Moors with apparent unconcern.' He was resigned to his fate. In fact, all the horsemen wanted to do was rob him, and they stole Park's cloak before allowing him to continue. After over three months in captivity, his Niger exploration thwarted, Park was free: 'It is impossible to describe the joy that arose in my mind, when I looked around and concluded that I was out of danger. I felt like one recovered from sickness; I breathed freer; I found unusual lightness in my limbs; even the Desert looked pleasant.'[22]

Park may have revived in spirits at being alone and free, but he had neither means of procuring food, nor prospect of finding water. In the heat of the day on 2 July 1796, Park – by now an embodiment of the explorer not as someone who conquers geography but as someone being slowly defeated by it – again resigned himself to his fate: 'here must the short span of my life come to an end'. As he lay prostrate Park was revived, however, by distant lightning and, following that, by a sand squall and then, mercifully, rain. Spreading his clothes on the ground to catch as much of the downpour as possible, he proceeded to suck the rainwater from them. Park then continued generally eastwards, sometimes heading north or east-south-east to avoid people. He was several times drawn to water by the croaking of frogs: what, in Pisania, had added to Africa's cacophonous unfamiliarity was now a sonorous clue to his survival.

On the morning of 5 July, Park entered the small town of Wawra. Again, Park was an object of curiosity: some thought he was an Arab, others that he was 'some Moorish Sultan'. There and in the neighbouring village of Wasiboo, he was fed and given corn for his horse. Again, he gathered fellow travellers: a party of eight Kaartans, like Park fleeing the Moors, were en route to Bambarra. Through the first half of July, Park and his companions continued eastwards. Park's horse was now so weak that Park walked behind it for much of each day's journey: this, and Park's appearance, made him 'the source of much merriment to the Bambarrans'. Even the slaves were ashamed to be seen in his company.

On 20 July, Park made his way to a small village near the town of Sego where he procured some victuals for himself and some corn for his horse, 'at the moderate price of a button'. In Sego, on the evening of 20 July, Park was told news that would have considerably revived his spirits – 'that I should see the Niger (which the Negroes call Joliba, or *the great water*), early the next day'. That night, Park lay awake in expectation – 'The thoughts of seeing the Niger in the morning, and the troublesome buzzing of musketoes [sic], prevented me shutting my eyes' – and he was ready, his horse saddled, before daybreak.[23]

On the morning of 21 July 1796, Park overtook the group of refugee Kaartans with whom he had earlier travelled. As they approached Sego, one of them called out, '*geo affilli* (see the water)'. Park recounts the moment of discovery thus: 'looking forwards, I saw with infinite pleasure the great object of my mission; the long sought for, majestic Niger, glittering to the morning sun, as broad as the Thames at Westminster, and flowing slowly *to the eastward.*' Park hastened to the river's edge and prayed in thanks 'to the Great Ruler of all things, for having thus far crowned my endeavours with success'.

Park further commented that he was not surprised by the direction of flow – 'although I had left Europe in great hesitation of the subject, and rather believed that it ran in the contrary direction' – because he had made frequent enquiries during his travels 'and received from Negroes of different nations, such clear and decisive assurances that its general course was *towards the rising sun*, as scarce left any doubt on my mind; and more especially as I knew that Major Houghton, had collected similar information, in the same manner'.[24]

In this simple act of observation, here tersely described, Park solved the first part of the Niger problem. His phrasing of the Niger's dimensions is interesting. In comparing it to the Thames at Westminster – words which he probably added later rather than thought of at once upon sighting the river – Park is giving his readers, and his sponsors in the African Association, a scale against which to judge a river they would never see. It is also misleading since the Niger that Park first encountered that July day was lower and slower and much less wide than it would be in flood in later months when the river near Sego could be nearly five miles across.

Park's admission that he left England in doubt over the direction of the Niger's flow and that he even supposed it ran east to west is puzzling. Since Rennell and Beaufoy were both convinced that it ran west to east, a view which Houghton's partial evidence had confirmed to their satisfaction, Park's remark suggests either that he had not paid attention to them before departure, or that he did not believe them. His admission about collecting information on the Niger as he travelled – from indigenous Africans but not from the Moors whom he held in unremitting contempt – points to a preparedness to trust indigenous testimony, as had Houghton, especially when it appeared to be consistent. What mattered most was the fact that Park had confirmed the direction of the Niger's flow from direct observation. But this would count for nothing unless he returned alive to report upon it. He had not yet got to Timbuktu as his instructions from the African Association required. Nor was Park any closer to knowing where the river ended.

In Sego, a town whose population he estimated at 30,000, Park waited to cross the Niger to pay his respects to Mansong, the king. He was told that he had to wait until he was summoned. So, Park sat, 'weary and dejected', until a woman took pity on him and gave him some food. As Park ate, he was serenaded by the female members of the woman's household as they spun cotton: 'The poor white man, faint and weary, came and sat under our tree.' Park was touched by the hospitality and the recital: 'to a person in my situation, the circumstance was affecting in the highest degree'. The next morning, he gave the woman two of the four remaining brass buttons on his waistcoat – 'the only recompence I could give her'.

After three days, word came from Mansong that Park should leave Sego, the king apparently apprehensive that he would be unable to protect Park from the Moors. The king provided Park with a guide and on 24 July the two entered the town of Sansanding. Anxious about the Moorish presence in the town, Park soon moved on, northwards and eastwards. On 28 July, he avoided attack by a 'large red lion', not by taking evasive action but by good fortune: his horse was too tired to run away, the lion too disinterested to attack.

The following day, his horse collapsed. Park was close to doing the same: 'I surveyed the poor animal, as he lay panting on the ground, with sympathetic emotion: for I could not suppress the sad apprehension, that I should myself, in a short time, lie down and perish in the same manner, of fatigue and hunger.' The next day he was taken by canoe to the fishing settlement of Morzan on the Niger's northern bank and, from there, to the larger town of Silla. At Silla, Park was treated well by the local Dooty, the chief magistrate, but, on the evening of 29 July 1796, as he weathered a further bout of fever, Park realised he had a decision to make:

> Worn down by sickness, exhausted with hunger and fatigue; half naked, and without any article of value, by which I might procure provisions, clothes, or lodging; I began to reflect seriously on my situation. I was now convinced, by painful experience, that the obstacles to my further progress were insurmountable. The tropical rains were already set in, with all their violence; the rice grounds and swamps, where every where overflowed; and in a few days

> more, travelling of every kind, unless by water, would be completely obstructed.

To push on risked all, 'for my discoveries would perish with me'. Before he left Silla, Park resolved to solicit as much information as he could concerning the further course of the Niger. He met only incomprehension: 'Of the further progress of this great river, and its final exit, all the natives with whom I conversed, seem to be entirely ignorant . . . They pay but little attention to the course of rivers, or the geography of countries.' On 30 July 1796, none the wiser concerning the second part of the Niger problem than when he set out, Park turned west.[25]

Homeward

As he turned back, Park was accompanied at first by the Dooty's brother who, like Park, was heading for Marraboo. There, on 31 July, Park was reunited with his horse, now somewhat recovered. As he moved west along the northern bank of the Niger, Park learned that rumours were circulating that he was a spy and because, in Sego, Mansong had not admitted Park into his presence, the Dooties of the different settlements felt at liberty to shun him. As he journeyed west, he was in consequence refused entry to several settlements – 'Every one seemed anxious to avoid me.' Park even considered the idea of swimming his horse across the Niger and heading south, towards the Cape Coast, before rejecting this plan in favour of continuing along the Niger's northern shore. Such food as he had – raw corn – he shared with his horse.

On the afternoon of 21 August, Park's fortunes took a turn for the better in the town of Marraboo, noted for its trade in salt. There he was well received by a local trader whose house 'was a sort of public inn for all travellers'. Park there shared a hut with seven other 'poor fellows'. He was similarly well treated at Bamako two days later where he made enquiries about the route ahead and the course of the Niger, only to be told that at the river's crossing point a little to the west, the ferry was not large enough to take his horse and that Park would have to wait some months for the river level to drop.

This was, Park wrote, 'an obstruction of a very serious nature'. There was, however, another more difficult route, over the hills, to Sibidiloo and Park took this in the company of a *Jilli kea*, a singing man, provided by the Dooty (upon arrival at a settlement, a singing man would extol the virtues of his would-be hosts in song and so hope to secure accommodation and food). As he travelled, Park again observed to the south 'some very distant mountains,' having previously sighted them from a hill near Marraboo and was told that they lay in a large and powerful kingdom called Kong. Park would not then have known how significant the mountains of Kong would later be in geographical enquiries about the Niger.[26]

Park left the village of Kooma, heading for Sibidiloo, on 25 August, in the company of two shepherds. Late that morning, he and his travelling companions were set upon by bandits, who Park initially mistook for elephant hunters, who claimed they had been sent by the king of the Foulahs with instructions to bring Park to him. As they entered the woods, Park was robbed. Park's hat was snatched from his head – a source of some concern since he kept his notes in his hat – they sliced a button off his waistcoat, and his pocket compass was stolen. Even his boots were inspected, one of which was held together, sole to upper, by a bridle rein. Park was stripped naked. And they took his horse.

After a moment in which the robbers debated whether to leave Park naked in the sun or afford him some protection, 'Humanity at last prevailed: they returned me the worst of the two shirts, and a pair of trowsers; and, as they went away, one of them threw back my hat, in the crown of which I kept my memorandum.' But it was all too much.

> After they were gone, I sat for some time, looking around me with amazement and terror. Which ever way I turned, nothing appeared but danger and difficulty. I saw myself in the midst of a vast wilderness, in the depth of the rainy season; naked and alone; surrounded by savage animals, and men still more savage. I was five hundred miles from the nearest European settlement. All these circumstances crowded at once on my recollection; and I confess that my spirits began to fail me. I considered my fate as certain, and that I had no alternative, but to lie down and perish.

Succour for the despairing Park came from two sources. One was 'the influence of religion' as he put it, for he felt himself 'still under the protecting eye of that Providence who has condescended to call himself the stranger's friend'. The other came via a plant: 'the extraordinary beauty of a small moss, in fructification, irresistibly caught my eye'. Here, too, Park sought solace in his faith: 'Can that Being (thought I), who planted, watered, and brought to perfection, in this obscure part of the world, a thing which appears of so small importance, look with unconcern upon the situation and sufferings of creatures formed after his own image? – surely not!'

It is possible of course, albeit rather unlikely, that despite his weakened state Park drew upon his botanical knowledge to at once identify the plant as a moss, even that at that moment of despair he had in mind James Dickson in whose company Park had identified mosses and other cryptograms in their native Scotland. Much more likely is that this moment of botanical epiphany was crafted later as Park prepared his notes for print. It may be, as one modern commentator unsympathetically argues, that Park 'enlists a sentimentalized physico-theology' to help resolve his gloom. This moment, the moss especially, would become a defining feature of Park's afterlife. Inspired by the moss, the still living Park rose and walked on, was reunited with the two shepherds and, at sunset, arrived at Sibidiloo, the frontier town of the kingdom of Manding.[27]

In Sibidiloo, Park was presented to the Mansa, the Dooty of the town, who promised to restore to Park what had been taken from him. Park waited two days for news of his belongings and horse. None being forthcoming, he pressed on westwards, arriving at the town of Wonda on 30 August. He resided there for nine days and was wracked with fever every day. Park's clothes and horse were brought back to him on 6 September, as was his pocket compass 'broken to pieces'. This, Park confessed, 'was a great loss, which I could not repair'. His compass had earlier attracted great interest, Ali and Park's Moorish captors in Benowm being curious to know why the iron needle always pointed to the Great Desert. To Ali, Park's answer – that the compass always pointed northwards to home, and to his mother: should she die, it would point to her grave – invested the device with magic qualities. In breaking his compass, Park's robbers perhaps thought to defeat its magic and, by implication, the power of the lone white traveller.

Having recently been reunited with it, Park nearly lost his horse, which fell into a well but was rescued. It was by now a mere skeleton. Park presented him in thanks to his landlord and continued on foot. Converting his half boots into sandals made walking easier but his progress was now impeded by a sprained ankle. He was given shelter and food for several days in a village called Kinyeto by Modi Lemina Taura, a trader. As Park headed for the town of Mansia, he was three times forced to lie down as he tried to walk up a hill toward the town, 'being very faint and sickly'. In the town of Kamalia, Park met the man who would provide the means to his survival and safe return to Pisania.

Karfa Taura, brother to Modi Lemina Taura with whom Park had stayed in Kinyeto, was a slave trader and *Bushreen*, a local term for a follower of Islam. At the time of Park's arrival in Kamalia, Karfa Taura was putting together a coffle of slaves to take to Gambia as soon as the rainy season ended. Upon their first meeting, Taura, in his tent with several *slatees* who hoped to join the coffle, passed Park an Arabic text and asked if he could read it. Park said no. Taura then produced another volume. Park was delighted to find that it was the *Book of Common Prayer*, in English, and proceeded to read aloud from it. Karfa Taura was equally delighted but for a different reason: some of the *slatees* present, who had seen Europeans at the coast, had doubts that Park was who he said he was. Park's skin was 'very yellow from sickness' – this and his 'long beard, ragged clothes, and extreme poverty' made him an improbable European to the *slatees*, who thought him 'some Arab in disguise'.[28] Park really was an unknown quantity: an ill white European thought by some to be an Arab in disguise. While the *slatees* distrusted what they saw, Karfa Taura placed trust in what he heard and was told.

Karfa Taura advised Park that further travel was not practicable until the dry season but that he would accompany Park to Gambia if, in return, Park would 'make him what return I [Park] thought proper'. Park responded by asking 'if the value of one prime slave would satisfy him'. Taura was satisfied at this response and immediately prepared Park's accommodation.[29]

It is difficult to know how to interpret this transactional moment. That it was founded on trust is clear: Karfa Taura trusted Park's word, that he, Taura, would be recompensed if and when they reached Gambia. Park, now so reduced in circumstances that words were all he had left to trade in, placed his trust in an Islamic slave trader he had only just met. It is possible, of course, that Park himself was that one prime slave (though hardly in prime condition) – that, in effect, Park enslaved himself to Karfa Taura until such time as he was returned to safety. Perhaps not for the first time and certainly not for the last, an ill explorer placed his trust in indigenous others. The success of the African Association's exploration of Africa here hung in the balance: it, and Park's life, was dependent upon Africans' agency.

For the next five weeks, Park was so weak he could hardly walk. Sometimes he would crawl out of the hut and sit in the open air. Mostly, he lay in his hut, shaking. But as the rains eased, so did Park's fever. Karfa Taura meanwhile was gathering slaves for his coffle, and on 19 December, he left Park in the care of Fankooma, the local schoolmaster, to go to Kancaba, a town on the banks of the Niger known as a slave market. The recovering Park used the time to gather information about the local region, attempting to address the Association's queries ill though he was. Taura returned on 24 January, with thirteen slaves taken as prisoners-of-war by the Bambarran army.

If Park had expected to be soon on his way, he was disappointed. The coffle did not depart until mid-April. It travelled at the slaves' pace, shackled as they were with chains either at the neck, or on both ankles, or by a chain or wooden bar connecting two slaves. The coffle was led by five or six singing men, followed by the free men, then the domestic slaves, then the free women, with Park to the rear. In this order, Park recalled of 23 April 1797, 'we travelled with uncommon expedition, through a woody but beautiful country, interspersed with a pleasing variety of hill and dale, and abounding with partridges, guinea-fowls, and deer, until sunset'. It is doubtful that the slaves in the coffle took the same view of the landscape and its abundant game. On most days, food was scarce and meals few. On a few days, the coffle changed route having heard that robbers lay in wait. In mid-May, the coffle was joined by a further party of slaves belonging to some Serawoollie traders.

The increased size of the group presented further difficulties, both in ensuring that all the slaves kept up and in providing for them. On several occasions, the large coffle was refused food and accommodation from village populations who had little but their own basic subsistence. On the morning of 4 June 1797, Park reached Medina, the capital of the kingdom of Woolli, which he had last seen in December 1795 and where Houghton's plans had started to fall apart years before. Later that day, they reached Jindey, a village to the west of Kootakunda, where Park had parted from Laidley and the Ainsley brothers. They were now only a short distance from Pisania. On 9 July, Park bade his farewell to the slaves in the coffle. Robert Ainsley met Park the following day and, two days later, Laidley returned from a trip downriver and 'received me with great joy and satisfaction, as one risen from the dead'.

Almost the first thing that Park did was make himself look like a European: 'I lost no time in resuming the English dress; and disrobing my chin of its venerable incumbrance.' To Karfa Taura, in shaving off his beard Park again transformed himself, this time 'from a man into a boy'. If he was surprised at Park's transformation, Taura was taken aback at the size of Laidley's schooner at anchor and at the scale of the European presence on the coast: 'Black men are nothing.' Park includes these words and other of his observations, he tells us, not to belittle Taura, but to demonstrate that he had 'a mind *above his condition*', and in the hope that his own account should be acceptable 'to such of my readers as love to contemplate human nature in all its varieties, and to trace its progress from rudeness to refinement'. This expression of Enlightenment sensitivity to ideas of moral improvement is unlikely to have been made on the spot.[30] Park settled his debt to Karfa Taura, paying him double the sum he had originally promised, and sent a present to Fankooma the schoolmaster.

By good fortune, an American slaving ship, the *Charlestown*, entered the Gambia river on 14 July in search of slaves and three days later, at Kaye on the Gambia, Park bade Laidley farewell, boarded this vessel now crammed with 130 slaves, and headed for Gorée.

Park was once again immobile however, keen though he was to be off. He and his American hosts did not leave Gorée until the beginning of October – a stay of over two months – as they waited for provisions. When the *Charlestown* finally sailed, Park found himself offering comfort to the slaves, both because he could speak their language and, with the ship's surgeon dead, by acting 'in a medical capacity in his room for the remainder of the voyage'. Of the 130 slaves held captive in the ship, three died on the Gambia, a further six or eight at Gorée, and eleven more on the Atlantic crossing. Being held in captivity was bad enough but the *Charlestown* was also unseaworthy being 'so extremely leaky, as to require constant exertion at the pumps'. Upon arrival in Antigua after the thirty-five-day voyage, the ship was declared unfit for sea. The slaves, who had saved the ship in their exertions, were there sold into labour.

Park stayed on Antigua for ten days before, on 24 November, he boarded the *Chesterfield* packet, which was returning to Britain from the Leeward Islands. Like Ledyard almost a decade earlier, Park charged the costs of his homeward voyage to Banks's account, presumptively informing Banks that he would have to pay the £40 upon Park's safe arrival in England. He arrived at Falmouth on 22 December, 'from whence I immediately set out for London: having been absent from England two years and seven months'.[31] On the morning of Christmas Day 1797, as he waited to report to Banks, Park met James Dickson in the gardens of the British Museum, it seems by chance. Both men were shocked and delighted in equal measure.

Perhaps the most remarkable thing about Park's journey of exploration was not that it was undertaken at all, or, even, that he solved the first part of the Niger problem by confirming what was suspected, but that he and his notes survived. Upon his return to Britain, Park had to ensure that his notes safely undertook a further daunting journey – the journey into print.

4

'The language of truth'

Park's Travels *(1799): The making of an explorer-author*

Like all books, exploration narratives are the result of collaboration: between the author and the subject in question; between the publisher, an editor or other reader, and the author; and between the book and its audience. In the field, note-taking was vital in exploration and to any future book about it, as a record of what one saw and was told, and as a guide to memory. But because an author's words – whether written on the spot in the field, or elsewhere and later, or both – may be revised by others with a view to the market as well as for their accuracy, the category 'author' seldom denotes a single voice or hand. Yet, only in being an author does one become an explorer. That is, the act of writing and the mediation of one's words by others produces both the character of the explorer-author and the nature and record of the exploration.[1]

Park hardly at all mentions writing during his travels. He several times and at others' request wrote a *saphie* – words, verse, or a prayer written from memory which, when worn about the recipient, was regarded as protection against evil spirits, the more potent for being written by a white stranger (and having a lock of that stranger's hair). He twice mentions that his notes were kept in the crown of his hat. Park made the time and found sufficient privacy to commit his experiences to paper: the fact that he was anxious about the possible loss of his notes during moments of personal danger indicates that he treasured them greatly and for obvious reasons. Yet Park nowhere reflects upon the process of writing in the field.

Only when he was back in London did Park have to consider the nature of authorship and then it was to think of it as a collaborative endeavour. While his travels in Africa distinguish Park as pioneering and intrepid, we know that only because his *Travels in the Interior Districts of Africa: Performed under the Patronage of the African Association, in the Years 1795, 1796, and*

1797 (1799) confirm him as author and explorer. His 1799 narrative was authored and assembled by himself and with the help of others, and his book, a post-exploration account, secured Park's reputation as an explorer after the event: Park both made his book and was made by it.

The making of Park's *Travels*

In the preface to his 1799 *Travels*, Park thanks his 'noble and honourable employers, the Members of the African Association' – to whom the book is dedicated by him, 'their faithful and humble servant'. He also makes clear there that the suggestion for 'an epitome, or abridgement of my travels' came from the Association's committee of management. Even as they praised Park for his 'perseverance and fortitude . . . prudence, ability and sagacity', they recognised that a full account would take time and that, for the benefit of members, an abstract was necessary to satisfy their curiosity and that of an eager public.[2]

Park was not called upon to abridge his own work. *The Abstract of Mr. Park's Account of his Travels and Discoveries* was produced for the Association's members from Park's notes by Bryan Edwards, secretary to the Association, in May 1798. It appeared in the Association's *Proceedings* that year. Almost all discussions of Park's work fail to recognise the importance of this *Abstract* (most fail to mention it), not simply as the first version of his book but in terms of the differences between the two works, the role that Edwards played in both, and for the fact that it was accompanied in the *Proceedings* for 1798 by *Geographical Illustrations of Mr. Park's Journey* drawn up by James Rennell. This too was a first version of what would appear in Park's 1799 *Travels*.

The *Abstract* is a short work: forty-eight pages in four chapters. The first chapter describes Park's journey as far as Jarra. The second concentrates upon Park's imprisonment by Ali in Benowm, omitting Park's inspection by the women there – 'A full recital of his sufferings . . . is reserved for the larger work' – with more promised on his escape. Chapter three focuses upon Park's confirmation of the west–east flow of the Niger, and the river's variant names: 'the Moors, in his hearing, uniformly called it the Nil il Abeed; the inhabitants of Sego, the Joliba'. Edwards was insistent that Park's 'ocular demonstration' had put the first part of the Niger problem beyond doubt and left the second part still to be solved: 'Thus, therefore, is all further question obviated concerning the existence and direction of this great river. Its termination still continues unknown.' The fourth chapter dealt, briefly, with Park's recovery under Karfa Taura's care with short descriptions of the economies of those west African kingdoms through which Park passed. There is little of Park's geographical or ethnographic commentaries, and nothing on slavery.[3]

Rennell's *Geographical Illustrations* is altogether different. It is a work of substantial scholarship, of 162 pages in seven chapters and with five maps. It is the result of prolonged enquiry, principally by Rennell but also by others whose knowledge of Africa and the Niger Rennell reviews. Consistent in its method with his own never-to-be-finished project on the geography of Asia, it is a work both of comparative analysis – 'the Ideas entertained by the Ancient Geographers, as well as the Moderns' – and of critical synthesis, in which he assesses others' textual accounts with the information received from Houghton, and Park's verbal testimony and notes. Rennell drew upon his 1791 *Elucidations of the African Geography*, based as it was on the communications of Houghton and Perkins Magra. To the subtitle of one of the five maps, 'Sketch of the Northern Part of Africa: Exhibiting the Geographical Information Collected by the African Association', which was first compiled by Rennell and printed in 1790 (see Figure 2.4), Rennell added the words 'Corrected 1793' – that is, by May 1798 he had incorporated in print the manuscript additions first discussed with Beaufoy five years earlier (Plate 4).

Rennell praises Park's achievements: 'The late journey of Mr. PARK, into the interior of WESTERN AFRICA, has brought to our knowledge more important facts respecting its Geography (both *moral* and *physical*), than have been collected by any former traveller.' Rennell knew, however, that Park's notes were not complete – 'A great part of Mr. Park's geographical memorandums are totally lost' – and that elements of what survived were faulty: 'Mr. Park was plundered of his sextant at Jarra, which accident of course put an end to his observations of latitude: and thus, unfortunately, left the remaining *half* (very nearly) of his geography in a state of uncertainty, as to parallel.'

In his use of the term 'moral', Rennell means the mores, customs and 'qualities of mind . . . of the Moors and the Negroes'. The term 'physical' used of geography in this context has two meanings. One meaning was literal – the topography and extent of a given region, including its mountains and the course of its rivers. The other is the sense in which contemporaries saw the physical environment as a determinant of human (moral) character: as Rennell put it, 'for that physical geography gives rise to habits, which often determine national character, must be allowed by every person, who is a diligent observer of mankind'. This view, then widely held, would become more forcefully argued from the later nineteenth century as the doctrine of environmental determinism.

Park's discoveries, claimed Rennell, 'give a new face to the physical geography of Western Africa' by proving, 'by the courses of the great rivers, and from other notices', that a belt of mountains ran west to east between ten and eleven degrees of latitude. As evidence for this mountain range, Rennell cited Abulfeda and others, and Park's sighting of the mountains of Kong. In

Classical reasoning, the continents were believed to be in balance one with another around the Mediterranean as the centre of the habitable world, the *ecumene*: Africa to balance Europe, mountains within Africa to partner those in Eurasia. One reason why Columbus headed west across the Atlantic was to 'prove' this idea of continental equivalence and balance – was there a western counterpart to the Eurasian landmass? The result, of course, changed the world dramatically.

As part of these arguments about geographical symmetry, the ancients reasoned that a great chain of mountains bisected Africa. These mountains were thought to stretch across the whole of Africa south of the Sahara. Those to the east were known as the Mountains of the Moon (*montes lunæ*). Those to the west were known as the Mountains of Kong. In west Africa, as he returned to Pisania, Park had seen mountains at a distance, to the south, in what he was told was the kingdom of Kong. For Rennell, as he reviewed Park's discoveries in relation to other evidence on the course of the rivers Senegal, Gambia, and Niger, the existence of a mountain range to the south helped explain the Niger's known west–east course and reinforced his own faith in the reliability of Classical authors. More importantly to Rennell's mind, the presence of this mountain range would prevent the Niger, whose later course and termination was still unknown, from flowing to the south.

Rennell was adamant on this question of a southern mountain range – 'seen by Mr. Park' as he twice puts it. Given others' evidence on the presence of inland lakes, he was equally adamant in his proposed solution to the remaining part of the Niger problem: 'On the whole, it can scarcely be doubted that the Joliba or Niger terminates in lakes, in the eastern quarter of Africa; and those lakes seem to be situated in Wangara and Ghana.' Rennell (and other Europeans at the time) was here confused in taking Wangara to be a place when the term referred to a scattered and semi-nomadic trading community. In proposing this view, Rennell returned to Herodotus, on whose work he considered himself the modern expert, who had argued that the Niger ran into an inland lake (see Figure 2.3).

But Rennell was wrong in reckoning the 'Mountains of Kong' justification for his argument concerning the Niger's later course and termination. What Park had seen was certainly hills but nothing approaching a mountain range and certainly not the Mountains of Kong – for the very good reason that they do not exist. Rennell was not the only one to err in this respect: the non-existent Mountains of Kong appear on numerous later maps and atlases and were not finally removed as geographical features until the early twentieth century.[4]

For several reasons, *Geographical Illustrations* is an important yet overlooked work. It is a detailed commentary upon the longitude, location, and geography of places in west Africa, based upon Park's observations and on

Rennell's reading of al-Idrisi, Abulfeda, d'Anville, James Bruce, and numerous others. In this combination of sources and critical method, it is consistent with Rennell's concerns for precision and accuracy. In its use of ancient and modern sources examined comparatively, it is indicative of new directions in 'modern' geography at that time.[5]

Yet Rennell – the 'geographical voice' of the African Association and a renowned geographical authority – got Africa's geography wrong. Because he incorporated, incorrectly, evidence from Park concerning the Mountains of Kong and because of his interest in the work of Herodotus, Rennell proposed a wholly erroneous view of the Niger's later route and termination. The consequences of Rennell's argument for later Niger explorers would be profound.

No less profound is the fact that Park's 1799 book about his exploration of the Niger was substantially shaped by others before he began to write it. Park's *Travels in the Interior Districts of Africa* was published on 5 April 1799 by George and William Nicol of London's Pall Mall and printed by London fine printer William Bulmer. It is more than a narrative of his Niger journey.[6] There is a frontispiece portrait of Park (Plate 7). Following a short preface, Park recounts his Niger journey in twenty-six chapters across 363 pages. Rennell's *Geographical Illustrations* is incorporated as a ninety-two-page Appendix. There is a short list of African and other terms provided for 'the reader's convenience', a seven-page vocabulary of the Mandingo language, and a short list of questions and answers thought 'useful in the West Indies' (presumably in speaking to those deported from west Africa).

There are three maps, two of which are folded to page size, and five engraved plates, three of which show scenes from his travels (one is Ali's camp at Benowm: Figure 3.1), and two of plant specimens. The plates are described in the book as 'J.C. Barrow delt. From a Sketch by M. Park'. No manuscript or preliminary version of the plates is known. Whether Park sketched them in the field or from memory upon his return is unclear. The maps are by Rennell based, in part, upon Park's hand-drawn sketch maps. There is a 'Song from Mr. Park's Travels', the words of which are by Georgiana Cavendish, the Duchess of Devonshire, set to music by the Italian composer and singing teacher, Giacomo Ferrari, whose one-act operetta, 'I due Svizzeri', premiered at London's King's Theatre on 14 May 1799, one month after Park's book appeared. The words are a variant upon those sung to Park by local women as he lay 'weary and dejected' in the village near Sego in July 1796. An explanatory postscript notes that the Duchess 'was pleased to think so highly of this simple and unpremeditated effusion, as to make a version of it with her own pen'.[7]

Of *Travels*' twenty-six chapters, twenty have Park's journey and his personal circumstances as their main subject. The six that do not, or are only partly about Park, are concerned, respectively, with descriptions of the peoples of west Africa and existing and possible future trade between west Africa and Britain (chapter II); the character of the Moors and a description of the 'Great Desert' (chapter XII); descriptions of the climate and of the Mandingo-speaking peoples of west Africa (chapter XX), which latter topic is also the subject of chapter XXI. Chapter XXII is entitled 'Observations concerning the State and Sources of Slavery in Africa' (which topic Park also addresses in chapter XXIV in recounting his passage as part of the coffle under Karfa Taura). Chapter XXIII is principally concerned with two of the commodities driving British and European interest in the region and with the river Niger as a means of access – gold and ivory.

Park's *Travels* differs from and enlarges upon the *Abstract* by offering for the first time ethnographic and geographical accounts of west African peoples based on first-hand encounter. In that sense, and almost regardless of its status as a pioneering travel narrative, Park's book is an important late Enlightenment text, written with considerable sympathy toward Africa's indigenous peoples and their modes of social organisation, and near total disdain for the Islamic Moors. The peoples bordering on the Gambia, 'though distributed into a great many distinct governments' as Park put it, could be divided into four great classes: the Feloops, the Jaloffs, the Foulahs, and the Mandingoes. The Feloops were 'of a gloomy disposition, and are supposed never to forgive an injury'. The Jaloffs, 'an active, powerful, and warlike race', inhabited the land between the Senegal river and the Mandingo states on the Gambia, were divided amongst themselves into several independent states, and were often at war with their neighbours. The Foulahs (or Pholeys as Park also termed them), were 'much attached to a pastoral life, and have introduced themselves into all the kingdoms on the windward coast, as herdsmen and husbandmen'. The Mandingoes made up the bulk of the population in the areas Park travelled through. For each of these peoples, Park presents a table of their language for the numerals from one to ten: in this, he was reflecting his own experiences while travelling in having to pay for items of food.[8] All of this was new and valuable information, most of which did not appear in Park's *Abstract*.

The key intermediary between Park's notes, the *Abstract*, and the published text was Bryan Edwards. Edwards was then Member of Parliament for Grampound in Cornwall having been a colonist and plantation owner in Jamaica for much of his life. His own principal works were his *History, Civil and Commercial, of the British Colonies in the West Indies* (1793), for which he was elected a Fellow of the Royal Society in 1794, and *An Historical Survey of the French Colony in the Island of St Domingo* [Haiti] (1797).[9]

Banks and Edwards knew one another through the Royal Society and the African Association. Edwards succeeded Banks as secretary to the Association in 1797, a position Banks had held temporarily following the death of Henry

Beaufoy in 1795. Given Edwards' experience and success as an author – his *West Indies* book went into several editions – Banks asked Edwards to assist Park in revising his notes and improving his style for what would become *Travels*. Authorship did not come easily to Park, but he improved with Edwards' guidance. 'Park goes on triumphantly', Edwards wrote to Banks: 'He improves on his style so much by practice, that his journal now requires but little correction; and some parts, which he has lately sent me, are equal to any thing in the English language.' Park wrote to his sister in September 1798 to enquire after her and attributing her silence to date 'to the hurry of harvest' and to report that he was 'very busy preparing my book for the press'.[10]

Written partly in Foulshiels and partly in London, Park's *Travels* thus emerged during the second half of 1798 and into early 1799, based upon Edwards' preliminary *Abstract* and from Edwards' tutelage of Park as author. At the first meeting of the African Association after the book's publication, on 25 May 1799, members noted that 'the public opinion on the merits of the work, called for acknowledgement, that it had been accomplished in a manner creditable to the author, and honourable to his employers'. Banks observed 'that the recital of discoveries and events throughout, bore every mark of veracity; that the remarks and observations interspersed, were intelligent and unassuming; and that the whole was conveyed in a stile [sic] unaffected, clear, and such as might distinctively and properly be termed "the language of truth".'

Park's authorial style reflected well upon himself. It also justified the Association's choice of him (since it was Banks, with Beaufoy, who had confirmed Park's appointment in August 1794, Banks's praise is self-directed): 'In such language, Mr Park's character bore a representation which did honour to him, and reflected credit on the choice of the Society [the African Association]. Strength to make exertions; constitution to endure fatigue; temper to conciliate; patience under insult; courage to undertake hazardous enterprise, when practicable; and judgement to set limits to his adventure, when difficulties were likely to become insurmountable, were every where exemplified in this book.'

Banks here illustrates how authorship produces the explorer: Park is depicted, after the event, as physically and morally able to undertake exploration even although its actual undertaking almost overwhelmed him. Banks also acknowledged Park's readiness to accept Edwards' assistance: 'In arrangement of the materials of this Journal [Park's book], and in the manner and stile with which it has been introduced to public notice and approbation, Mr. Park professed his acknowledgements, to the advice and critical taste of the worthy Secretary of the Society. The literary character of Mr Edwards, stood too high in the learned world, to require testimony; but the candour of the avowal was honourable to Mr Park, and vouched his general accuracy, and laid claim to further confidence.'[11]

Banks's intent in thus praising Park, Edwards, the African Association (and himself) was to establish the veracity of the book by virtue of the moral quality of its author, from Park's conduct in Africa and, once returned, his gracious admission of Edwards' involvement. Banks's position within the Royal Society and his many scientific connections, in London and across Europe, meant that he was in a position to confer further patronage upon Park and, in doing so, affirm his own standing. In stressing the moral qualities of Park the author and the physical capacities of Park the explorer, Banks was imparting authority and credibility to the book, even as he recognised Edwards' role. He was bestowing upon Park's words what has been called 'epistemological decorum' – the association between social status, the content and truth claims of published narratives, the moral basis to authorship, and the justification for the narrative to be seen as authoritative.[12]

These remarks apply equally to the maps in Park's *Travels*. The first map, commonly folded to the front of the text between the Duchess of Devonshire's song and the start of the text, shows 'The Route of Mr Mungo Park', compiled, as the subtitle notes, 'from Mr Park's Observations, Notes, & Sketches, by J. Rennell' (Plate 5). Park's route on the outward leg of his journey is shown in the solid red line; his return journey by the broken red line. To the foot of the map, south of the 12° line of latitude, Rennell has written 'A Chain of Great Mountains extends along these Parallels'. This map appears in Rennell's *Geographical Illustrations* which accompanied Edwards' *Abstract*.

The second map, commonly placed after Park's text and before Rennell's Appendix in the 1799 *Travels* (his 1798 *Geographical Illustrations*), was entitled 'A Map shewing the Progress of Discovery & Improvement in the Geography of North Africa' (Figure 4.1). The detail evident in Rennell's earlier maps of Houghton's route and the presumed course of the Niger is absent from this map. The Niger is shown, labelled the 'Neel Abeed', north of the 15° line of latitude. To the south, a line of stylised hills running west to east – the 'Mountains of Kong' – seem to mark the boundary between what is known and named and what not. The third map showed lines of magnetic variation and was used by Rennell to elaborate upon matters of longitudinal accuracy in his *Geographical Illustrations*.

Park recognised Edwards' work on his words, and acknowledged Rennell's on his maps: 'Major Rennell was pleased also to add, not only a Map of my route, constructed in conformity to my own observations and sketches (when freed from those errors, which the Major's superior knowledge, and distinguished accuracy in geographical researches, enabled him to discover

Figure 4.1 James Rennell's 1798 map of Africa was based on information received from William Browne (for north Africa), from James Bruce (for Abyssinia) and from other sources for the west coast and the interior, including Houghton and Park (*cf.* Figure 2.4 and Plate 4). Rennell here shows the Mountains of Kong and regional descriptors and place names for kingdoms around the Niger. Source: J. Rennell, 'A Map showing the Progress of Discovery & Improvement in the Geography of North Africa', in *Proceedings of the Association for Promoting the Discovery of the Interior Parts of Africa* (London, 1798). Reproduced by kind permission of the National Library of Scotland.

A MAP,
shewing the Progress of
DISCOVERY & IMPROVEMENT,
in the GEOGRAPHY of
NORTH AFRICA:
Compiled by J.Rennell,
1798.
Geographic Miles
British Miles
The double line of dots shews
Mr. Park's Route in the West,
Mr. Bruce's in the East.
SEA
SYRIA
PALESTINE
Cyprus
Aleppo
Damascus
Jerusalem
Cairo
Alexandria
Desert of Barca
Desert of Lybia
Medina
ARABIAN GULF
RED SEA
YEMEN
Mocha
Bab. Al Mandeb
NUBIA
ABYSSINIA
SENNAAR
KORDOFAN
DARFOOR
BORNOU, or KANEM
KAWAR, or
KUKU
NIGRITIA
KOROROFAH
GINGIRO
ADEL
FATIGAR
MACHIDAS
MARACATOS
INDIAN OCEAN
Melinda
Division of Maps
Library of Congress

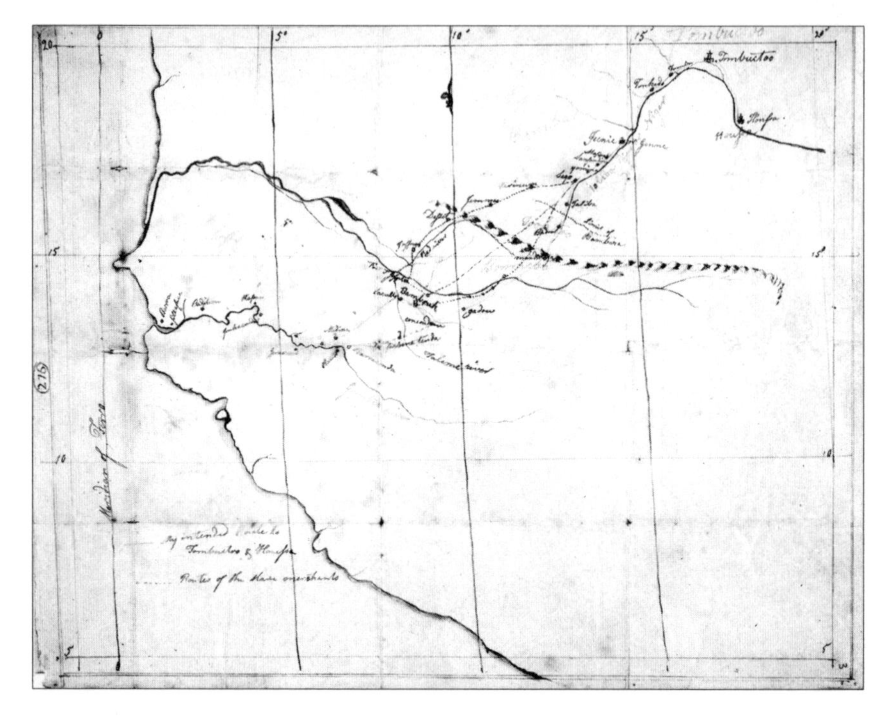

Figure 4.2 Mungo Park's manuscript map of his routes in west Africa on his first African journey. Note Park's indication of 'My intended route to Tombuctoo & Hausa' and the 'Routes of the slave merchants'. The 'Joliba or Niger' is shown to the top right-hand corner. Reproduced by kind permission of the National Library of Scotland.

and correct), but also a General Map, shewing the progress of discovery, and improvement in the Geography of North Africa.'[13]

The map of his travels (Plate 5) was based on two manuscript maps by Park (Figures 4.2 and 4.3). It is not clear when and where Park drew up these maps. Their form and content suggest that one (Figure 4.3) is a refinement of the other (Figure 4.2): several names are shown more distinctly ('Gambia River', 'Bambarra', and the Niger, here named 'Joliba'). Neither is it clear what the 'errors' are to which Park refers. He appears to have used Ferro as his base prime meridian and not Greenwich (which Rennell used), despite receiving instructions from Nevil Maskelyne, the Astronomer Royal, through whose *Nautical Almanac* Greenwich was then being advanced as the principal prime meridian.[14] Park was diligent in taking longitudinal calculations which, given the trigonometry involved, would have demanded that he stop

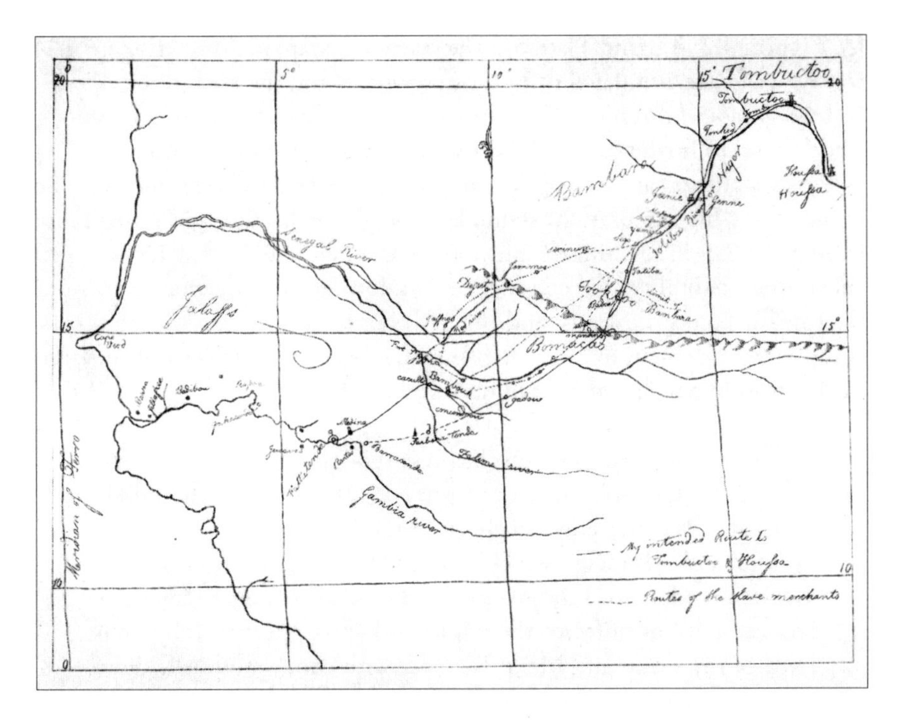

Figure 4.3 In this second manuscript map from his first African journey, Park has included lines of latitude and longitude and has more clearly marked the course of the Niger (*cf.* Figure 4.2). Reproduced by kind permission of the National Library of Scotland.

long enough to render his sightings and figures into positions and time from his known start point (Figure 4.4).

It is possible that Park produced these sketch maps in the field and kept them hidden in his hat. Given their detail and content, it is more likely that he produced them under Rennell's guidance sometime in 1798. Park's comment about Rennell's 'superior knowledge and distinguished accuracy' should be interpreted as a reference to Rennell's longitudinal calculations and accurate positioning of places that distinguish his *Geographical Illustrations* and to the fact that he had been accumulating evidence on the mapping of Africa for at least a decade.

Taken together, the evidence suggests that Park's text was worked up at least twice between its making in the field and its publication (and remember that Park lost some of his original notes in Africa): from Park's notes to Edwards' *Abstract* to drafts worked on by Park and Edwards under Banks's instructions. Park's manuscript maps shaped Rennell's map of Park's route (*cf.* Figures 4.2, 4.3 and Plate 5). The 'General Map' of Africa (Figure 4.1) was Rennell's, accumulated from various sources and inserted in the *Abstract* and *Geographical Illustrations* to correct Park's account, to provide a guide to readers, and to further establish Park's credibility as author-explorer.

Park's status as an author, which had to be clear in what and how he wrote about his exploration, stemmed from his epistemological credibility (seeing the Niger for himself and returning safely), and from his personal and moral credibility (his conduct in London and his acceptance there of the roles of Banks, Edwards, and Rennell). Park had to acknowledge others' authorial involvement or face criticism either over his narrative style or from the institution which had funded his travels:

> Availing myself therefore, on the present occasion, of assistance like this, it is impossible that I can present myself before the Public, without expressing how deeply and gratefully sensible I am of the honour and advantage which I derive from the labours of those Gentlemen; for Mr. Edwards has kindly permitted me to incorporate, as occasion differed, the whole of his narrative into different parts of my work; and Major Rennell, with equal good will, allows me not only to embellish and elucidate my Travels, with the Maps before mentioned, but also to subjoin his Geographical Illustrations *entire.*[15]

For these several reasons, *Travels in the Interior Districts of Africa* is not entirely Park's book. Others may also have thought this. In November 1803, in the wake of Edwards' death, his replacement as secretary to the African Association was Sir William Young, who completed a revised edition of Edwards' *West Indies* after the author's death. Young raised questions about his predecessor's role in Park's work, and referred in doing so to 'the judicious compilation and elegant recital of the travels of Mungo Park'. Park took issue with this wording and its implications regarding his authorship.

Although the correspondence between the two does not survive in full, what remains suggests that Young meant his words to refer to Edwards' *Abstract*, not Park's 1799 *Travels*. Park was nevertheless much miffed by Young's words. He requested that Young's explanation should be inserted in copies of the *Abstract*, or as *Travels* was given out to review, but this appears not to have happened. Young's wording may have been careless, and he certainly expressed himself ready to 'fully render justice to your [Park's] high and well earned reputation'.[16] Their exchanges appear to have ended by late May 1804. This is evidence, albeit partial, of Park's sensitivity regarding his authorship. Further evidence to suggest that what was published

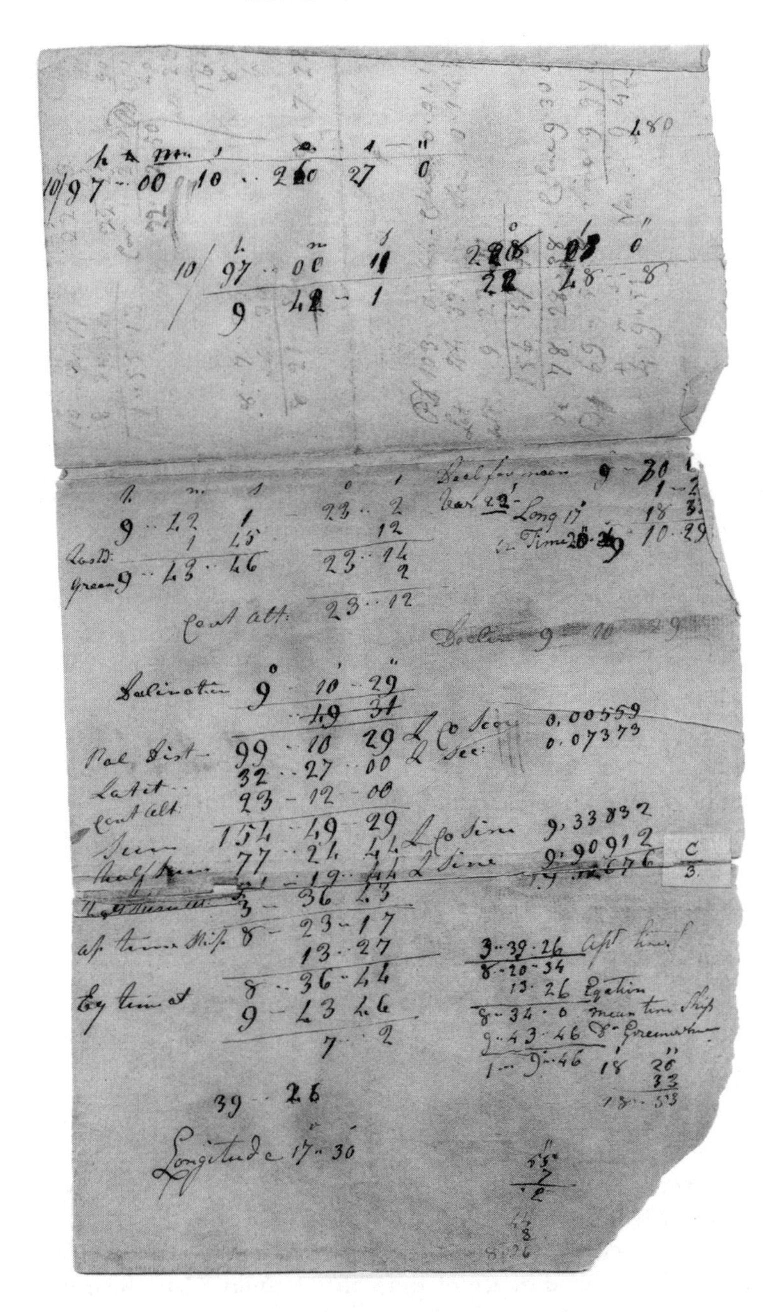

Figure 4.4 A surviving fragment of Park's longitudinal calculations. Determining one's position by using trigonometry to calculate longitude from a known prime meridian was a time-consuming but vital element in exploration. [rgsLMSP4-8] © Royal Geographical Society (with IBG).

was not a full account of Park's exploration of west Africa comes from Park himself.

In the preface to his *Travels*, Park stresses the truthfulness of his narrative: 'As a composition, it has nothing to recommend it but *truth*. It is a plain unvarnished tale, without pretensions of any kind, except that it claims to enlarge, in some degree, the circle of African geography.' It may be that Park is here paraphrasing Shakespeare's *Othello* – 'I will a round unvarnish'd tale deliver / Of my whole course of love' – more than he is following either the Royal Society's instructions for clarity in prose, Edwards' textual interventions, or Banks's faith in his narrative as the 'language of truth'. Park's declaration is consistent with the trope of the modest author, that is, his claims to have enlarged contemporary understanding of Africa's geography strike true because of his modesty, his personal credibility.[17]

Yet, what Park claimed in writing in 1799 to be the truth may not have been the whole truth. Five years later, in private conversation with Sir Walter Scott, Adam Ferguson, and Dugald Stewart at Hallyards, Ferguson's house near Peebles in the Scottish Borders, and as he was deliberating upon his future, Park admitted that his published work was only a partial account. Scott later recounted the circumstances in correspondence with John Whishaw, who was then preparing the biographical account of Park that would appear in Park's second and posthumous book, *Journal of a Mission* (1815): 'Mr Park himself was prepared, and with a judicious caution . . . [to omit] the relation of many real incidents and adventures, which he feared might shake the probability of his narrative in the public estimation. This fact has been proved beyond doubt, by the testimony of many of his intimate friends and relatives, to whom . . . he freely admitted many singular anecdotes and particulars, which he scrupled to admit to the jealous eye of the critical public.'

Reasonably enough, Scott asked Park why he had demurred over telling a fuller story: 'His answer was to this effect: "Whatever I have had to tell where the information contained appeared of importance to the public, I have told it boldly; leaving it to my readers to give it such faith as they may be disposed to bestow. But I will not shock their credulity, or render my travels more marvellous, by producing anecdotes which however true, can only relate to my own personal escapes and adventures".' Unfortunately, Scott could not remember what he and his companions were told by Park: 'I wish I could remember the anecdotes to which the above declaration alluded – But I feel no confidence at this distance of time, that I could relate them with any accuracy & will not do my deceased friend the injustice to produce them imperfectly.'[18]

If one reason for Scott's forgetfulness was the passage of time, another may have been the fact that he, Park, Ferguson, and Stewart were the worse for drink: 'I have often endeavour'd to recollect the passages you mention,' Scott informed Whishaw, 'but they were communicated near the close of an evening of conviviality & although I am positively certain of the scope of the conversation I cannot at this distance of time rely on my memory as to the particular narrative which led to it.'[19]

Park's private admission and Scott's testimony about it has significant implications. The text that is central to Park's reputation is, by its author's own admission, partial. This admission qualifies one modern interpretation of Park's work as a narrative 'of personal experience and adventure' since the author chose, or so he told his friends, to leave out instances of just such experiences. Park's authenticity in terms of the geographical claims in his *Travels* ought not to be cast into doubt by this revelation. His role as the sole author of that book, and his claim that 'It has nothing to recommend it but *truth*' must be.[20]

The audience for Park's *Travels*

Park's *Travels* was successful for all that it may not have been the full story. Park secured 1,000 guineas (£1,050) for the first edition. As early as 7 May 1799, by which date the first edition had sold out, George Nicol the bookseller (and bookseller to the African Association by virtue of his friendship with Banks) was urging a further printing: 'If we allow it to be long out of print, some fresh Tub will be thrown to the Whale, and poor Mungo will be forgotten.' Three further editions appeared in 1799, and by January 1800, the profits from these editions had reached over £2,000.

The success of Park's book, it has been argued, owed much to Park's credibility as an ordinary man who overcame the rigours of geography and to a frontispiece illustration (Plate 7) which 'helped establish Park's persona as a credible, even heroic figure whose stories of distant adventures could be believed'.[21] It is generally easier in book history to know the facts of production – numbers of copies printed, price, the financial rewards to the author, and so on – than it is to know those associated with the book's reception: who bought it, read it, and why. But we can get a sense of who bought Park's book and, presumably, read it from the second London edition which has an eight-page list of 336 subscribers.

The subscribers were mainly London-based men and women of letters and political and scientific influence: Sir Joseph Banks, the natural philosopher Henry Cavendish whose sister's song was included in the book, hydrographer Alexander Dalrymple, Bryan Edwards, James Rennell, and many more. John Barrow of the Admiralty, then completing a book on his southern Africa travels in 1797 and 1798, was a subscriber: Barrow would come to play a major role in coordinating the Niger's exploration after Park. James Anderson of *The Bee* also subscribed. So did George Chalmers the historian and antiquary, the Duke of Buccleuch, landlord to the Parks' farm at Foulshiels, Lord Palmerston, and the MP and abolitionist William Wilberforce. Subscription or local libraries and book societies are also listed: Bristol, Brompton, Bury St Edmunds, the county library in Cornwall, Deal, Exeter, Halifax, Kendal, two different libraries in Leeds, Macclesfield, Stamford, Trowbridge and Warminster Book Society, Repton Book Society, Rochester Book Society, Sittingbourne, and, in Yarmouth, the Old Book Society and the Monthly Book Society. The Literary Society of Newcastle subscribed as, in London, did the Society for the Encouragement of the Arts.[22]

In Edinburgh, publishers and booksellers John Bell and John Bradfute sold Park's *Travels* to other booksellers throughout Scotland – in Aberdeen, Alloa, Elgin, Glasgow, Kirkcudbright, and Peebles amongst others – as well as to individuals across Scotland, in England, and overseas. One buyer, the advocate James Ferguson, later returned his copy. In Edinburgh, Professor Daniel Rutherford purchased a copy for the university in mid-May 1799, perhaps to read about his former student's achievements. In February 1807, their stocks running low, Bell and Bradfute placed an order with Nicol in London for a further twenty-four copies. The purchase price varied depending on the edition and format, either quarto or octavo, with customers who bought the work in boards, that is, unbound so that they might bind it as they chose, paying a lower sum.

On 6 January 1801, Mungo Park, then living in Peebles, bought a work of African travels from Bell and Bradfute. The book was Christian Friedrich Damberger's *Travels Through the Interior of Africa from the Cape of Good Hope to Morocco* (1801). In this work, Damberger (his real name is thought to be Zacharias Taurinius) recounted his journey between 1781 and 1797 from southern Africa, through eastern Africa, before turning west to reach Timbuktu and thence across the Sahara. Published in German, French, and English editions, it was an immensely popular work with seven English editions in its first year. It was also, however, a fake, an amalgam of the accounts of authentic explorers, Park's included. Perhaps because he found his own words misused or considered the whole thing far-fetched, Park later returned his copy to Bell and Bradfute.[23]

Park sold well in the United States too. The first American edition, published in Philadelphia in 1800, has a list of 137 subscribers. Many were prominent figures in the early Republic. Benjamin Rush, for example, was Professor of Medicine at the University of Pennsylvania, some-time Treasurer to the US Mint and, like Park, a former student of medicine at Edinburgh. Samuel Stanhope Smith, the pioneering ethnologist and president of Princeton, was a subscriber. So was Jedidiah Morse, America's leading geographer. Several of the American subscribers were booksellers and bought multiple copies: Thomas Dobson, for instance, secured thirty copies; Messrs Thomas and James Swords of New York took 100 copies; William Wilkinson of Rhode Island a dozen.[24]

This picture of Park's *Travels* as having international reach and significance – as a travel narrative if not always for its importance vis-à-vis the Niger problem – is both reinforced and complicated by the fact that it was translated into several European languages: two German translations in 1799, with French, Italian, Swedish, Dutch, and Danish editions as well. Translation is more than the original in another language: it can be used to promote particular interpretations of the original. Like the French, the Danes had reason to be interested in Park's claims given their colonial interests on the Guinea Coast. In 1799, the Danish priest Hans Christian Monrad, who in 1822 would publish a major history of Danish colonialism in west Africa, published his *Efterretninger om det indre Afrika*, an abstract of Houghton's and Park's travels. This was intended, as its subtitle had it, 'as a forerunner to the latter's more extensive travel account'. The first Danish edition of Park's *Travels* appeared in 1800. The Swedish edition of Park's book, by botanist and theologian Samuel Ödmann in 1800, was based not on the English language original but, in abridged form, upon the German edition published in Berlin in 1799.

The original English version was later bowdlerised to suit particular needs. In 1812 – at about the time the circumstances behind Park's death on the Niger were becoming more widely understood – the anonymous author of *A New System of Geography* produced a book based on Park's *Travels* and those of John Barrow, hoping to capitalise on public interest in Africa's exploration. Once his words were in print, Park had little control over them.[25]

The reception of Park's *Travels*

Knowing what readers thought of Park's book is more difficult. In London, his book was read by members of the self-help Spitalfields Mathematical Society amongst others, and borrowers from subscribing libraries and book societies in Britain and in the United States no doubt read it to be similarly educated and entertained. Bryan Edwards' copy survives, as do those of several contemporaries: Banks, Dugald Stewart, the Duke of Buccleuch, Henry Cavendish, and Sir Walter Scott. But none has marginalia or other reader's marks from which we may know their motivation for reading Park's *Travels* or their reactions on doing so.

James Rennell read Park to correct himself. Notes in his hand in his copy reveal that Rennell made several amendments, principally to the Appendix, his own *Geographical Illustrations*: on the accuracy of Ptolemaic placenames; adjustments to Park's compass bearings and, by implication, questioning the accuracy of Park's route map (Plate 5). One marginal amendment refers to Park's second Niger journey: 'Mr Park actually found 17° 2/3 W variation near Sego, in 1805.' Park's longitudinal positionings published in 1799 were being added to, after his death, by the person who had helped produce them in life.[26]

Park was the subject of public notice – the explorer as celebrity, feted for his accomplishments by a British public weary perhaps of war news – before his book was. In January 1798, several London papers pronounced upon Park and his travels, not always correctly. Most people who met him found him shy, even dull, and certainly diffident, but this may say more about their expectations than his personable qualities. By mid-1799, reviews of his *Travels* (and, in a few instances, of Edwards' 1798 *Abstract*) reveal what others thought of his achievements. Several themes recur: the importance of Park's travels and comments in revising Europeans' view of Africa and Africa's peoples, his narrative style, and the geographical truth of his account – without, necessarily, reference being made to the Niger – and, because it was then high on the political agenda and in the public's consciousness, the matter of slavery.

The *Edinburgh Magazine* considered that Park had extended the limits of geographical knowledge: 'a considerable portion of Africa is now known, which hitherto has been impervious to every traveller, and to no one has the world been so much obliged as to this gentleman'. The *Anti-Jacobin Review and Magazine* praised Park for his comments on African peoples, seeing in them recognition of the shared qualities of human nature, something 'essentially the same in all climates and situations'. Park was praised too for the utility of his findings, his own conduct under duress, and the style of his book: 'From the character of the author, and the verisimilitude of the narration, we entertain no doubt that the whole and every part of what he alledges [sic] is true.' To the reviewer of the monthly *New London Review* in July 1799, Park's 'simple and perspicuous' style was well adapted to the subject and 'shews [sic] that the writer was rather ambitious to establish a character for exactness and authenticity, than to gain praise for well-turned periods at the expence of either'. Rennell was likewise the object of praise, more so indeed than Park:

> It is needless to say that the geographical illustrations of Mr. Park's journey, by Major Rennel [sic], are a most important addition to the work. While they afford an opportunity to the Major of displaying his consummate knowledge of the subject, they considerably enhance the value of Mr. Park's publication, and render it

> as complete as existing information will permit. Both ancient and modern authorities are here brought forward to elucidate the geography of the interior of Africa; and we are proud that Great Britain has at last produced a man who is qualified not only to rival, but eclipse the fame of a D'Anville and a De Lisle.

To the same reviewer, Park offered 'judicious reflections on the slave trade, and the misery it occasions', the reviewer also noting that this was not an issue on which Europeans were to be solely blamed, 'as slavery is hereditary in Africa'.[27]

Park was aware of public concern over slavery years before he encountered slaves, slavery, and the slave trade at first hand. As a contributor and a subscriber to *The Bee*, he would have read James Anderson's short pieces in 1791 and 1792 critical of slavery and the slave trade. At home in the Borders, he would have known about, and perhaps taken part in, the widespread agitation throughout Scotland in 1792 against the slave trade: petitions were signed across the country. In Selkirk, one leading abolitionist was the Rev. Dr George Lawson, minister of the Secessionist Church to which the Parks belonged. Park would have known of Lawson's views on the matter (which included abstaining from sugar in protest at its production through enslaved labour) and of the petition in Selkirk. Yet there is no evidence to show more exactly what Park thought about slavery and the slave trade, what he might have read on the matter, or with whom he may have spoken about it. As Duffill notes, 'if he did hold abolitionist opinions, he certainly kept quiet about them, and at no stage in his career did he ever publicly denounce the Atlantic slave trade or slavery.'

Park's 'judicious remarks' on the subject form the shortest chapter of his *Travels*. It combines historical enquiry into the causes of slavery with descriptions of its prevalence in west Africa. In recounting the experiences of his return to Pisania, Park notes the sufferings of slaves in Karfa Taura's coffle, but he is not condemnatory of their enslavement. Recounting his time on the *Charlestown*, Park admits to helping the slaves as they lay ill but does not reflect upon the condition of slavery, or muse upon their enslavement and the mortality of slaves as a matter of morality or, in the language of Andrew Duncan, his one-time teacher, of medical jurisprudence.[28] In his book, Park identified what he termed 'two distinct classes' of slaves – those born into slavery because their mothers were already enslaved, and those born free, 'but who afterwards, by whatever means, became slaves'. Slaves of the second sort became so from one or other of four causes: captivity, famine, insolvency, or crime.

In offering his brief account of the causes of slavery, Park wrote as a historical-cum-political commentator, an enlightened reviewer of social iniquity. In this, and in other remarks on slavery's form and origins, he is consistent in his tone and mode of enquiry. His *Travels* is the first detailed and largely sympathetic account of the lives and customs of the peoples of the African interior, an account which shows there was little difference in the moral codes binding the society of black Africa compared with those of white Europe. But as an individual of moral conscience who encountered slavery and the slave trade during his travels and who might have offered comments about their cessation, Park concluded his chapter equivocally, even evasively:

> How far it [slavery] is maintained and supported by the slave traffic, which, for two hundred years, the nations of Europe have carried on with the natives of the Coast, it is neither within my province, nor in my power, to explain. If my sentiments should be required concerning the effect which a discontinuance of that commerce would produce on the manners of the natives, I should have no hesitation in observing, that, in the present unenlightened state of their minds, my opinion is, the effect would neither be so extensive or beneficial, as many wise and worthy persons fondly expect.[29]

These observations were taken up by contemporaries who opposed its abolition as being supportive of the Atlantic slave trade. Park's biographers differ in their judgement of them, and of Park. For Duffill, Park temporised over the issue of the slave trade, 'primarily because he did not wish to alienate those powerful individuals and groups that had an interest in the slave trade and slavery'. In short, Park demurred from public condemnation for reasons of possible personal advancement: 'his equivocal stance over the issue of slavery and the slave trade in his book was derived from his interest in keeping open the possibility of future employment in West African exploration'.[30] This is possible, but in 1798 and 1799 as Park wrote his book, Park was by no means certain that he would again head for Africa.

Lupton, by contrast, sees Bryan Edwards' influence as important, a view first articulated by John Whishaw, Park's first biographer, in 1815. For Whishaw and for Lupton, Park the proven explorer but inexperienced author deferred overly to Edwards. Although the owner of enslaved peoples on his plantations in Jamaica, Edwards was probably a 'moderate' in his views over the slave trade: believing that it should be abolished, but over time, and that abolition was a matter for the British Parliament, not for the autonomous West Indian Assemblies. Many of Edwards' remarks upon slavery in his 1793 *History* are similar in tone to Park's in 1799: he comments upon the history of slavery in the Caribbean (but not its ending), and upon the origins of the slave trade (which, he admits, was 'very wicked'). Edwards recognised the moral and physical degradation produced by slavery – 'so degrading is the nature of slavery, that fortitude of mind is lost as free agency is restrained' – but he forbore from calling for an end to it.[31]

We will never know the exact nature of Edwards' involvement with Park's words. By his own admission, we do know that Mungo Park's *Travels* was not entirely complete, not wholly his own. His book is not a simple narrative of exploration, nor was it read as such. Edwards influenced its style and contents, and in its geographical accuracy, it owed much to Rennell. These facts do not diminish Park's truth claims, his achievements in Africa, or his 1799 text about them. But in the book's wake, and as he found himself a successful explorer, Park faced a problem encountered by many explorer-authors – what next?

5

'The enlargement of our geographical knowledge'

Indecision and complications, 1799–1805

Whatever others thought of Park's achievements and of his book, Sir Joseph Banks was in no doubt. He declared his view to those present at the African Association's meeting in May 1799: 'We have already by Mr. Park's means opened a Gate into the Interior of Africa, into which it is easy to every Nation to enter and to extend its commerce and Discovery from the West to the Eastern side of that immense continent.' While Banks recognised the commercial possibilities attendant upon west Africa's continued exploration – a view he and Beaufoy had articulated a decade earlier – the words 'easy for every Nation to enter' were a warning given French ambitions in west Africa.

In the five or so years after his book's publication, Park's personal ambitions were more and more caught up in these matters: with others' plans for him – plans which were not always coincident with his own – with the effects of international conflict, the politics of colonialism and exploration, and with different theories about the Niger. And even as Park worked to write his book upon his return – indeed, even before he had returned from his African travels – the African Association had sent another man into the field.

Park's shadow: Friedrich Hornemann and the Niger, 1796–*c.*1801

In May 1796 Joseph Banks had engaged a German theology student, Friedrich Hornemann, on behalf of the African Association. Banks had written to Park in mid-September 1795 – a letter which Park never got – expressing his hopes that by the time Park received it, 'you will no doubt have Returnd [sic] from a Perilous Journey'.[1]

Far from completing it, Park would not begin his journey for a further ten weeks being then still in Pisania recovering from fever. Communication between west Africa and Britain was uncertain at the best of times. Banks received only one letter from Park, the rest being lost or seized by the French ships who frequently intercepted correspondence between Britain and west Africa. By appointing Hornemann, Banks may simply have been trying to maximise the chances of success and so enhance the Association's reputation in the absence of information from or about Park.

Friedrich Hornemann was not an obvious candidate for African exploration. He had matriculated at the Georg-August-Universität in Göttingen in 1791. The university was then the centre of enlightened ideals in the German-speaking territories, its liberal curriculum making progressive education more possible there than elsewhere. Hornemann was a student of theology, but he never got on with religion and spent much of his time studying Africa's geography.

In 1795 and with a view to contacting the African Association, Hornemann introduced himself to Johann Friedrich Blumenbach, the Göttingen-based comparative anatomist and ethnologist on which subjects he corresponded with Banks. Blumenbach shared Banks's African interests but from a different direction: his ethnological work was especially critical of contemporaries then arguing for the innate inferiority of Africans. When Blumenbach communicated the possibility of Hornemann's involvement to him, Banks was warm to the idea. The proposal that Hornemann should engage in African exploration was not formally discussed by the Association until May 1796. As Beaufoy had of Ledyard, so Blumenbach endorsed Hornemann's physical and moral capacities: 'He is a young robust man of an athletic bodily constitution, infatigable [sic] but always preserving great care and good precaution for preserving his health.' Perhaps mindful of what had happened to Ledyard, Hornemann thought it sensible that, rather than head for Africa too soon, he should stay in Göttingen, there to learn natural history from Blumenbach, to learn Arabic and how to sketch, to be taught by the university's mathematicians and astronomers, and 'to acquire some necessary practical knowledge of Domestic Medicine and Surgery'.

If to his mind these were the accomplishments necessary for Africa's exploration this was because Hornemann had already planned what he wanted to do. At the May 1796 meeting, the Association's committee members discussed the 'Proposals from Mr. Hornemann' which accompanied Blumenbach's letter of recommendation. Hornemann outlined the need for financial prudence – welcome news to the Association given the Willis debacle – and his plan to travel to Cairo and from there to Murzuq in the Fezzan, there to speak with the two shereefs from whom Simon Lucas had solicited information. Once in Murzuq, his plan was to despatch his journals via Mesurata and Tripoli to London, for which purpose he would need 'an address to some Trusty person either at Mesurata or Tripoli'. From Murzuq, he planned to travel south to Caina (today Katsina in northern Nigeria). In a list of 'the chief points I shall attend to' – the proper manner of travelling in Africa, the conduct of travellers, 'Observations on the Manners of the Natives', and so on – he identified 'Researches concerning the course of the Niger'.[2]

Hornemann was directed to come to London before departing for Cairo. Banks paid his fare as he had for Ledyard. In London, he was given clear final instructions: acquire Arabic and such other languages as might be useful; engage with strangers from the African interior and others likely to be of help but in no circumstances was he to disclose his purpose to them. Because Park was exploring the upper Niger, Hornemann was instructed to turn his attention to the river's middle reaches around the Hausa kingdom. Britain being at war with France, Banks arranged freedom of movement for Hornemann with the French authorities and a reception in Paris from France's leading natural philosophers. Hornemann was in Paris by July 1796. There he met, among others, a Turkish merchant from Tripoli named Sidi Mohammed d'Ghies who, in addition to being a trader, had diplomatic connections. He later became the Bashaw's foreign minister. Forgetting his instructions about not disclosing his mission to others – an indiscretion which greatly angered Banks – Hornemann asked for and was given advice about Saharan travel and customs.

Hornemann was delayed by a lack of ships and by French mastery of the Mediterranean but was in Cairo by early October. Like Park in Pisania, and Ledyard before him, Hornemann was then unable to leave Cairo. Plague thwarted his first attempt, he had financial difficulties, and, in the wake of Bonaparte's seizure of Alexandria in July 1798, Egyptian animosity towards Europeans meant he lived under house arrest for several weeks. Ironically, Napoleon's victory over the Egyptians at the Battle of the Pyramids in July 1798 – and Nelson's subsequent destruction of the French fleet at the Battle of the Nile in early August 1798 which ended French naval control of the Mediterranean – led to an upturn in Hornemann's fortunes. Bonaparte, now confined to Egypt, was well disposed to historical and geographical enquiries. The multi-volume *Description de l'Égypte* published between 1809 and 1829 would be proof of this: the publication rendered Egypt one of the closest studied and best-mapped countries of the Near East.[3]

Like Park, Hornemann was ready to travel but restricted by circumstances beyond his control. Unlike Park, Hornemann travelled with another European, Joseph Frendenburgh, a German who had converted to Islam, spoke Turkish, and spoke and read Arabic. Hornemann was further unlike Park in adopting disguise. He was the first of many Niger explorers to do so: 'It is a new plan to travel as a Mohamedan in these Countries . . . but dangerous only for the beginning: – afterwards it is safer and better.' By pretending to be a Moslem trader and adopting Arab dress and the Turkish name Jussuf, Hornemann aimed not only to travel more easily within Islamic Africa but also to do better than Park. To Hornemann, Park would have accomplished more had he not dressed as a European: 'I found his account very true, but shall discover, travelling as a Mahomedan, what he could not do in his character.'[4]

Disguise in geographical exploration is a complex issue, whether in the adoption of local dress, in using another language, by pretending to be a travelling physician, or, as was common, employing scientific instruments by subterfuge, either to measure distance or calculate longitude, or to hide one's writing in the field. Several of those who followed Hornemann and Park did so in disguise. Disguise in its basic sense was a means to ensure safe travel by deceit. However much it was a matter of practicality and safety, disguise raises further questions of trust and truthfulness in the explorer-author: can we trust the word of persons who have gathered what they know by pretending to be what they are not?[5]

Hornemann and Frendenburgh left Cairo in a Fezzan-bound caravan in early September 1798 and arrived in Murzuq on 17 November. They were several times questioned about their identity. To judge from a letter to Bryan Edwards written on the eve of his departure, Hornemann knew that any display of scientific instruments would arouse suspicion: 'In respect to my astronomical instruments, I shall take care never to be discovered in the act of observation; should those instruments, however, attract notice, the answer is ready, "they are articles for sale".'

Frendenburgh died of fever within days of arriving. After several weeks' residence, rather than remain in Murzuq and in the absence of the expected coffle to take him south toward the kingdom of Bornu and the Niger, Hornemann headed north to Tripoli, arriving there in August 1799. He there renewed acquaintance with Sidi Mohammed d'Ghies and, in October 1799, met Simon Lucas, the British consul having been absent on leave when Hornemann first arrived in Tripoli. Both men were helpful and Hornemann left his journals in Lucas's care: they became the basis to his book.[6] Hornemann returned to Murzuq on 20 January 1800. In a letter to Banks of 6 April, Hornemann wrote that he was that evening intending to join a caravan to Bornu. It was the last letter received from him.

Rumours concerning Hornemann's travels and his demise were reported to the Association in 1804, 1805, and 1809, but his fate was not confirmed until much later, by the Bashaw in Tripoli and by the later Niger explorers Joseph Ritchie and George Francis Lyon, and by Hugh Clapperton later still. It is likely that Hornemann died from dysentery, early in 1801, in Nupe, south of

Bornu. Hornemann made several interesting archaeological and ethnological observations during his travels, which Rennell discussed in an essay of 1802 entitled 'Geographical Illustrations on Mr. Hornemann's route', the German's work being compared with that of the Englishman William Browne.

Modern historians of exploration have generally been kinder to Hornemann than his achievements merit. To Bovill, Hornemann was resolute and determined, but 'not of the stuff that makes a great explorer'. Sattin considers Hornemann's trans-Saharan journey a 'magnificent achievement', his death a tragedy for 'the cause of African geography'. For Hallett, Hornemann's 'extraordinary exploit' could hardly have failed to have influenced the course of European contact with western Africa – 'had he lived'.[7] This qualification is all important, of course, since it applies equally to all those who died in search of the Niger.

Politics, exploration, and personal ambition, 1798–1805

In the little over seven years between returning from Africa in December 1797 and departing for it in January 1805, Mungo Park entertained the possibility – or it was entertained for him – of exploring New South Wales, farming in Scotland, or moving to the Caribbean. Travel to China was also mooted. That none of these things came to pass, and that Park spent part of this time as a doctor in Peebles, was the result of political circumstances over which he had no control, and of changes in his personal life, including in his relationship with Banks.

Mungo Park left London for Scotland only in June 1798, six months after returning from Africa, and following the publication of his *Abstract* and the African Association's meeting in late May. At that May 1798 meeting, Banks raised with Park the possibility of exploring New South Wales, Britain's Australian colony established in 1788. Banks had in mind the parallel appointment of Matthew Flinders, then on HMS *Norfolk* exploring the coast of Van Diemen's Land (modern Tasmania), with Flinders to lead the nautical work and Park the terrestrial exploration. Park was keen on the idea but, anxious to see his family after a long absence, he returned to Scotland before arrangements were finalised.

To judge from what Banks wrote in May 1798 to John King, the Under-Secretary of State for the Colonies, Banks and Park reached agreement over the latter's future remuneration even before Park returned home: 10 shillings a day with rations on top, with outfitting costs for instruments and fire-arms and presents estimated at a further £100. Park returned to London in early September, his *Travels* only part-complete, to benefit from Edwards' and Rennell's assistance and, with Banks as intermediary, to meet with John

King and his namesake, Philip Gidley King, the lieutenant governor of New South Wales, then in London but preparing to return. At their meeting on 9 September, Banks's letter to John King could not be found. When asked what he thought Park's remuneration should be, John King suggested between 6s and 10s a day, with no outfit money. Park bridled at this and withdrew from the plan.

An exchange of letters then followed. Their content indicates, on Park's part, a hardening of attitude and on Banks's, increasing irritation at Park's stance. On 14 September, Park apologised to Banks 'that any misunderstanding should have taken place respecting my present voyage to New South Wales' and indicated that the proposal was his idea rather more than Banks's: 'it was a plan of my own proposing and it was the hight [sic] of my ambition to have accomplished it with honour'. But, he continued, the wages 'appeared too small'. Further, the fact that Governor King was planning to set sail within the next ten days made it 'impossible for me to make the necessary preparations'.

In a letter of 20 September, Park appeared to excuse Banks for his part in the matter but was insistent that he would not head for New South Wales: 'tho I have every reason to be satisfied with the part you have taken in the business, I feel that those triffling [sic] misunderstandings have considerably damped that enthusiasm which prompts to, and is necessary to ensure the success of such an enterprise.' Park continued in rather aggrieved tones: 'as I have no wish to deceive nor disapoint [sic] my patrons by engaging in an enterprise which from the circumstances I mentioned to you cannot be perfectly agreeable – I must therefore want some other opportunity, when perhaps I may be better reconciled to the situation.'[8]

This was too much for Banks. His reply the following day suggests that he thought Park was keeping something from him and that Park was showing a side to his character not previously evident:

> You, who are acquainted with the real Motives of your now declining an Engagement, which till your absence from London you always appeared to me to Solicit with Eagerness, are alone able to determine upon the propriety of your Conduct. I shall offer no arguments on the subject; but I cannot help observing that, as in all my former transactions with you I never noticed you to be more than properly attentive to pecuniary considerations, I am astonished to find your Enthusiasm dampt [sic] by the small difference between the Pay you had taught yourself to expect and that really offer'd to you; because real Enthusiasm, which you certainly possess, is very little sway'd by pecuniary motives.

By chiding him thus Banks perhaps hoped to sting Park into changing his mind. Park's refusal to take further the possibility of exploring New South Wales – a plan which he suggested was his in the first instance – certainly made things awkward for Banks. In his letter of 21 September to Robert Ross, Under-Secretary of State, Banks termed Park 'this fickle Scotsman' for not 'going to New Holland'.

Further exchanges followed. On 24 September, Ross wrote to Banks to inform him that John King had raised the salary to 12s 6d and expressed himself 'much obliged to you for Park's final determination, which I really hope will be in the affirmative'. The following day, Banks wrote to Park to inform him that the Secretary of State [for the Colonies] had authorised Banks to raise the pay. Equipment would be in addition: 'Respecting outfit, I will undertake that it shall be sufficient to furnish you with all necessary Instruments, and also with arms and a sufficiency of Presents for the Natives.' With this, Banks hoped the matter resolved: 'Misunderstanding must, however, be now at an end, as the Pay offered is somewhat more than the highest rate that has been originally thought of: if any other kind of misunderstanding remains, it will be necessary for you to state the nature of it.' He closed his letter with an appeal to Park: 'My advice to you, however, is, that, if you decline the Voyage in consequence of some new View which has opened itself to you since it was proposed to you, that you say so openly, and do not any longer attempt to defend your Conduct by the Plea of unexplained misunderstanding, which necessarily lays upon me all the blame it takes off from you.'

Park's reply came the next day. From what he could recall, the initial sum mentioned had been 12s per day and £100 for outfit, but, upon his return to London from Foulshiels, Park found this reduced to 10s a day 'and no outfit'. Monetary matters alone, 'however contemptible in themselves', were nevertheless 'a good criterion by which to estimate the importance or inutility of any office or pursuit'. What Park in his mind 'had painted as a Voyage of some Importance', government officials saw differently. While largely absolving Banks of blame, Park ended with his 'final determination': 'I am sorry to observe, that, after having wrote to my Relations what has happened, and dropped every kind of preparation for the Voyage, it will be very inconvenient for me to accept of the terms offered, even as they now stand.'[9]

There the matter closed, for Park anyway. Banks and government officials knew that the Pacific world lay largely unstudied from a European perspective – Banks from personal experience having voyaged there with Cook – and that for the New South Wales colony to be successful, its natural wealth needed to be understood. But Park's refusal meant that Banks had to look elsewhere.

The man Banks turned to, Robert Brown, was like Park in several ways.

Born in Montrose in 1773, Brown had matriculated as a medical student at Edinburgh, taken John Walker's class in natural history as his interests swung towards botany, left the university in 1793 without completing his medical studies, and undertaken plant hunting trips to the Scottish Highlands. His interest in cryptograms brought him into contact with James Dickson. It is possible that Park and Brown knew one another, either through Dickson or Banks, or by reputation through London's socio-scientific circles.

After considerable delay, and the departure of Governor King to New South Wales, Brown sailed south with Flinders on the *Investigator* in July 1801. Brown spent almost four years botanising in Tasmania and Australia. He spent much of the rest of his life becoming a world-leading botanist and dealing with the floristic diversity he collected there (despite losing many of his specimens in a shipwreck). Like Park, Brown owed his patronage to Banks, and he returned the support shown by providing Banks and the Royal Society with a range of botanical information.[10]

Flinders' *Investigator* voyage and the circumnavigation of the Australian continent was a key moment in Europe's encounter with the Pacific world. For Flinders – who is credited with proposing the name 'Australia' to replace 'New Holland' and 'Terra Australis' – it came at some personal cost: captured and held prisoner by the French on Mauritius for six years, he wrote up his discoveries while confined.[11] Whatever Park thought as he learned of Brown's achievements and Flinders' imprisonment – about opportunities missed, reputations made, authorship under duress – he did not commit to paper.

After the difficult autumn of 1798, and the publication of his book in April 1799, Park returned to Foulshiels in May 1799. One good reason to return was Allison Anderson. It is possible that Banks knew of Park's feelings for her, and that this was Park's reason for declining the New South Wales opportunity – even that Banks was told this by James Dickson. Mungo Park and Allison Anderson – 'My lovely Allie' as he called her – were married on 2 August 1799. He was twenty-eight, an explorer-author of renown, tall, good-looking (Plate 7), and beginning to recover from his African travels (though he never really did, being plagued by indigestion and stomach pains). She was nineteen and later described by one of Park's biographers, rather unfairly, as 'amiable in disposition, with no special mental endowments, and if anything somewhat frivolous and pleasure-loving'.

The couple settled in Foulshiels, but Park did not settle to a Borders life. By early 1800 he was back in London. What opportunities if any had come his way in the intervening months is not certain. In a letter of 10 January 1800 to his wife, Park wrote that he had dined with Nicol, his publisher, and had met with Banks who assured him 'that he was always ready to serve me'. Any awkwardness resulting from their exchanges in the previous autumn was overlooked, the events forgiven if not completely forgotten. Despite this, Park was personally unsettled: 'I have been constantly running about since I came here but I am far from happy as my mind is always anxious about my Allie.'

In this letter, Park mentions 'the China appointment'. Whatever was being mooted – no details survive to tell us – Banks 'was afraid it would not answer my expectations'. Banks urged Park to 'bestir myself among my freinds [sic] as soon as they come to town at the meeting of parliament . . . on the 21st Jany'. Between 1792 and 1794, Britain had made formal diplomatic contact with China through the Macartney Mission named after its leader, George Macartney. John Barrow was involved in this before he ventured to southern Africa. The Macartney Mission was a failure, largely as the result of misunderstandings over trade policies and slights over diplomatic protocol, despite British efforts and gifts of the best scientific instruments. Whether Park hoped to be involved in a follow-up mission is not known (there would be one, but not until 1816). There being nothing then available as a government appointment, Park returned to Foulshiels in the spring of 1800.[12]

Mungo and Allison Park became parents on 28 May 1800, naming their son Mungo (after Park's father and grandfather). Throughout 1800, the Parks lived off the royalties from his 1799 book. One senses that Park cast his eyes south more than he looked for employment in Scotland. On several occasions, hopes were raised of the possibility of an appointment, either from Banks or from the government, but nothing came of it. Bryan Edwards may have offered him prospects in the Caribbean, but he died in July 1800. That month, the capture by the French of the British-held island of Gorée off the coast of Senegal briefly raised hopes of returning to west Africa, but this never materialised. The recapture of the island by the British in April 1801 would probably have rendered any such role short-lived. We can reasonably suppose that Park helped with the family farm, and perhaps assisted Thomas Anderson, his father-in-law, but there are no records of his doing the first and, with Alexander Anderson's return to Selkirk after his medical studies in Edinburgh, there was no realistic possibility of the second as a career path.

Park returned to London in early 1801, from where he wrote to Allison to say that the change of government ministers 'has thrown everything into confusion'.[13] In February, he passed an examination of the Royal College of Surgeons of London, a qualification necessary to practise as a physician. It may be, by then, that Park was thinking of utilising his medical training, possibly as a regimental surgeon, and had had conversations over taking up the lease of a farm under the Duke of Buccleuch. There are also signs that he had more distant horizons in mind. In a letter to his wife on 12 March 1801, he wrote 'I am happy to know that you will go to New South Wales with me, my sweet wife.' Given the activities of Brown and Flinders, it is unlikely that this possibility of a colonial life was a renewal of the earlier opportunity for exploration. It is, more likely, a simple expression of an intention to emigrate.

By October 1801, any plans the Parks had for a life in Australia had come to naught. That month, Park wrote to Banks and noted how 'I left London, as you may easily suppose, a little down hearted: the romantic village which my fancy had erected on the shores of New Holland, as a habitation for myself and family, had completely disappeared; and I journeyed towards my native country with the painful but not degrading reflexion that I must henceforth eat my bread by the sweat of my brow.'

He further admitted that 'On my arrival in Scotland, it was my wish to occupy a farm; but the high price of cattle and the enormous rents which landowners everywhere expected made it rather a dangerous speculation.' Rather than farming, Park took up doctoring. To judge from the closing lines of his letter to Banks, he did so with some reluctance: 'In the meantime I hope that my friends will not relax in their endeavours to serve me. A country surgeon is at best a laborious employment; and I will gladly hang up the lancet and plaister laidle whenever I can obtain a more eligible situation.'[14]

The Parks – by mid-September a family of four, their daughter Elizabeth was born on 9 September 1801 – settled in the Borders town of Peebles. Doctoring in small town and rural Scotland may not have been Park's first choice, but his reputation afforded him access to the local nobility and gentry including the Earl of Dalkeith, heir to the Duke of Buccleuch, Sir John Hay, Colonel John Murray, and Sir James Montgomery, the last of whom enrolled Park into the Tweeddale Yeomanry, a local militia created during the Napoleonic Wars in the event of invasion. Park later wrote a patriotic song in their honour.

It was in nearby Hallyards, Adam Ferguson's home, that Park confessed to Walter Scott, Ferguson, and Dugald Stewart over the incompleteness of his *Travels*. A second son, Thomas, was born in March 1803. To judge from later correspondence, Park still hoped for a government appointment. He would have learned of others' interests in the Niger – of Hornemann's travels, and of Rennell's meeting with William Browne on 31 December 1801 where Browne was quizzed over his travels and thoughts on the Niger, conversations which informed Rennell's 1802 essay on Hornemann. He may have wondered about Hornemann or learnt of a Mr Watt and a Mr Winterbottom, English merchants, who in 1794 had tried to head for the interior from Senegal but without success. But Park heard nothing of substance from Banks or his government contacts for almost two years.[15]

Politics and the militarisation of west African exploration, 1803–1805

Park may not have heard from him, but Banks still had the exploration and commerce of west Africa as a central concern. In a 1799 memorandum to the Board of Trade written following the May meeting of the African Association, Banks proposed that Britain should annexe the west African coastline north of Sierra Leone to about 350 miles north of Gorée, the better to protect British interests and to reduce slavery and the slave trade. The proposal was never actioned.[16]

War with France was draining the nation's reserves. William Pitt's government fell in March 1801, brought down by public discontent at the conflict, leaving confusion in its wake, a fact Park transmitted to his wife. Henry Addington's government created the new post of Secretary of State for War, incorporating the responsibilities of the previous Secretary of State for the Colonies. The result of this, over time, was a growing association in the British political consciousness between African exploration and military intervention.

The first occupant of the new post was Robert Hobart, Earl of Buckinghamshire and, from 1793 to 1804, Lord Hobart. His under-secretary – in effect his only employee – was John Sullivan, who had Indian experience and whose patron was George Macartney. Sullivan understood Banks's concerns. In a letter of 13 October 1802, prompted by Banks's further urgings on the matter, Sullivan wrote to Banks over plans for a small-scale exploratory mission but one with a clear military nature:

> our little Flotilla for African Discoveries should proceed as soon as possible and I have every reason to think every thing at the Admiralty will be ready in ten days . . . I have gathered much from Moore's book, from Park and from the publications of your Association, but this only makes me the more anxious to have the lights of your mind upon the subject . . . The present [expedition] will consist of two small Frigates, some Gun Brigs and small craft, to be augmented by as great a number of large canoes and African navigators as Captn. Beaver who is to command may think proper to employ.[17]

The expedition never sailed. Plans made and delayed, or never realised, would be a common feature of the next eighteen months. But from late 1803, they at last involved Park.

Early in October, Banks wrote to Park to say that he had been 'requested by Lord Hobart to desire your attendance immediately at his Lordship's

Office in Downing Street'. Banks was not authorised to say more but, he advised Park, 'I have every reason to believe that the expedition which is on foot . . . will suit your wishes, and that proper terms of engagement will be offered to you.' Lupton has argued that this summons to Park was connected to a proposal that a Colonel Charles Stevenson should raise a corps of men to go to west Africa to attack French interests there. For that reason, it is unlikely – which Lupton claims – that Park's mission was 'exploration pure and simple'. Returning home after seeing Lord Hobart, Park wrote to Banks on 20 October: 'I found that his Lordship wished me to take another trip into the Centre of Africa, with a view to discover the termination of the Niger, that a guard of 25 soldiers would be allowed me, and that I should have 10(s) per Day for subsistence, and £200 per annum during my stay in Africa.'[18]

Park spent the remainder of 1803 in Peebles but was in London again by early January 1804. Park's letter of 5 January to his brother-in-law, Alexander Anderson – who he hoped would come too – makes clear the military and commercial nature of the Hobart–Stevenson proposal: 'I arrived here last night, and this morning had an interview with Sir Joseph and Mr Sullivan, and find that it is the wish of Government to open a communication from the Gambia to the Niger for the purpose of trade, and they do not wish me on any account to proceed further than I shall find it safe and proper. 200 pick't men embark this day at Falmouth for the expedition.' Letters between Sullivan and Stevenson in early 1804 reveal that Stevenson's plan was more concerned with a military intervention from the Gambia to the Niger for commercial purposes than it was with the geographical question of the termination of the Niger. For Park, the prospect of a second west African journey was nevertheless now clearer and closer.

Even so, the political and military circumstances were not propitious. War with France had renewed following the cessation brokered by the Treaty of Amiens between March 1802 and May 1803. Addington's government fell in May 1804. Gorée twice changed hands, from the British to the French and back again, between December 1803 and March 1804. Although Park and Anderson were prepared to leave with only a few days' notice, the troops, ships, and funds needed for west Africa's exploration were simply not available. As Stevenson understood, and as Park knew from personal experience, journeys to west Africa had to be timed to ensure travel in the dry season, or risk disaster.

Park, downhearted, returned to Peebles in early March 1804. Disappointed not to have left already but in preparation for when he did, Park brought with him Sidi Ombark Bouby, a Moroccan who had arrived in London as an interpreter to the Egyptian Ambassador. He stayed in Peebles for over two months, teaching Park Arabic. William Pitt returned as prime minister in mid-May 1804, with John Jeffreys Pratt, Lord Camden, as the new Under-Secretary of State for War and the Colonies.

On 26 September 1804, Park met with Camden in London. Park wrote to Alexander Anderson that day: 'I had a long conversation with his Lordship, he said that if I would undertake to go on an Embassy or rather a Journey of discovery to the Niger (we should be absent from England about 18 months) a sufficient force would be allowed to me and if you went with me he said Government would have no difficulty in allowing you one thousand Pounds and four thousand to me to be settled on my wife and family in case of my Discease [sic].'

Camden also wrote that day to William Pitt, to inform him of the meeting with Park:

> Mr Parke [sic] has just been with me. He is inclined to attempt the expedition proposed for the sum I mentioned – viz. £4,000 for himself and £1,000 for his Friend [Alexander Anderson] – An expedition upon the scale stated in Mr. Sullivan's memoire is *now* quite out of the Question and it is therefore to be determined in what manner a Journey of discovery and of Enquiry for commercial purposes can best be attempted. Mr. Parke seems to think that he shall be able to travel with less suspicion and therefore with more Effect, if he was only accompanied by 2 or 3 Persons on whom he could depend. He has however promised to consider and give his Sentiments in writing.

The differences expressed over the intended expedition – not previously the subject of comment in studies of Park's travels – are noteworthy. If we take these two letters at face value, Camden was prepared to allow 'a sufficient force' to accompany Park: that is, the government would support exploration with troops and materiel. Park, it would appear, expressed the view that he could travel better with only two or three companions: 'less suspicion . . . more effect'. Whatever form the mission was to take, Camden passed responsibility for it to Banks, writing to him on 28 September to that effect and stressing that 'it is material Mr. Park should sail by the Middle of October'.

Park laid out his 'Sentiments' regarding the expedition in a lengthy memoir delivered to Camden on 4 October 1804. The Memoir was structured under three headings: the objects of his journey; the means necessary to accomplish it; the manner of its undertaking. The first was clearly stated: 'The Objects which Mr. Park would constantly keep in view are, *the extension of British Commerce, and the enlargement of our Geographical Knowledge*' (the emphasis appears in the original). Questions of commerce centred around trade routes, the value of the merchandise (with gold as a priority), and profit margins.

Park proposed four subjects for 'his geographical researches'. Ascertaining the correct latitude and longitude of the different places visited in getting to the Niger was the first: he may have been mindful of Rennell's comments in his *Geographical Illustrations*. The second and third concerned the Niger: '2dly. To ascertain, if possible, the termination of that river. 3dly. To make as accurate a survey of the river as his situation and circumstances will admit of.' The last – 'to give a description of the different kingdoms on or near the banks of the river, with an account of the manners and customs of the inhabitants' – echoed the instructions he had received from the African Association almost a decade earlier, and Houghton before him.

The 'Means necessary for accomplishing the journey' shows that Park changed his mind on the nature of the expedition, perhaps to meet Camden's approval, perhaps from reflecting upon the task ahead. Now, he favoured travel with a large party: thirty European soldiers, six European carpenters, '15 or 20 Goree Negroes, most of them artificers'. Asses and mules would be bought en route in Cape Verde. Each man was to have a 'Musquito veil', a hat with a broad brim, a greatcoat for sleeping, a gun and bayonet, and a pair of pistols. Carpenters' tools and other articles were to be provided for building two boats, each forty feet in length, to navigate the Niger. Then came the merchandise and the necessary presents: the list included blue, white, and scarlet cloths, amber (to the value of £150), coral, garnets, gold beads, five double-barrelled guns, five swords with belts, small mirrors, scissors, spectacles, and much else.

From Cape Verde, Park planned to land at Gorée, then to make for Fattatenda 500 miles up the Gambia river where he would disembark having secured the permission of the King of Woolli. From there, he would 'proceed on his journey to the Niger'. From his own experience, he recognised the need 'to use conciliatory measures on every occasion'. First on horseback and then by boat, he proposed to follow the Niger, after surveying Lake Dibbie, to the kingdom of Wangara. If the river should end there – a reference to Rennell's belief that it did just that – 'Mr. Park would feel his situation extremely critical; he would however be guided by his distance from the coast, by the character of the surrounding nations, and by the existing circumstances of his situation.' Returning westwards was thought impossible, northwards equally so, and any route to Abyssinia he considered 'extremely dangerous'. The only remaining route that promised a successful exit was towards the Gulf of Guinea.[19]

Written out thus, it all seems so straightforward. Plans were laid that the expedition should sail as part of a Caribbean convoy, then divert south to Africa. But word was received that the French presence in Senegal was greater than thought. No further troops could be spared, or other resources diverted, for a larger undertaking. On 20 October 1804, Camden postponed the expedition. Disappointed though he would have been, Park maintained a state of readiness all the same: Sidi Ombark Bouby continued with his Arabic instruction; stores were purchased in expectation of the expedition's renewal. Like Banks, Park was anxious about the effect of delay on the arrival time in west Africa's dry season. Permission finally came, in late December 1804. Camden issued Park with his formal instructions on 2 January 1805, essentially reiterating the substance of Park's October 1804 memoir.

In his memoir to Camden, Park had expressed his view concerning the Niger's later course and likely end. Park believed that the Niger flowed into the Congo; that is, that the Niger river and the Congo river were one and the same. In this, he was at odds with Rennell, and both Rennell and Park differed from those who believed the Niger connected with the Nile. These views were not the only ones concerning the Niger's end.

Ideas concerning the Niger

By 1804, four hypotheses were extant concerning where the Niger ended – what contemporaries called 'this most doubtful and obscure problem in modern geography'. One was that the Niger terminated in the Nile. This view, held by Hornemann and a few others, was rejected by most contemporaries, stridently by Rennell, chiefly on the grounds of the difference in height above mean sea level of the two river systems – that is, Africa's topography made it unlikely that the Niger could maintain its flow across the continent (an estimated 2,300 miles) and join the Nile. Of all the hypotheses concerning the Niger's termination, 'that which supposes it to be a branch of the Nile, is the most unfounded, and the least consistent with acknowledged facts'.

The second hypothesis, 'supported by the greatest authorities' both ancient and modern, was that the Niger 'has an inland termination somewhere in the eastern part of Africa, probably in Wangara or in Ghana: and that it is partly discharged into inland lakes, which have no communication with the sea, and partly spread over a wide extent of level country, and lost in sands or evaporated by the heat of the sun.' This was Rennell's belief. His argument was based on 'the best informed writers of antiquity on the geography of Africa' (principally Herodotus), contemporary testimony, and his interpretation of Park's words on the Mountains of Kong.

The third idea, to which Park held – that the Niger terminated in the river Congo – 'is entirely a recent conjecture, adopted by Park in consequence of the information and suggestions of Mr. Maxwell, an experienced African trader'. George Maxwell hailed from near Longtown in the Scottish Borders. In July 1804, he had written to William Keir of Milnholm near Longtown, a mutual friend of his and Park, expressing his view that the Niger ended in the Congo, and asked that Keir forward his letter to Park. In October 1804, Maxwell wrote to Park directly, noting that he, Maxwell, had planned to explore the Congo before now but had been put off 'by the war of 1793' which prevented him from carrying what he called 'six supernumerary boats' on board his merchant ship, to be crewed, had his plan worked, by thirty or forty 'black rowers' to be hired in west Africa. Maxwell's argument was simple but it was neither wholly accurate nor consistent: that the volume of the Niger's flow was great, that the source of the Congo's equally great flow was somewhere 'far to the northward', that the idea of the Niger losing itself in the desert 'rests wholly upon the authority of the Romans', and that the Congo was of such magnitude that 'whether the Congo be the outlet of the Niger or not, it certainly offers the best opening for exploring the interior of Africa of any scheme that has ever yet been attempted'.

To moderns now familiar with the geography of west Africa, Maxwell's ideas seem far-fetched. The Congo and the Niger are separated by several hundred miles at their nearest point: the Congo's watershed lies in central Africa, the Niger's in the mountains of west Africa. But to contemporaries, they illustrate exactly that level of ignorance that so vexed Henry Beaufoy in 1788: Africa's coastal margins were known, but its interior was still not understood by Europeans.

Maxwell's reasoning powerfully shaped Park's 1804 memoir. What Park called 'The total ignorance of all the inhabitants of North Africa respecting the termination of that river' (the Niger) was reason enough to suppose it flowed elsewhere than across Africa. He drew from Browne's work and from Hornemann's, the latter having argued that, as the Niger flowed eastward into Bornu, it took the name '*Zad*'. '*Zad*', understood Park, was the name of the mouth of the Congo and for that river 650 miles inland. Furthermore, the volume of water discharged by the Congo was, Park reasoned, so great that it 'cannot be accounted for on any other principle, but that it is the termination of the Niger'. And the floods of the Congo agreed in their timing with the floods of the Niger.[20]

Park arrived at this reasoning in several ways. It was a mixture of textual exegesis (of Browne and Hornemann especially). It derived from memory of his own experiences when, in captivity, he had tried and failed to ascertain the Niger's course from indigenous sources and travelling merchants. And it may also be described as blind faith in Maxwell's unsubstantiated argument over the comparable volumes of water in the two rivers. To those who held a different view – most strongly Rennell – what denied the Niger–Congo idea legitimacy was 'That it supposes the course of the Niger to lie through the vast chain of the *Kong* Mountains (anciently *Montes Lunæ*), the great central belt of Africa.' This was not the first time that Park's views contradicted those of Rennell. Upon encountering the Niger and confirming by 'ocular demonstration' that it ran west–east, Park confessed that he had thought it flowed the other way (and this after receiving guidance, based on Houghton's evidence, from Rennell and Beaufoy). Had Park known of the fourth hypothesis, he might not have travelled at all.[21]

This was the view advanced by a German geographical commentator and civil servant, Christian Gottlieb Reichard. Reichard was born in 1758 and studied law in Leipzig. In January 1783, he took up a post, which he held until his death in 1837, as a civil administrator in Löwenstein, in Baden-Württemberg. He conducted his civic affairs so efficiently that he had time for his real love – geography and map-making. His reputation as a map-maker and surveyor soon spread: he was offered employment by Napoleon as an engineer-cartographer and, later, by the Russian Court, but declined both. Apart from short journeys within the German-speaking territories, Reichard undertook all his research while resident in Löwenstein: he was the epitome of an 'armchair geographer' – what the French called a *geographe de cabinet*.

Reichard's work on the Niger appeared in German geographical periodicals in 1802 and, at greater length, in 1803. In his 1803 article, he included a map on which he showed the course of the Niger flowing east to an inland lake (Figure 5.1). Unlike d'Anville and Rennell, however, he made the Niger continue south and south-westwards to reach the Atlantic (*cf.* Figure 4.3). Reichard showed the Mountains of Kong – *Gebrige Kong* – but, unlike Rennell, he did not depict the range as continuous. The title of Reichard's 1803 map – *Verkleinerte* [reduced] *Probe-Charte von C. G. Reichard's Atlas des Ganzen Erdkreises* – indicates that it was a working draft of the map of Africa which appears in Reichard's *Atlas Des Ganzen Erdkreises in der Central Projection*.[22]

In his 1802 article, Reichard considered and dismissed the Niger–Congo hypothesis. In his 1803 work, he focused on the hydrography of the Niger's lower reaches and on the river's probable volume as indicated by comments in travellers' accounts. From these, he estimated the river's capacity and size. He also compared descriptions of the Niger with what was known of the deltas of other major rivers, such as the Ganges, the Nile, and the Mississippi. He realised that the landforms, soils, and the form of river systems along Africa's Guinea coast all pointed to the existence of a large delta, which, in his estimation, could only be the Niger. He was right. In effect, Reichard solved the Niger problem by analogy, by sitting still and reasoning hard, and by placing trust in the testimony of others. He never saw the object of his

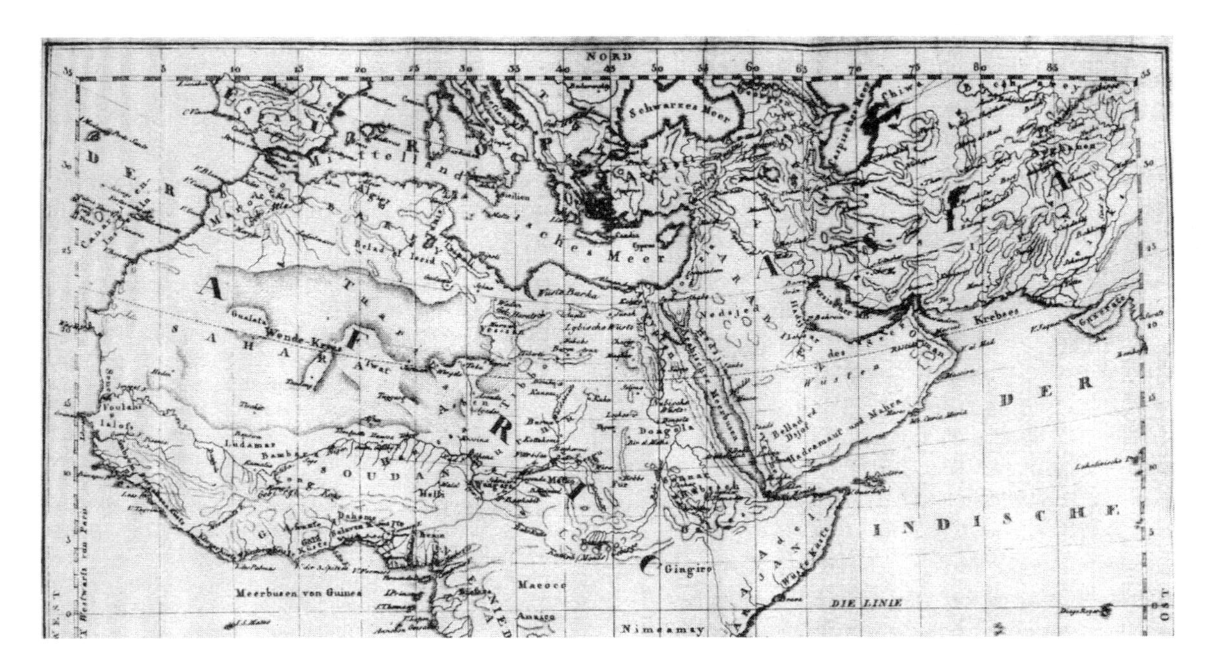

Figure 5.1 Christian Gottlieb Reichard shows the Niger rising in the mountains of west Africa and flowing in an east-north-easterly direction before swinging south and then west to empty into the Atlantic in the Gulf of Guinea. Source: C. G. Reichard, 'Probe-Charte', in *Allgemeine Geographische Ephemeriden* 12 (1803), between pp. 264 and 265. Reproduced by permission of the Syndics of Cambridge University Library, Q690. D. 3. 12.

enquiries. Yet he must be regarded as the first European to answer the second part of a 2,000-year-old geographical conundrum.

Reichard's solution may be understood in several ways. It can be interpreted as a problem in the mobility of print culture; that is, how widely were Reichard's geographical claims known and, if known, believed? Simply, a speculative armchair geographer got right what several explorers in the field did not. But seeking explanation only in the uneven circulation of geographical knowledge in print does not do justice to the wider context. For even if Rennell, Banks, Camden, or Park had known of Reichard's claim – it is likely that Rennell did know of it – they would have rejected it given Rennell's belief that the Niger flowed eastwards into an inland lake, given Park's belief in the Congo hypothesis, and, quite possibly, given Reichard's status as a stay-at-home German town clerk. Later commentators doubted it given the supposed existence of the Mountains of Kong. Whether Reichard was known about and was believed or not, Park would in all probability have returned to the Niger to confirm its geography for himself: seeing things for yourself and reporting those facts to others was what explorers did.[23]

Park closed his 1804 memoir by noting the importance of his forthcoming expedition: 'Considered in a commercial point of view, it is second only to the discovery of the Cape of Good Hope; and in a geographical point of view, it is certainly the greatest discovery that remains to be made in this world.'[24] Contemporaries would have agreed with his geographical claim. Yet, by the time Park penned those words and had set off for Africa in January 1805 to determine their truth, a German town clerk had solved the remaining part of the Niger problem without even leaving home.

6

'Once more saw the Niger rolling'

To the Niger again, 1805–1806

Mungo Park's second journey to the Niger was very different from his first. Park too was different. On his first trip, he had travelled as a lone white man on behalf of the African Association with only two African companions. He had gone as something approaching a volunteer adventurer – albeit one formally contracted by Joseph Banks – was but three years out of university, unmarried, and had only a few weeks' experience of foreign travel and tropical climates. On his second, he was a commissioned officer of the Royal African Corps responsible for over forty men, none of them African, and African servants: he was, in effect, a political-cum-commercial agent of the British government in charge of a militarised expedition. As he set off for west Africa for the second time, Park was married, a father of four – he would never see his third son and fourth child, Archibald, born in December 1804 – a celebrated author, a confidante of Britain's leading natural philosophers, geographers, and politicians, and with first-hand experience of the regions he was travelling to.

His first journey was personally punitive but geographically successful. His second would be a slowly unfolding disaster with almost no lasting geographical consequences and two mysteries at its end. And even as Park departed in January 1805 on his second journey, another man was in Africa on behalf of the African Association.

The best laid plans

The fact that Park identified the 'Means necessary for accomplishing the journey' in his October 1804 memoir says much about his forethought and, perhaps, about his own belief that things would come to pass despite others' prevarication. Park had an itemised list of the necessary scientific instruments ready by January 1804, several of which he had already bought. He continued his Arabic lessons with Sidi Ombark Bouby in London in further readiness. 'Om Bark', as he was called in a letter from Andrew Anderson to his brother Thomas in Selkirk in December 1804, was reported to be missing the Park family. London life was rather different from Peebles. On one occasion, as they left Park's London lodgings after a language lesson, Sidi Ombark and Anderson were pursued by a prostitute: 'If you had but seen him and one of the Sporting Girls that attack us every night when we are coming from Mr Parks, she ran after him a long way calling &c.'[1] If Park knew of his Arab tutor's evening adventures, he does not say.

Despite his pre-planning, Park was anxious – about what lay ahead, at the scale of the venture perhaps, certainly at leaving his wife and family, and at his responsibilities for others in Africa. Others also had their concerns: Banks that it was too dangerous, Rennell that Park was wrong-headed with respect to the Niger. And everyone was aware of the question of timing: when to leave to ensure travel in the dry season.

Park wrote to his wife within days of completing his October 1804 memoir. He had not the smallest doubt, he assured her, that the trip would be accomplished 'in one year from the time we leave England' and that 'unless the King of Bondou consent to receive and protect us we shall return immediately to Goree'. Whether or not Park really intended to withdraw if such support was not forthcoming, we will never know. These words, and their original emphasis, may reflect Park's awareness of the need to secure the support of local rulers when travelling through west Africa. Equally, they may simply have been to reassure his wife.

The couple exchanged letters between London and Foulshiels throughout that autumn and into the new year. By late October, Park had been to see Rennell in Brighton for two days where he was doubtless cautioned that his and Maxwell's Niger–Congo theory was wrong. He called twice on Lord Camden but, as he told Allison, for the most part nothing of consequence had happened: the present ministers had a good deal of the 'uncertainty' that characterised their predecessors. He expressed his anxieties over her health given her pregnancy and promised to send some money. On 14 November, he wrote to say that only the day before he had purchased £600 worth of beads for the expedition – even although it was not yet sanctioned officially – and noted how he expected to return in a year when 'Alexr. [Alexander Anderson] and I will then have both of us made something handsome in the money way.'

By the end of that month, no vessel had been identified for them nor had a sailing time been fixed, but he had received one year's salary, the remainder of what was agreed to come upon departure. Promising to send Allison £100, Park hoped that she wanted nothing for 'health and comfort'. On

Plate 1. Newark Tower. The Tower stands in the valley of the Yarrow Water very close to Foulshiels, Park's birthplace. Built in the fifteenth century, the tower was the site of a massacre of women and children after the Battle of Philiphaugh in September 1645 and the setting for Sir Walter Scott's *The Lay of the Last Minstrel*. Photo courtesy of A. Withers.

Plate 2. Park's draughtsmanship and eye for natural detail is evident in his 1797 drawing of this Sumatran fish species, *Balistapus undulatus*, the orange-lined triggerfish. © The Trustees of the Natural History Museum, London.

Plate 3. Mungo Park's drawing of the African clematis *Clematis Thunbergii St* [Steudel]. This, with other botanical and zoological sketches, including a turtle (see Figure 6.2), was drawn when Park was in Pisania, recovering from fever, before he set out on his first African travels. Reproduced with permission from the Library Collection at the Royal Botanic Garden, Edinburgh.

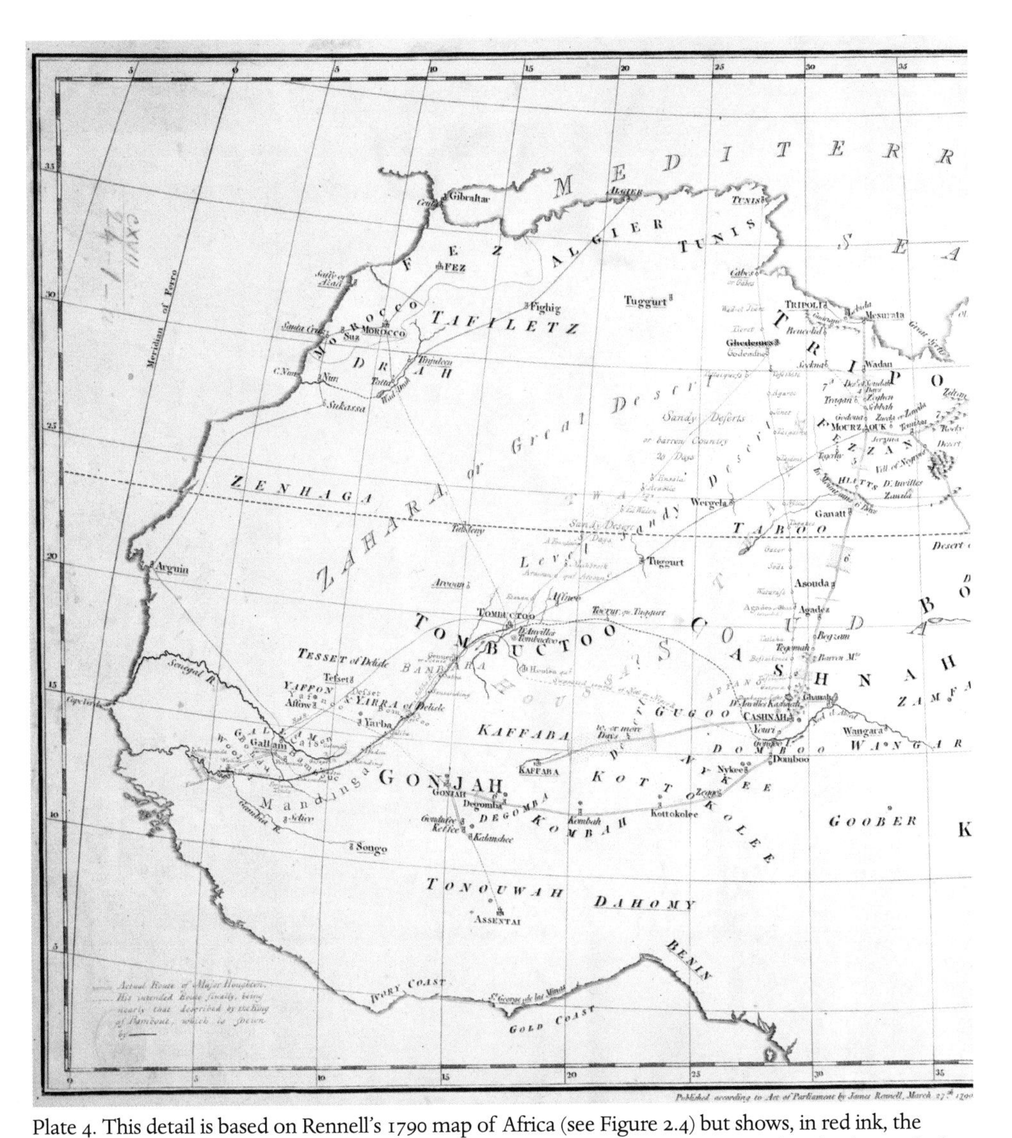

Plate 4. This detail is based on Rennell's 1790 map of Africa (see Figure 2.4) but shows, in red ink, the addition of further information on north Africa's geography and travel distances, and in the thin pecked line in red ink, the 'supposed course of the Niger'. It was compiled by James Rennell and Henry Beaufoy as an aggregative or working record of new information as they received it between *c.*1793 and *c.*1798.

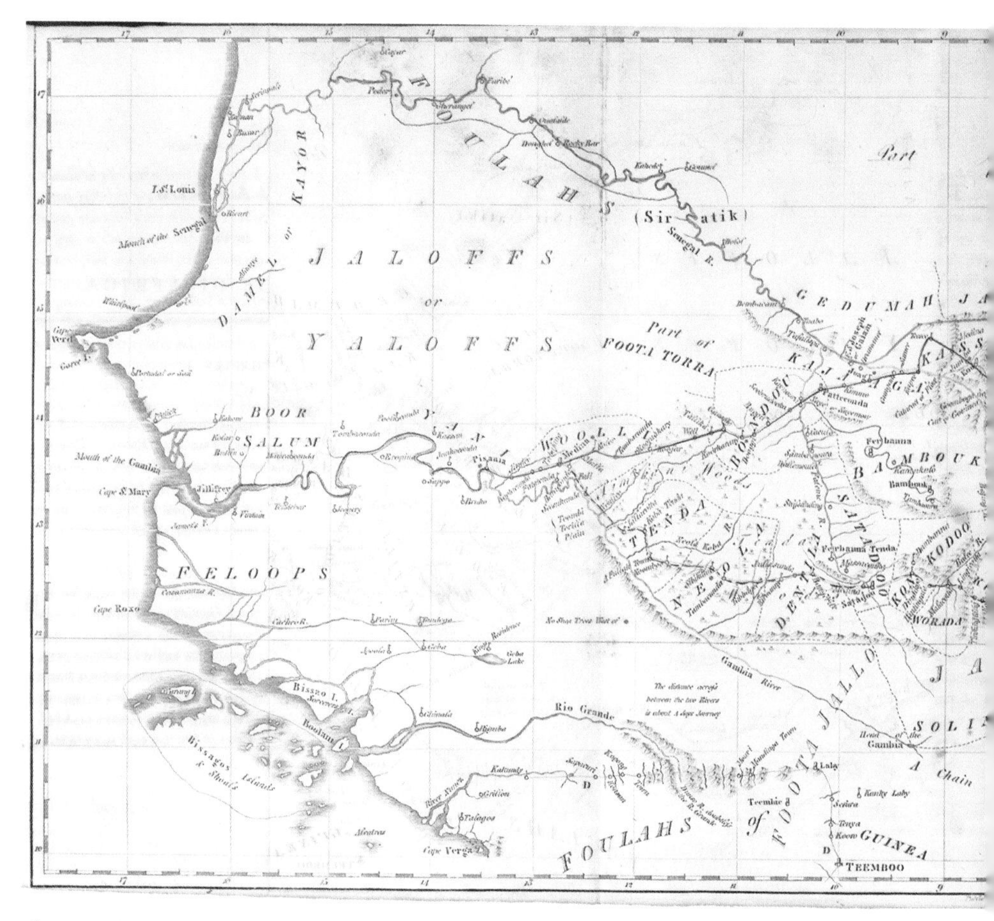

Plate 5. Park's routes on his first Niger journey, 1795–7. Sego is where he encountered 'the majestic Niger . . . flowing slowly *to the eastward*'. Source: M. Park, *Travels in the Interior Districts of Africa* (London, 1799). Reproduced by kind permission of the National Library of Scotland.

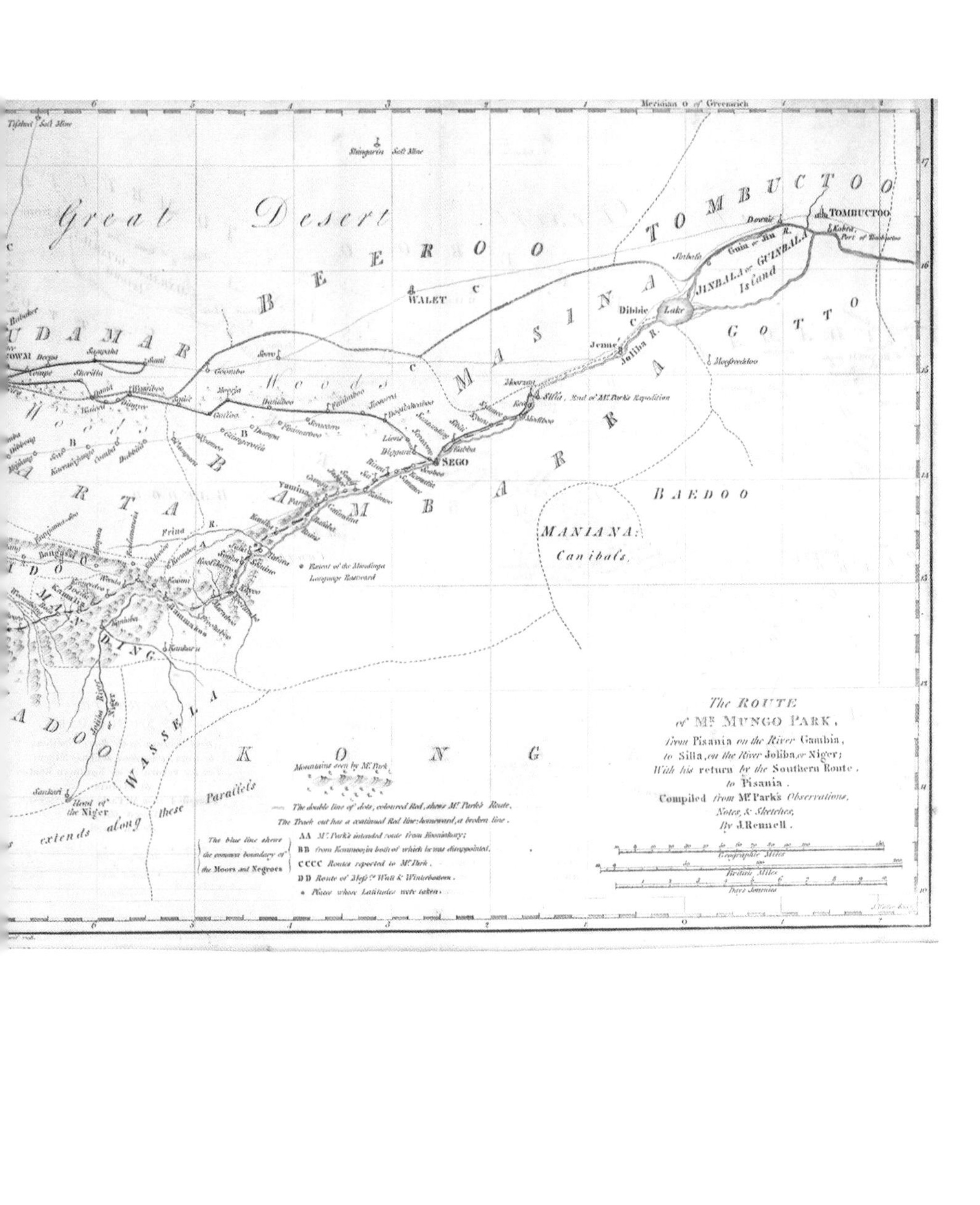

The ROUTE
of Mr. MUNGO PARK,
from Pisania on the River Gambia,
to Silla, on the River Joliba, or Niger;
With his return by the Southern Route,
to Pisania.
Compiled from Mr. Park's Observations,
Notes, & Sketches,
By J. Rennell.
Geographic Miles
British Miles
Days Journies
Meridian 0 of Greenwich
Great Desert
BEEROO
TOMBUCTOO
TOMBUCTOO
Kabra, Port of Tombuctoo
WALET
MASINA
JINBALA or GUINBALA Island
Dibbie
Lake
Jenne
Joliba R.
Silla, End of Mr. Park's Expedition
SEGO
GOTTO
BAMBARRA
BAEDOO
MANIANA: Canibals.
KAARTA
LUDAMAR
WASSELA
KONG
Head of the Niger
Mountains seen by Mr. Park
The double line of dots, coloured Red, shews Mr. Park's Route.
The Track out has a continued Red line: homeward, a broken line.
The blue line shews the common boundary of the Moors and Negroes
AA Mr. Park's intended route from Kooniakary;
BB from Kemmoo; in both of which he was disappointed.
CCCC Routes reported to Mr. Park.
DD Route of Messrs. Watt & Winterbottom.
Places whose Latitudes were taken.
Extent of the Mandinga Language Eastward

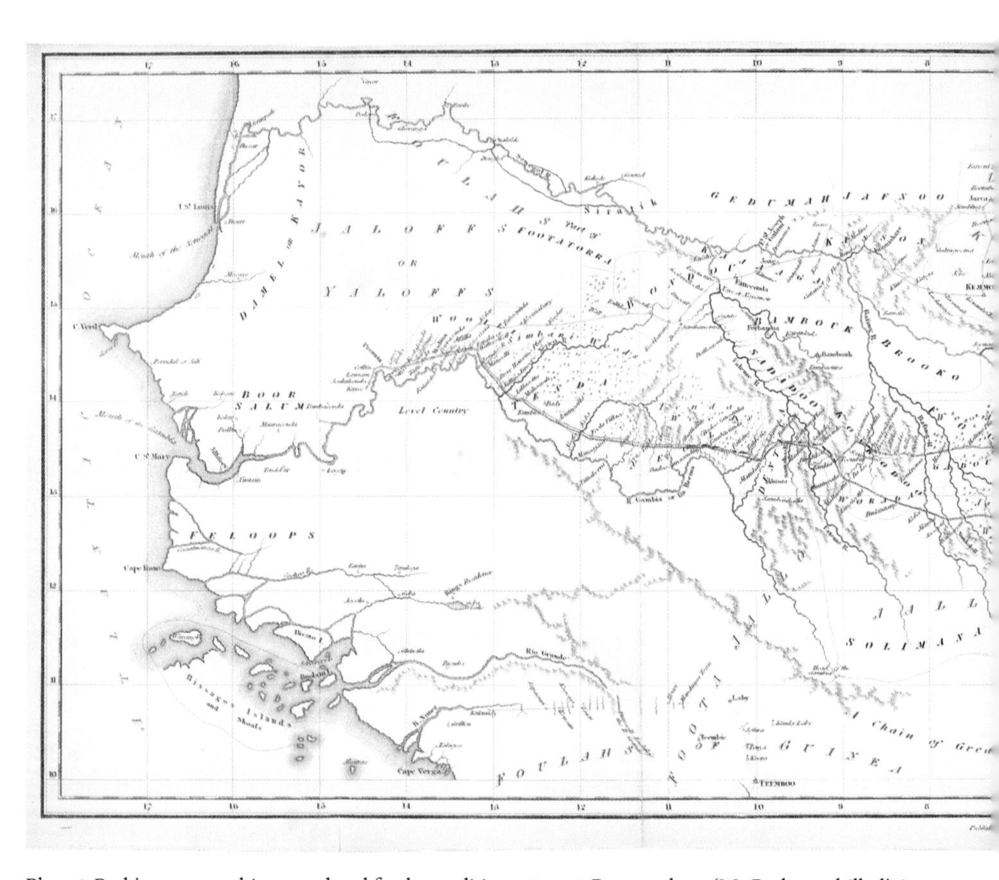

Plate 6. Park's routes on his second and fatal expedition, 1805–6. Bussa, where 'Mr Park was killed', is shown to the far right of the map as the Joliba or Niger is shown flowing eastwards into an unknown African interior. Source: M. Park, *The Journal of a Mission to the Interior of Africa* (London, 1815). Reproduced by kind permission of the National Library of Scotland.

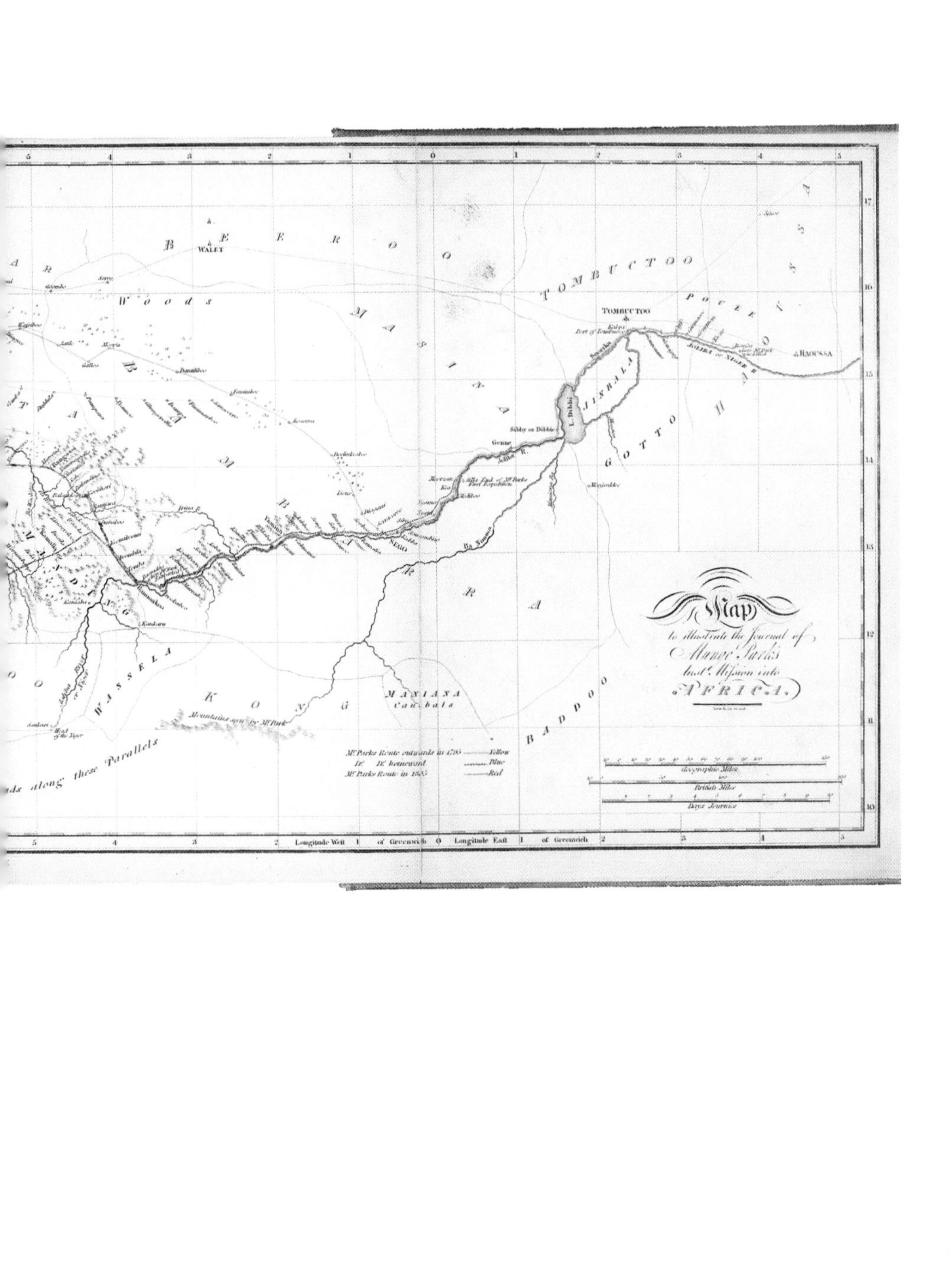

Map
to illustrate the Journal of
Mungo Park's
last Mission into
AFRICA.
Mr. Parks Route outwards in 1795 ——— Yellow
Do. Do. homeward ——— Blue
Mr. Parks Route in 1805 ——— Red
Geographic Miles
British Miles
Days Journies
Longitude West 1 of Greenwich 0 Longitude East 1 of Greenwich
TOMBUCTOO
SEGO
L. Dibbie
JINBALA
GOTTO
BADDOO
MANIANA
Canibals
WASSELA
KONG
WALET
Woods
Mountains seen by Mr. Park
...ds along these Parallels

Plate 7. Mungo Park from Thomas Edridge's engraving as he appears in the frontispiece of Park's 1799 *Travels*. African women who came to inspect Park attributed his white skin and prominent nose to his being immersed in milk as a child, his mother holding him by the nose and so pinching it as she dunked him. Source: M. Park, *Travels in the Interior Districts of Africa* (London, 1799). Reproduced by kind permission of the National Library of Scotland.

10 December, Park wrote to say that the expedition 'moves forward' but no departure date had been set. He was, as he put it, 'on the fidgits' until her 'business' (her pregnancy with Archibald), was over. On Boxing Day, he sent her £10, with the promise of more to come, and thanked his wife for the lock of her hair which she had sent him. On 3 January 1805, he wrote to reassure his wife over her anxieties expressed about the dangers of the expedition – 'the unhealthiness of the Climate is the only thing we have to fear'. The plan was to be on the coast for only a short time and since this was planned for the healthy season, he hoped that they 'would escape sickness entirely'. He was, he wrote, 'sometimes almost at the crying when I think about you'.[2]

On 28 January, his departure only three days away, Park wrote again to his wife. His letter begins with expressions of concern and of love: 'My Beloved Wife, I have waited with great anxiety for these two days for another letter from you, ah my sweet Allie I never knew till now how much I loved you.' Park asked that his best wishes be passed to Thomas Anderson, his father-in-law – 'tell him I shall write him before we sail'. He did, there expressing his anxieties about 'Allie' to Anderson and styling himself 'your affectionate son'. His letter to his wife ended with notice of his new authority. The day before, he had been made a captain and her brother, Alexander, a lieutenant: 'we are not obliged to wear any uniforms but have these Commissions to keep the soldiers in proper order.'[3]

These exchanges afford insight into Park's state of mind and that of his wife. They also clear him of that charge of callousness which some biographers have aimed at him over his leaving his wife and family and heading for London, and for preferring travelling in Africa over doctoring in Scotland. They show Park's confidence about what lay ahead, misplaced though it turned out to be. Both anxious and excited about going, he was certainly a changed man.

Sometime in January 1805, Park had his portrait painted by Thomas Rowlandson (Figure 6.1). Rowlandson was noted for his satirical representations, and it is possible that he depicted Park in that light. There is, certainly, a marked contrast between the formal, rather elegant Park in Edridge's engraving which is the frontispiece to his 1799 *Travels* (Plate 7), and the rather careworn figure painted by Rowlandson. Allison also had her portrait painted, sometime in 1804 and perhaps, with Mungo also being painted then, as early as 1798. Mungo Park carried a picture of her, doubtless an engraving of the original – done 'in fine stile' as Park put it – and it was, he told her, his constant companion.[4]

The timetable for the trip was set by Banks. His memorandum in response to Park's October 1804 memoir was prepared for Camden in January 1805 following talks between the two on 18 December 1804. It begins 'I am aware that Mr. Park's expedition is one of the most hazardous a man can undertake; but I cannot agree with those who think it too hazardous to be attempted.' For Park's plan to be successful, timing was everything. It was essential, argued Banks, that Park leave 'on or before the 8th of January', that date giving him six months 'from the time he embarks on the sea to that on which he means to embark on the River Niger'.

This six-month period Banks divided into three: two months from departure to Fatteconda, including a stop in Cape Verde to buy asses and at Gorée to recruit men; two months from there to the Niger, assuming a rate of ten miles a day over fifty days of travelling and with ten days of preparation; and 'two months more for building the Boats, Felling Trees, sawing Planks &c &c &c' by which means Park was to navigate the river to its end. Banks also stressed how important it was that the commander of the British garrison at Gorée should cooperate with Park and help him choose 'such men as excel in Bodily strength & ability to bear fatigue & heat'. In a short paragraph at the end of his memorandum headed 'Difficulties', Banks noted 'Tho' inured to the hot climates at Goree, the garrison have never been subjected to the miasmata, which produces Typhus symptoms on the Banks of the Gambia. Time must be allowed for the Seasoning Fever to be got over, & a certain loss by that Fever incurred before the Journey commences.'[5]

While Park's words to his wife over 'the unhealthiness of the Climate' and his own earlier experiences demonstrate his awareness of the dangers of travelling in west Africa in the wrong season, Banks's to Camden suggest that a certain loss of life was reckoned necessary as Europeans 'seasoned' there. Park nowhere voiced concerns in his correspondence about the schedule set for him. Neither did he about the fact that the African Association had put another man in the field as the government's expedition was being planned, postponed, and altered before it was finally approved.

Henry Nicholls had made himself known to the African Association in 1804 and was interviewed by William Young, James Rennell, and others in June that year. Perhaps cautious about his abilities – Nicholls was at first told to go away, read Park's *Travels*, and reflect upon the dangers of the enterprise – the Association was prepared to fund Nicholls only for a year in Africa and then only if he was able to travel more than 500 miles inland.

Nicholls was sent to Newtown, a trading settlement in Old Calabar, with a view to developing commercial links with local *slatees* and, in time, a route into the interior. He arrived on 17 January 1805 – two weeks before Park left for Africa – the letter confirming his arrival reached London only in August that year. Two other letters of Nicholls's survive: he soon met local rulers, settled in well, and was 'in good spirits, no obstacle appearing to prevent

Figure 6.1 Mungo Park, by Thomas Rowlandson, *c.*1805.
Reproduced courtesy of the National Portrait Gallery, London.

me from eventually accomplishing my mission'. However, two months after arriving, Nicholls was dead from fever. The Association only learnt of his death in 1807.

Nicholls has been afforded little recognition in the history of the Niger's exploration, probably rightly. Yet his brief residence in Old Calabar has a significant element of irony about it. Had Nicholls explored the local region further, he might now be more widely recognised. The town lay close to the deltaic river systems formed as the Niger breaks up into smaller rivers and enters the sea: Nicholls had even noticed on arrival how the Old Calabar river emptied itself into the Gulf of Guinea. Further exploration might have allowed him to put two and two together. As one commentator has concisely phrased it, 'Nicholls's starting-point was in fact his destination'.[6]

Park's second Niger journey

Park departed from Kayee towards Pisania on 27 April 1805, the date at which his *Journal of a Mission* (1815) begins. We know his movements and timings before then from letters to Banks and others.[7]

The expedition's ships, the *Crescent*, the transport on which Park sailed, and its escort the *Eugenie*, anchored in Porto Praya Bay at St Iago in Cape Verde on 8 March, having made slow progress since departing Portsmouth. Park there oversaw the purchase of asses, and of grain as feed, haggling with the island's governor over prices which he thought too high. The ships left for Gorée on 20 March, forty-four asses having been embarked the day before. The ships arrived at Gorée one week later, having lost two asses on the first night after they had slipped on deck: many others had 'their legs much cut with kicking each other'. A small island – only about forty-five acres, most of which was fortified or given over to buildings to hold slaves – Gorée occupied an important strategic position near the Senegal coast and the Cap Vert peninsula, a fact which was reflected in the many times it switched hands between the British and the French, and before them, the Portuguese.

At Gorée, Park found the British garrison had 'suffered much by sickness' with only 195 rank and file men fit for duty. Park offered double pay to those who would come on the expedition, and a discharge at the end of it: unsurprisingly perhaps, 'in two days the whole garrison offered themselves, as did also Lieut John Martin' [this is usually spelt 'Martyn', the spelling I use here]. On 6 April, the selected soldiers and two sailors embarked: they, like the ship's carpenters who had joined at Portsmouth, were also on double pay.

In total, the party numbered four officers (Mungo Park, Alexander Anderson, John Martyn, and George Scott, the draftsman and, like Park and Anderson, a Borders man), forty-one soldiers, seamen and carpenters, Isaaco the appointed guide, his wife and child, and several of Isaaco's people. On 7 April, both vessels grounded in the river Gambia, on a sand spit 'not laid down in any chart', but they incurred no damage. As the expedition made its way up the Gambia, gifts were given to local rulers. As circumstances permitted, temperature readings were taken: '100 in the shade at 3 oclock' [pm] as the ships anchored off Devil's Point on 10 April; '123 at noon' two days later off Seahorse Island; '78' the recorded figure 'of the air at 7 in the Even[ing]' four days later.[8]

On the morning of 15 April, the ships arrived at Kayee. The days following were spent purchasing more asses, taking longitudinal readings by which to correct and regulate the chronometer and, from the soldiers, entertaining a local king to a display of precision drilling and a fusillade. On the evening of 26 April, Park wrote to Thomas Anderson: 'We set off tomorrow morning for the interior, with the most flattering prospect of finishing of our Expedition in the course of six months with honour to ourselves and benefit to Mankind.' He enquired after the welfare of 'my dear Allie'. In a separate enclosure, Park laid out his will and what to do with monies accrued 'In case it should please the Almighty to take me to himself.' He wrote to Allison, noting there that he had not looked at her picture for six days 'for fear it would make a coward of me'.[9]

In a separate letter to Banks, Park noted his plans for departure. For several reasons, these included retracing his earlier route: 'Mr Ainsleys vessel is to carry our goods to Woolli . . . Mr. Anderson goes with her: the soldiers & myself travel by land . . . I propose to travel by the Southern route on the same road as I returned from the interior. I do this chiefly on account of the Abundance of water & the smallness of the Kingdoms I shall pass.'

Everyone was expectant, but Park was reflective, more so than in letters to his wife and father-in-law:

> The Troops are all in high health: we have lost one man by sickness, but he was unwell when we left Goree: they are all in high spirits & promise to be useful men: if I can find an opportunity, I will write from Woolli. I hope to find means of sending accounts from the interior: nothing shall be wanting on my part to compleat [sic] the plan laid down to me before I left England: if I succeed, I shall be the happiest man on Earth. I will enjoy, even in my last moments, the pleasing reflection that I have done my utmost to execute an undertaking worthy of the best of Governments.[10]

With a salute from the ships' guns, the expedition got underway at 10.00 a.m. on 27 April. They arrived at Pisania the following day. Two days later, Ainsley's schooner arrived with their baggage and the rice. Park had earlier paid his respects to Seniora Camilla, who he had met on his first travels: she was 'much surprised to see me again attempting a journey into the interior of the country'. When the expedition departed Pisania, the asses were each marked and numbered with red paint and divided among the six groups into which the whole party was split: George Scott and one of Isaaco's men led the way, with Martyn in the middle, and Park and Anderson to the rear. This positioning allowed Park to keep the asses moving, to assist those lagging, and to help when, as often happened, animals shed their load or refused to continue unless goaded.

The narrative of the first few days has a repetitive tone: dealing with truculent animals, fatigue from the heat, searching for water, and presenting gifts to local rulers some of whom, on seeing the size of the party, sought additional tribute before allowing it to proceed. Park had time to observe cotton being dyed with indigo, his notes on the process incorporating the botanical names for the plants involved. As if in an instructional note to himself, his description of the process ends abruptly: 'To return to the narrative'. His narrative in the first few weeks reads as a record of making and breaking camp, securing provisions, and, increasingly, of illness among the men under his command.

At dusk on 15 May, as they were unloading the asses for the evening, one of the soldiers, John Walters, 'fell down in an epileptic fit, and expired in about an hour after'. Perhaps mindful of Rennell's observations on the accuracy of his longitudinal readings from his first journey and perhaps because he was charged with surveying the course of the Niger, Park was at pains to record his calculations regularly and to establish their correctness. Observations taken on the evening of 16 May, for example, showed his watch to be fast, by three minutes and thirty-five seconds. 'It is difficult to account for such a difference in the rate of going in the watch in the course of one month', noted Park, 'but the excessive heat and the motion of riding may perhaps have contributed to it; for I think my observation of the immersion was correct.' Not for the first time, anxieties about the reliability of scientific instruments – could they be trusted to give accurate readings? – affected how explorers thought of their conduct and position in the field.

In a letter of 26 April to Banks, Park wrote that they were proceeding 'with as much success as I could reasonably expect'.[11] He apologised for his lack of attention to botanical and natural historical matters given the demands of the expedition, but he had sent a few specimens back in a trunk 'which I hope will come safe'. He also sent on some Mandingo cloth dyed with indigo but had not been able to procure any Bondou frankincense. He asked that Banks pass on his compliments to Rennell 'and tell him that I hope to [be] able to correct my former Errors'. By this we must presume he meant his longitudinal calculations rather than his belief in the Niger–Congo theory.

Park adopted a more positive tone in his letters, to Banks and to family, than he did in his notebook: where he recorded instrumental errors in his notebook and pondered upon the reasons for them, he told Banks that 'The watch goes on so correctly that I will measure Africa by feet and Inches'.[12] Park wrote on the move: his letter of 27 April to Banks, which continued the theme of economic botany from the day before, was 'Written in great haste in an asses saddle, the morning we left Kayay [Kayee].' More usually, he would have tried to find some time at the end of each day to record his thoughts and the day's events.

These and other remarks reveal exploration – if we may take Park's second journey as an exemplar – to be a slow, even rather repetitive, process routinely performed day after day after day: securing food and water, making and striking camp, maintaining discipline, undertaking astronomical measurements and calculating longitude while encouraging one's fellow travellers, and, not least, finding the time to write about events in ways that would be useful when, later and elsewhere, one came to construct a narrative about them. Exploration required a certain regimen, for one's fellows, and for oneself. For Isaaco, it could be assisted by propitious offerings: on the morning of 15 May, as the party left the village of Kussai, he slaughtered a black ram on the road as a good omen, the animal then being offered to the slaves in Kussai 'that they should then pray for the expedition's success'.

Their first difficulty came on 20 May, at the village of Tendico or Tambico, whose local ruler, the Farbana of Jallacotta, was a subject chief under the rule of the King of Woolli. The tribute gifts initially offered were refused, being deemed inadequate. The refusal, and the fact that Park's party was met by twenty-six armed men, alarmed Park sufficiently to warrant his sending a note back to Martyn to 'be ready for action at a moment's notice'. That evening, as Isaaco's horse was being led to water, Farbana's people seized it. As Isaaco enquired into their actions, they seized him, stripped him of his weapons, tied him to a tree, and flogged him. Park tried but failed to take two locals as hostage for Isaaco: 'I was,' he wrote, 'a little puzzled how to act.'

The incident prompted the first expression from Park of the possibility of an inappropriate and excessive response: 'We could have gone in the night and burnt the town. By this we should have killed a great many innocent people, and most probably should not have recovered our guide.' Following discussion with Anderson and Martyn, caution prevailed. The following morning Isaaco was released. His weapons were not returned.[13]

In the light of later circumstances, it is possible to see in this moment the first expression of a rather different Park. Anxious about the expedition's progress, the accuracy of his instrumental readings, and the welfare of the men under his command, he was, it seems, ready to countenance the use of force to secure progress, even if that meant not paying proper attention to the demands of local rulers, several of whom, in his view, were asking too much in their demands over tribute. This question of tribute and safe passage is a recurrent theme in his second journey – and of its end.

On 26 May, a swarm of African bees attacked the party, having been disturbed by some of Isaaco's men looking for honey. 'The bees came out in immense number,' Park recorded, 'and attacked men and beasts at the same time.' The asses galloped up the valley to escape. Horses and men were badly stung. Left unattended, the cooking fires spread and set fire to the huts built for shelter: 'In fact, for half an hour the bees seemed to have completely put an end to our journey.' Yet, in his letter to Allison on 29 May,

Park said nothing of this. There, he reckoned they were halfway through the trip and had achieved this 'without the smallest accident or unpleasant circumstance'.[14]

Neither did he mention it in his letter – a 'hasty scrawl . . . only meant to let you know I am still alive and going forward in my journey' – to Banks the day before: 'we have had as prosperous a journey thus far as I could have expected'. Judging by his latitudinal and astronomical observations, his own earlier map (Plate 5) was in error, the course of the Gambia river being too much to the south. Park further noted that, 'I find that my former Journeys on foot were underrated. Some of them surprise myself when I trace the same road on horseback.'[15] The map that appears in Park's *Journal of a Mission* shows his routes in 1796 and in 1805 (Plate 6).

Early in June, in the town of Julifunda, the expedition was halted for three days while Park dealt with what he clearly felt were the extortionate tribute demands of the local chief, Mansa Kussan. Mansa Kussan initially appeared perfectly satisfied with the amber, scarlet cloths, and coral which Park paid in tribute, but he later changed his mind and refused to let the expedition proceed further unless substantially more was paid. Should Park not pay more, Mansa Kussan 'would do his utmost to plunder us in the woods'. Park passed the king a little more amber and scarlet cloth but held fast at that. The chief relented only when Park, who considered he had paid significantly over the odds to secure safe passage, stated firmly that he would give nothing more – and certainly not the requested gunpowder and flints – and that if the party should be obstructed, 'we would do our utmost to defend ourselves'. Progress was slow, hampered by lack of water, by tending to one of the carpenters who died on the evening of 8 June, and by the over-burdened and fatigued asses. And the rains were coming: 'a heavy tornado with much thunder and lightning' on 8 June presaged worse.

The fortunes of the expedition changed dramatically once the rains arrived:

> The tornado which took place on our arrival [at the small town of Shrondo on 10 June], had an instant effect on the health of the soldiers, and proved to us, to be the *beginning of sorrow*. I had proudly flattered myself that we should reach the Niger with a very moderate loss: we had two men sick of the dysentery; one of them recovered completely on the march, and the other would doubtless have recovered, had he not been wet by the rain at Baniserile. But now the rain had set in, and I trembled to think that we were only halfway through our journey. The rain had not commenced three minutes before many of the soldiers were affected with vomiting; others fell asleep, and seemed as if half intoxicated. I felt a strong

> inclination to sleep during the storm; and as soon as it was over I fell asleep on the wet ground, although I used every exertion to keep myself awake. The soldiers likewise fell asleep on the wet bundles.[16]

By mid-June, as the expedition neared the village of Fankia, it had become strung out, often over several miles. The thickness of the vegetation in places meant that the expedition was never fully in sight. Park rode and walked back and forth, tending sick soldiers, walking at their faltering pace, putting them on his horse when they could walk no more. He confessed himself 'very sick, having been feverish all night'. As they left Fankia, their route took them on a steep and rocky ascent through a pass in the Tambaura mountains. Park describes their progress vividly: 'The number of asses exceeding the drivers, presented a dreadful scene of confusion in this rocky staircase; loaded asses tumbling over the rocks, sick soldiers unable to walk, black fellows stealing; in fact it certainly was *uphill work* with us at this place.' Amber and beads were stolen from Anderson's horse, which he had lent one of the ill soldiers, but to Park's relief, the pocket sextant and artificial horizon were not taken. As they left the village of Dindikoo on the morning of 13 June, progress was hindered as one group of asses with their drivers took the wrong road. Park confessed himself 'Very uneasy about our situation: half the people being either sick of the fever or unable to use great exertion, and fatigued in driving the asses.' And 'to my great mortification' as he put it, the ass which carried the telescope had not appeared. Anderson was sent back with one of the soldiers to find it, but after retracing their route for five miles, was unable to find it.

Word of the expedition preceded it, and this caused difficulties on more than one occasion. In mid-June, near the village of Gimbia, with Park at the rear bringing up some asses which had shed their loads, he was alerted by noises ahead. The cause of the tumult 'was, as usual, the love of money. The villagers had heard that the white men were to pass; that they were very sickly, and unable to make any resistance, or to defend the immense wealth in their possession.' On 23 June, after the party had arrived at Gimbia, locals started to unload the asses and unbridle the horses. What appeared at first to be acts of assistance was soon revealed as theft: 'everything seemed going into confusion'. The soldiers loaded their muskets and fixed their bayonets. This made the locals back off and Park laughed at the group of local men, armed with bows and arrows, who stood threateningly, watching proceedings. Park informed the headman that 'we did not come to make war; but if any person made war on us, we would defend ourselves to the last'.[17]

Such comments, and similar ones about being continually robbed, should not lead us to suppose that the Park of the second expedition was less empathetic towards Africans than he had been earlier. There is, rather, a sense that the size and nature of the expedition was from the Africans' perspective a potential source of immediate wealth, in contrast to the poor pickings to be had from a lone white man. The expedition regularly advertised its size and nature. As it straggled in wooded or hilly country, or as its members lost sight of one another, muskets were fired as signals: what was a form of direction finding by sound might easily be misinterpreted.

On 2 July, another of the soldiers, Roger McMillan, was left behind, to Park's obvious regret. McMillan, wrote Park, 'had grown old in the service of his country'. He had been a soldier for thirty-one years, twelve times a corporal, nine times a sergeant, 'but an unfortunate attachment to the *bottle* always returned him to the ranks'. By then, Anderson and Scott were sick with fever, and Park pronounced himself 'very sickly, having lifted up and reloaded a great many asses on the road'. Many of the soldiers were too ill even to stay on their mounts. William Alston, one of the two sailors from HMS *Squirrel* who had joined at Gorée, begged to be left behind: Park did so, leaving him a loaded pistol and some cartridges in the crown of his hat.

As Isaaco led six asses across a river on 4 July, he was attacked by a crocodile and dragged under. With great presence of mind, Isaaco thrust his finger into its eye, at which the crocodile let him go. Before Isaaco could reach the far bank, he was attacked again. Again, he poked it in the eyes, and the crocodile finally relented. Isaaco was 'very much lacerated' with deep wounds to his thighs. Park bandaged the wounds with slips of adhesive plaster and sent Isaaco ahead to the nearest village to recover. As he did so, Park was astonished to see Alston, the sailor he had left behind in the woods the previous night, walking naked towards him, having been stripped of his clothes during the night.[18]

Isaaco was a good guide and a valued member of the party but, on one occasion, he accidentally struck two of the soldiers as the party fended off thieves. This nearly cost him his life: one of the soldiers drew his bayonet, making to stab Isaaco, but Alexander Anderson stopped him in time. Park reproved Isaaco for his actions 'in the strongest manner' but Isaaco 'went off in a *pet* with his people, leaving us to find our way across the river in the best manner we could'. The rigours of exploration often made personal relationships fraught, but Isaaco and Park were soon reconciled.[19]

Sattin has observed that 'There is a nightmarish quality to Park's writing once the rains start'.[20] This is true inasmuch as Park records how travelling became more difficult and the depredations of thieves more frequent and more brazen. Yet even when he and others were constantly inconvenienced by being soaked and having to dry goods and clothing, or in having to re-load the pack animals and ward off threats, Park found the time to make and to record longitudinal calculations and to comment on local customs, methods of gold panning, and the physical environment. As they left Sullo in late June and 'travelled through a country beautiful beyond imagination', Park and his companions briefly halted at a rock formation which looked to them like a ruined Gothic abbey – 'A faithful description of this place would certainly be deemed a fiction' – and a dome of red granite without so much as a blade of grass: 'never saw such a hill in my life'.[21]

Park's *Journal* is not the polished narrative of his *Travels*, but something more akin to a field notebook, with entries for most days but of different length and register. Sometimes, clipped entries suffice for a whole day – 'July 7th. – Dressed Isaaco's wounds: they looked remarkably well' – with longer descriptive passages detailing several events and written when Park had time free from his other responsibilities. He did not neglect his longitudinal observations. Resting in Secoba on 26 June, and somewhat restored by having purchased supplies of chickens and milk, Park checked the satellites of Jupiter and the copy of Maskelyne's *Nautical Almanac* which he had carried with him to arrive at their longitude. His watch was running forty-eight seconds slow. By late July, it was in greater error, being two minutes and thirty-four seconds slow. Arduous though it was, travelling demanded such calculations in order to fix their position and so, in time, plot positions on a map and correct earlier maps.

From mid to late July onwards, the themes of ill and dying soldiers, the constant thieving, the demands of tribute, and Park and Anderson's resort to the use of force or to the threat of force, became the dominant themes in his *Journal*. Near Seransang on 18 July, Park shot and wounded a thief who had made off with his great coat: Park left him 'bleeding amongst the branches of the tree'. On 27 July, shortly after leaving the village of Bangassi, three soldiers lay down under a tree, refusing to go further. Park later sent the men a note, with three strings of amber to buy provisions 'till you arrive at the Niger'. They never did. On 6 August, two more soldiers and one of the sailors fell behind, one of the soldiers appearing later. Four more men were lost by 11 August. Alexander Anderson was 'still in a dangerous way, being unable to walk or sit upright'. On 12 August, as he helped the ailing Anderson into his saddle, both men heard deep growls from the bushes nearby. Wondering what animal it was, and proceeding only 100 yards further, Park 'was not a little surprised to see three lions coming towards us'. 'I was afraid', he wrote, that if his musket should misfire, 'we should all be devoured by them.' Instead of retreating, Park advanced: 'I therefore let go of the bridle, and walked forwards to meet them.' At this, the lions stopped. Park fired at one, at which all three retreated into the bushes.[22]

Some respite from the rigours and dangers of travel came on 15 August in the town of Doombila where Park again met Karfa Taura, now resident in Boori where he had heard reports 'of a coffle of white people . . . conducted by a person of the name of Park, who spoke Mandingo'. 'You may judge of the pleasure I felt,' recorded Park, 'on seeing my old benefactor.' And on the afternoon of 19 August, Park crested the brow of a hill and '*once more saw the Niger* rolling its immense stream along the plain!'[23] He had hoped to reach it by 27 June.

Park's observations at this point evidence a certain pride in the achievement, and a somewhat unrealistic view of the expedition's progress and its likely success:

> After the fatiguing march which we had experienced, the sight of this river was no doubt pleasant, as it promised an end to, or to be at least an alleviation of our toils. But when I reflected that three-fourths of the soldiers had died on their march, and that in addition to our weakly state we had no carpenters to build the boats, in which we proposed to prosecute our discoveries; the prospect appeared somewhat gloomy. It however afforded me peculiar pleasure, when I reflected that in conducting a party of *Europeans*, with immense baggage, through an extent of more than five hundred miles, I had always been able to preserve the most friendly terms with the natives. In fact, this journey plainly demonstrates, 1st. that with common prudence any quantity of merchandize may be transported from the Gambia to the Niger, without danger of being robbed by the natives: 2dly, that if this journey be performed in the dry season, one may calculate on losing not more than three or at most four men out of fifty.

As the party headed for Bambakoo from Toniba (on which short journey three more soldiers died), Park gave a succinct and more accurate count of the expedition's mortality: 'Of thirty-four soldiers and four carpenters, who left the Gambia, only six soldiers and one carpenter reached the Niger.'[24]

Travelling on the Niger was made easier by the river being in flood. The party's canoes passed easily over numerous rapids. As they headed downstream, Park was too ill to shoot at a large elephant (whether for sport or for their safety is not recorded). Hippopotami, he reported, could be frightened away by the sound of a musket. One of the canoe men speared a turtle 'of the same species as the one I formerly saw, and made a drawing on in Gambia' (Figure 6.2). As they had turtle and rice for supper, the rains continued 'with great violence all night.'

Late in August, the expedition halted at the town of Marraboo where Park prepared to send ahead tribute goods for Mansong, the King of Sego, the gifts to be negotiated between Isaaco on Park's behalf, and Modibinne, Mansong's prime minister. The gifts included a 'handsome silver plated tureen', two double-barrelled guns and two pairs of pistols, all silver mounted, and lengths of coloured cloth. Park retained the guns and some gunpowder to be given 'as soon as I heard accounts that Mansong would befriend us'. Isaaco departed, with his wife and the goods, on 28 August. Park, who had been suffering from dysentery since arriving in Marraboo, was probably too ill to notice. He took extreme measures to alleviate his condition: 'as I found my strength was failing very fast, I resolved to charge myself with mercury. I accordingly took calomel till it affected my mouth to such a degree, that I could not speak or sleep for six days.'[25]

Figure 6.2 Mungo Park sketched this picture of a turtle while he was recovering from fever in Pisania in late 1795. The outlines above the animal are Park's attempts to depict the dimensions of the turtle's shell. Note, too, the ship (not to scale) to the far left. Reproduced with permission from the Library Collection at the Royal Botanic Garden Edinburgh.

Park waited in Marraboo for days for news about Isaaco. When he was told, on 8 August, by Bookari, Mansong's singing man, that he could proceed to Sego, departure was delayed a further four days given Bookari's fondness for the local beef and beer. Despite claims from Bookari that Mansong was pleased with the gifts sent with Isaaco, Isaaco informed Park when the two were reunited on 19 September that 'Mansong had never yet seen any of them'. Mansong was prepared to allow Park permission to proceed but he used Modibinne as a go-between to quiz Park over his motives. Speaking in Bambarran, Park told him that he was 'the white man who nine years ago came into Bambarra'. He used Modibinne to flatter Mansong: his earlier generous conduct towards Park 'has made his name much respected in the land of the white people'.

Being further encouraged to disclose his purpose in returning to Africa, Park stressed the trading opportunities and emphasised how direct contact would be cheaper than current networks. In doing so, he disclosed his thinking on the course of the Niger: 'For this purpose, if Mansong will permit me to pass, I propose sailing down the Joliba [the Niger] to the place where it mixes with the salt water; and if I find no rocks or danger in the way, the white men's small vessels will come up and trade at Sego, if Mansong wishes it.' Park urged Modibinne to disclose this to no one save Mansong and his son – 'for if the Moors should hear of it, I shall certainly be murdered before I reach the salt water'.[26] While it is not clear from these words if Park intended to reach the 'salt water' by way of the Congo, the suggestion that the Niger reached the sea rather than emptied into an inland lake again placed him at odds with Rennell.

Mansong gave permission without ever meeting Park but promised to send canoes to facilitate his travels. Park and the remaining men then moved to Sansanding as the preferred location from which to proceed downriver. By 8 October, the promised canoes not having arrived, Park turned merchant and shopkeeper to secure the necessary resources. He opened a stall in the local market selling his trade goods to secure sufficient cowries to buy two canoes: Park even produced a price list giving the exchange rates. For over a week, he was so busy that 'I was sometimes forced to employ *three tellers at once* to count my cash'. But when the canoes finally arrived on 16 October, one was part rotten, as was its intended replacement. Park set about joining the two better halves together. Over eighteen days with the assistance of Private Abraham Bolton, the two men '*changed* the Bambarra canoe into *His Majesty's schooner Joliba*; the length forty feet, breadth six feet; being flat bottomed, draws only one foot water when loaded'.[27]

Any sense of achievement at this was dashed with Alexander Anderson's death early on the morning of 28 October. Park ends his entry for that day with reflections upon his 'dear friend' and, after burying Anderson, a note on his emotions: 'I then felt myself, as if left a second time lonely and friendless amidst the wilds of Africa.' There were no further entries until 14 November. By then, the canoe was nearly ready for departure. Park was anxious for Isaaco's return from Mansong's court so that he could entrust his guide with his notes. Isaaco returned the following day, as Park was completing the awnings on the canoe, with the news that Mansong was anxious that Park should depart soon, lest the Moors catch news of him. On 16 November, Park began his journal thus: 'All ready and we sail to-morrow morning, or evening.'[28]

Park's notes end with remarks upon local names, upon the accuracy of his observations using Troughton's pocket sextant which he had bought twenty-one months earlier, and a sketch map of the course of the Niger derived from a local who had seven times been to Timbuktu (Figure 6.3). Later, and in

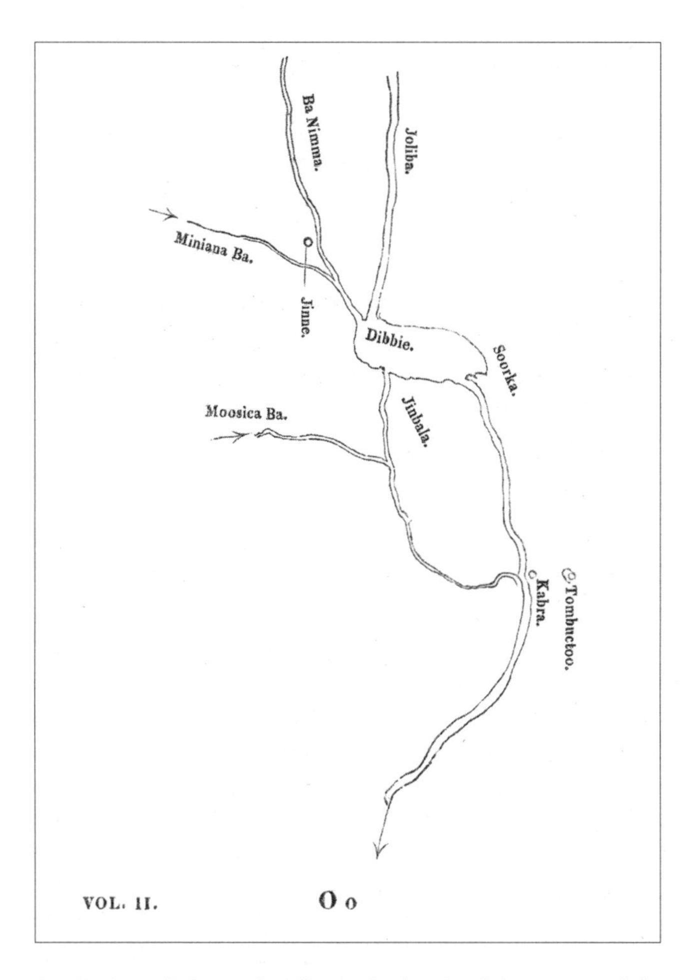

Figure 6.3 Park took down this 'facsimile sketch of the course of the Niger' in November 1805 from an African trader who had been to Timbuktu on seven occasions and was soon to depart for the town again. Source: M. Park, *The Journal of a Mission to the Interior of Africa* (London, 1815). Reproduced by kind permission of the National Library of Scotland.

others' hands, these notes, calculations, and sketches would become *Journal of a Mission*.

The end of the expedition

What we know of the expedition's final days comes from letters which post-date Park's last notes as they appear in the *Journal* and from Isaaco who gathered evidence from Amadi Fatouma, Park's guide in the expedition's later stages.

Park penned numerous letters in the days before departing from Sansanding. His last surviving letter to his wife was written on 19 November 1805. There he reported upon the deaths of Alexander Anderson and George Scott, referring her for details to the letter he had written to Thomas Anderson three days earlier. Park was, he told his wife, in good health, the rains were over and 'the healthy season has commenced, so that there is no danger of sickness and I still have a sufficient force to protect me from any insult in sailing down the river to the sea'. In addition to Fatouma, Park had seven companions: Lieutenant Martyn and three privates, one of whom was 'deranged in his mind and three slaves'. In a letter of the same day to the trader Thomas Dean at Kayee, Park praised Isaaco and asked Dean to pay Isaaco what was due him.[29]

Park's letter to Banks of 16 November is detailed and business-like, understandably different in tone from letters to his wife and family. After an initial note of thanks to his patron – 'I should be wanting in gratitude, if I did not embrace every opportunity of informing you how I have succeeded in this Enterprise' – Park laid out his 'future views':

> It is my intention to keep the middle of the river, and make the best use I can of winds & Currents till I reach the termination of this mysterious stream. I have hired a guide to go with me to Kashna [Amadi Fatouma]: he is a native of Kapon, but one of the greatest travellers in this part of Africa having visited Miniana, Kong, Badoo, Gatti, and Cape Coise Castle to the South, and Timbuctoo, Housa, Nyffe, Kashna and Bornou towards the east. He says the Niger after it passes Kashna runs directly to the right hand or the South: he never heard of any person who had seen its termination, and is certain that it does not end anywhere in the Vicinity of Kashna or Bornou, having resided some time in both these Kingdoms. He says our voyage to Kashna will occupy two months, that we touch on the Moors no where but at Timbuctoo; the north bank of the river in all other places being inhabited by a race of people resembling

the Moors in colour, called Surrka, Makinga, & Twarick [Tuareg], according to the different kingdoms they inhabit.

Park expected that 'we will reach the sea in three months from this; and, if we are lucky enough to find a vessel, we shall lose no time on the Coast; but at all events, you will likely hear from me, as I mean to write from Kashna by my guide, and endeavour to hire some of the merchants to carry a letter to the North from that place'. He ended his letter by asking Banks that he be remembered 'to my friend Major Rennel [sic]'.[30] Again, Rennell may have been in his mind as Park admitted to Banks his hopes of reaching the sea.

Park was more direct in his letter of 17 November to Lord Camden: 'With the assistance of one of the soldiers I have changed a large canoe into a tolerably good schooner, on board of which I this day hoisted the British flag, and shall set sail to the east with the fixed resolution to discover the termination of the Niger or perish in the attempt. I have heard nothing that I can depend on respecting the remote course of this mighty stream; but I am more and more inclined to think that it can end no where but in the sea.'[31]

With these letters, all reliable information concerning the expedition and Park's fate comes to an end.

In the following year, word reached British settlements on the coast that Park and his companions had been killed but this information came 'upon no distinct authority'. As the rumours increased, all without substantiating evidence, Lieutenant-Colonel Charles Maxwell, the governor of Senegal, obtained the permission of the British government to send someone to determine the truth. For this, he engaged Isaaco, Park's guide for much of the expedition. Isaaco was absent from Senegal undertaking this task between January 1810 and September 1811.

Isaaco's journal in search of the facts of Park's death is principally a record of Isaaco's travels and the events which befell him and his companions en route: gifts in tribute to local kings; gifts to him of livestock for food; an attack by African bees which killed one of his asses – 'stifled by the bees getting into its nostrils and one of my men almost dead by their stings' – and meeting emissaries in order to facilitate safe passage. In early August 1810, in the village of Sannanba, he encountered his sister and one of his wives who he had left behind while guiding Park. They had been told that Park was dead, one informant having been to the place where Park was reported to have died. Early in October, Isaaco arrived at Sansanding. There, he met Amadi Fatouma. 'I demanded of him, a faithful account of what had happened to Mr. Park. On seeing me, and hearing me mention Mr. Park, he began to weep; and his first words were, "They are all dead".'[32]

The following day, Amadi Fatouma told Isaaco what had happened.

7

'To do justice to Mr Park's memory'

Park's death and his Journal of a Mission *(1815)*

Park's second Niger expedition did not have a successor for ten years – not until John Peddie and Thomas Campbell headed for the Niger in 1815 and James Hingston Tuckey set out in February 1816 to trace the course of the Congo and to determine whether, in fact, it was the Niger.

Africa's exploration nevertheless continued to be of interest in the intervening decade. The African Association put further men in the field, despite the ageing Banks's frailties and downturns in the Association's finances, but as individuals, not as leaders of an expedition. From the government's viewpoint, the Niger remained an interesting geographical problem and a latent commercial opportunity, but ships and men could not be spared given the long-drawn out war with France and the campaign against the slave trade. There was, in truth, little political will for further expensive exploration after 1806 and before 1815.

The decade between Park's second Niger venture and the expeditions of 1815 and 1816 was marked too by deep uncertainty over Park's fate and that of his expedition. One feature of this was how to treat the testimony of slaves and of Africans regarding Park's death. Park's death made the Niger's exploration again a two-part mystery notwithstanding Reichard's solution and others' competing theories. To 'where did the river end?' was now added 'how and where had Park died?' Answers of a sort would first come from Isaaco's report as it appeared in *The Journal of a Mission* (1815), Park's posthumous work. This book contains, as the first credible account of Park's death, what Isaaco was told by the weeping Amadi Fatouma in August 1810.

The death of Mungo Park

Amadi Fatouma's account of Park's death and the end of the expedition takes up just seven printed pages in Isaaco's report to Governor Charles Maxwell in *The Journal of a Mission*. It reads as a record of close encounters, cultural transgression and, ultimately, death and disaster.

Amadi Fatouma reported that, on departing Sansanding in late November 1805, the expedition first landed at Silla, where Park had turned for home nine years earlier. As they proceeded down the Niger, canoes from local settlements came out to meet them, either to seek tribute or because they were simply curious at the passage of so large a vessel crewed in part by white men. On most occasions, their attempts were seen off by force, each man having fifteen muskets, 'always in order and ready for action'. At one point, sixty canoes came after them: the expedition fired upon them and 'killed a great number of men'.

Amadi Fatouma reportedly protested to Martyn at the slaughter only to be threatened by him. Although Park's initial intention was not to stop, the expedition halted several times to take on food and water. At Yaour in the Hausa kingdom (what is today Yauri, in western Nigeria), Park and Amadi Fatouma parted company since the agreement made between them had been for Fatouma to guide Park only as far as that kingdom. At that point, Park had travelled about 1,500 miles of the river's total length of 2,600 miles and was only about 600 miles from the Atlantic.

Park met the local ruler but only briefly. The party continued downriver after Yaour, Amadi Fatouma recounted, without leaving sufficient tribute gifts or seeking formal permission to travel. Thinking himself duped by Park and by Amadi Fatouma, the king clapped the latter in irons and, the next morning, 'sent an army to a village called Boussa near the river side'. Bussa, as it is more commonly spelt, is about forty miles downstream from Yauri. Fatouma then described what happened:

> There is before this village a rock across the whole breadth of the river. One part of the rocks is very high; there is a large opening in that rock in the form of a door, which is the only passage for the water to pass through; the tide current is here very strong. The army went and took possession of the top of this opening. Mr. Park came there after the army had posted itself; he nevertheless attempted to pass. The people began to attack him, throwing lances, pikes, arrows and stones. Mr. Park defended himself for a long time; two of his slaves at the stern of the canoe were killed; they threw every thing they had in the canoe into the river, and kept firing; but being overpowered by numbers and fatigue, and unable to keep up the canoe against the current, and no probability of escaping, Mr. Park took hold of one of the white men, and jumped into the water; Martyn did the same, and they were drowned in the stream in attempting to escape.

The sole survivor, a slave, was taken to the king along with the canoe and a sword belt, the only remaining trace of its occupants. It was from this slave that Amadi Fatouma, once released from captivity, learnt the manner of Park's death. Although no date could be given with certainty, Fatouma recounted that 'Mr. Park died four months after his departure from Sansanding,' which would place his death as sometime in March or early April 1806.[1]

Isaaco's report to Maxwell was in Arabic as was Amadi Fatouma's testimony to him. Initially, doubts were raised about these materials – because they were written by Africans, in Arabic, because neither Isaaco nor Amadi Fatoumi was with Park at the time of his death, and, of Amadi Fatouma's account because it was translated by someone with only moderate knowledge of the Arabian dialect in question. Doubts attached to Isaaco who some felt had left Park's company only to serve his own interests. They were additionally directed at Amadi Fatouma because of his background, his failure to stay with Park, and his preparedness to believe a slave.

Isaaco saw no reason to doubt Amadi Fatouma: he was 'a good, honest, and upright man, I had placed him with Mr. Park; what he related being on his oath'. Park had good reason to trust Isaaco: writing on 19 November 1805 to the merchant Thomas Dean at Kayee, Park had remarked 'any information you may wish for respecting our Journey he [Isaaco] will readily give it and his information may be depended on for I never found him guilty of telling lies.' Whishaw in his account of Park's life notes that the evidence in combination was considered credible: 'Upon the whole there seem to be no reasonable ground of doubt with regard to the fact either of Park's death or of its having happened in the manner described in Isaaco's Journal. The first of these may be considered as morally certain, the latter as highly probable.'[2]

The anonymous reviewer in the *Edinburgh Review* was highly complementary of Whishaw's management of Park's papers – 'it could not have been entrusted to better hands, whether for diligence, accuracy, or ability' – but entertained some doubts over the nature of Amadi Fatouma's account and how it was imparted to Isaaco: 'whether this was communicated in writing, or was only taken down by Isaaco from oral conference, we are not strictly informed.' The reviewer's suspicions rested upon the fact that Amadi Fatouma's account seemed too detailed to have been given verbatim to Isaaco. Why, continued the reviewer, should Amadi Fatouma have committed the information he gleaned from the surviving slave to paper, if indeed he did. Weighing the evidence, the reviewer came down in favour of oral transmission a fact which left other questions hanging: 'Upon the whole, we rather incline to the supposition that Amadi *told* it to Isaaco – in which case, its particularity seems highly injurious to its credit. There seems moreover a suspicious anxiety to account for his leaving Park.'[3]

These facts are essentially all there is concerning Park's death. They raise important questions about the credibility of oral testimony, trust, the role of translation in exploration narratives, as well as about exploration as fieldwork and personal observation by reliable – that is, white European – witnesses. For Whishaw, happy though he was with the general tenor of the account, the details concerning the expedition's end and Park's death were left in 'great obscurity' both from a lack of precision in the narrative, 'but principally because the particulars related, depend altogether upon the unsupported testimony of a slave . . . from whom the information was obtained at the interval of three months after the transaction. It is obvious that no reliance can be placed on a narrative resting upon such authority; and we must be content to remain in ignorance of the precise circumstances of Park's melancholy fate.'[4]

Doubts over the facts of Park's death, how they had been secured, and from whom, had been expressed years before. On 10 July 1806, the *Charleston Courier* in South Carolina carried a short report which stated that after having been shown around the town of Sego by the local king, Park and his remaining companions had been murdered there: 'This intelligence is furnished us by a gentleman recently from the Rio Pongus, who received the information from traders from the interior country, and on whom reliance might be placed.' While the location of Park's death is inaccurate, this evidence is interesting for the credence placed upon the word of traders – slavers working between west Africa and the southern American states: slavers might be trusted, but not the enslaved.[5] *The Times* quoted parts of this account on 1 September 1806 and, on 20 January 1807, reasserted claims that Park had died in Sego but without further supporting evidence.

Over the course of the next few years, *The Times* ran a series of brief notices on Park: that he was still alive (as asserted on 12 January 1808, and 19 May, 14 June, and 8 November in 1810); that he was dead (4 October 1811); and that he might still be alive (4 April 1812). *The Times* was not alone. On 1 January 1808, the *Edinburgh Courant* noted, also without proof, 'that the enterprising Mr Mungo Park is still alive, although the sole survivor of his party, and has gone from Africa for the West Indies'.[6] In April 1808, Parliament granted his widow the sum of £3,236 10s the sum owed to her from the £4,000 Park had negotiated in the event of his death on the second expedition. It seems parliamentarians were convinced enough of Park's death, probably with Banks's approval, to make good on the financial settlement Park had agreed with Camden.

Contemporary interest in Park's death survived on rumour until Isaaco's and Amadi Fatouma's more solid evidence. Rumour was at moments lent fresh weight by further partial evidence from west Africa. In November

1814, Lieutenant Thomas Perronet Thompson, the former governor of Sierra Leone, forwarded a letter to Adam Park, Mungo's brother, that had come into his possession via the Gorée garrison. This letter, dated July 1808, concerned an earlier undated letter from Lieutenant Martyn to a fellow soldier at Gorée. Martyn's letter, written at Sansanding as the expedition prepared to set sail down the Niger, is noteworthy on two points. It illustrates Martyn's belligerent and unpleasant character: 'Whitbread's beer is nothing to what we get here, as I feel by my head this morning having been drinking all night with a Moor and ended by giving him an excellent thrashing.' And it makes clear Park's view concerning the Niger problem as they set off from Sansanding: 'I [Thompson] forgot to say that he [Martyn] adds "Capt. Park had no doubt that this river is the Congo River".' Thompson closed his communication to Adam Park by noting that no further intelligence had been received from these voyagers for three years.[7]

Determining the precise circumstances of Park's death became a defining concern for later Niger explorers, hoping not only to resolve the remaining element of the Niger problem by first-hand observation. As they searched for the remains of Park's expedition, they sought also to rescue the truth of his death from being the 'unsupported testimony of a slave', testimony initially relayed by word of mouth and, latterly, from one Arab to another and in translation. That the facts were uncertain was a matter not just of what was known, but of who was a credible source and who not, and in what form information was relayed to others.

Park's death continued to fascinate long after the Niger problem had been solved. In February 1921, the secretary of the Royal Geographical Society received a letter from an H. S. N. Edwardes who, in October 1920, had visited Bussa and gone down the rapids where Park was drowned. With the aid of a sketch plan scribbled on the back of an envelope (Figure 7.1), Edwardes provided a possible explanation of Park's death:

> At the head of the rapids the main channel bears away to the left & the thunder of the water over the rocks can be heard for miles. The water spills over in the channels at the same point in the right bank. The only possible canoe passage requires a sharp swing to the right close under the bank of the mainland & then a turn to the left down the small channel. With skilful canoe-men the passage can then be made.
>
> To a stranger coming down the middle channel, X, straight ahead, appears the obvious course, it certainly did to me. The natives shouting & gesticulating on the mainland were probably trying to guide him to the right channel. Whether he thought them hostile,

> or simply did not hear what they said owing to the thunder of the rapids, we shall never know. In either case, without local pilots or men accustomed to the passage, he would probably take the middle channel, and be smashed up & drowned. Had he by any chance got through he could hardly have survived the much more formidable Potachi and Burru [?] rapids lower down.[8]

It is possible, of course, that what Park and Martyn took for threatening gestures were attempts at signalling the canoe to take another course, based on local knowledge over which rapids offered safe passage, especially when the water levels were low. But this neither corroborates nor explains what Amadi Fatouma was told about the explorers having lances, arrows, and stones thrown at them, and their leaping into the Niger in attempting to evade that attack. Such an attack was, after all, understandable. By ignoring the required toll in gifts or in offering them too little, and in killing men who came after them on the river, Park irreparably damaged relations with indigenous leaders.

As explorers in the years following sought clues about Park's death, so modern biographers have debated it. For Sanche de Gramont, 'Park did not

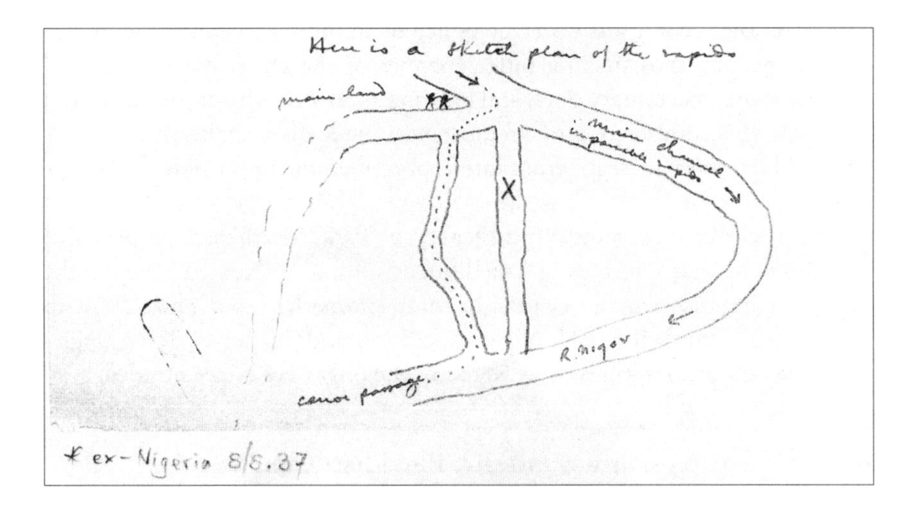

Figure 7.1 H. S. N. Edwardes's sketch map of the Malali rapids at Bussa at the point where Park drowned in early 1806. It is possible that the locals were trying to warn Park and his companions of the 'impossible rapids' – or that they were directing 'HMS Joliba' through a different stretch of the Niger in order to attack it more easily. [rgs550488] © Royal Geographical Society (with IBG).

meet his death in the way that Ahmadi [sic] described.' De Gramont identifies three 'major implausibilities' in the latter's account: Bussa was a separate kingdom and its king would not have permitted the passage of Yauri troops in the ways recounted; Park could swim; and when, in 1857, Lieutenant John Glover travelled to Bussa and there recovered an eighteenth-century nautical manual consisting chiefly of logarithm tables which belonged to Park, the volume showed no signs of water damage despite claims that 'every thing they had in the canoe' had been thrown in the river.[9]

The reality is that we shall never know the truth of what happened to cause the death of Mungo Park and the remaining members of his second Niger expedition. Hard enough to discern the circumstances of his death, it is impossible now to check descriptions of the site against the geography of Bussa and the rapids: the town of Bussa, and the rapids where Park came to grief, were flooded in 1968 by the waters of Lake Kainji as that section of the Niger was altered by the construction of the Kainji Dam.

Park's death was basically due to two tragic circumstances:

> First the initial delay, which had cost him the loss of most of his men in the rains between the Gambia and the Niger, and but for which he would have reached Bussa early in the year when the river would have been high enough for his boat to pass without fear of interception. The second was his crazy policy of shunning all contact with the people, thus making bitter enemies of the chiefs by denying them their customary dues, and of firing at anyone who approached him. This combination of circumstances made disaster almost inevitable so soon as escape from interception became impossible.[10]

The scientific and geographical legacy of Park's death had its principal expression in that continuing expeditionary impulse directed towards the Niger, an impulse which began – or, more properly, was renewed – with Peddie and Campbell's 1815 Niger expedition and Tuckey's 1816 Congo–Niger expedition. However, they were not the only ones Niger minded.

In Park's footsteps, unsuccessfully: Burckhardt and Salt

Several other men had turned their attention to the Niger years before Isaaco reported upon Park's death, and the African Association had a further candidate within a year of its receiving news of the death of Henry Nicholls in 1807.

Jean-Louis Burckhardt, a Swiss, had come to London in July 1806 to seek a career in the diplomatic service, unsuccessfully as it turned out. Like

Hornemann, Burckhardt carried a letter of introduction to Banks from Blumenbach (Burckhardt had studied at Leipzig and at Göttingen). Through Banks's patronage, Burckhardt presented his plans to the Association's committee of management on 21 March 1808. Burckhardt planned to travel to Tripoli by way of Malta – the political situation in Egypt being still too unsettled – there to solicit information about north African affairs. Upon arrival in Tripoli, he hoped to garner information from merchants and traders travelling between Murzuq and Timbuktu, 'or to any of the considerable towns on the Banks of the Niger' and to do so by adopting 'the character of a Moorish merchant'.

As had Hornemann and Frendenburgh, Burckhardt took the view that disguise would more readily allow safer travel and easier access to information, even to the Islamic world overall, than if he appeared as a Christian asking questions about Moorish trade. Even before he received the Association's final instructions, Burckhardt went to considerable effort to ensure that upon arrival in Africa he should be taken for what he was not. He attended lectures in London and in Cambridge in Arabic and in 'those branches of science which were most necessary in the situation wherein he was about to be placed', namely chemistry, astronomy, mineralogy, medicine, and surgery. He talked with William Browne and James Rennell amongst others. He grew his beard 'and assumed the Oriental dress'. Between his studies in 1808, in what in England was an especially long and hot summer, he adopted a regimen of what we might describe as 'pre-exploration hardening' in preparation for the rigours to come: 'long journeys on foot, bareheaded, in the heat of the sun, sleeping upon the ground, and living upon vegetables and water'. In Malta, Burckhardt adopted the name Ibrahim Ibn Abdullah.[11]

Because he was 'entering Africa on the Northern Side' as the Association's instructions put it, Burckhardt was instructed to head for Syria after leaving Malta, and to stay there for two years. Such residence, he was advised, 'will we trust enable you then to pursue your Journey under the same advantages which a Native would possess[.] on your arrival at Cairo let your appearance in that City be such as to ensure your good reception amongst the Mohammedans of the Maggrebin Caravan in whose company you are to traverse the desert to Moorzouck [Murzuq]'.

Burckhardt did not leave England until March 1809. He arrived in Aleppo seven months later and spent almost three years in Syria, making several journeys in the Near East. He continued to adopt his alias in Syria despite 'finding no necessity for such a disguise at Aleppo', there wearing Turkish dress, 'as is often assumed in Syria by English travellers, less for the sake of concealment than to avoid occasional insult'.[12] He did not arrive in Cairo until September 1812. From there, rather than wait for a caravan to Fezzan and head south towards the Niger, Burckhardt undertook journeys across the Nubian desert, down the Red Sea and to Mecca, where he made the pilgrimage.

Burckhardt's knowledge of Islamic culture 'enabled him to assume the Mussulman character with such success, that he resided at Mekka [sic], during the whole time of the pilgrimage, without the smallest suspicion having arisen as to his real character'.[13] He returned to Cairo in 1815 and, following further delays with caravans to Fezzan, prepared in 1817 to join a party of pilgrims to travel there. But like Ledyard, Burckhardt never left Cairo. He died there, of dysentery, in October 1817.

Burckhardt's Syrian and Nubian travels formed the basis to two important travel narratives, the second of which was seen into print on behalf of the African Association by William Martin Leake, a future Association secretary and himself a traveller in the Middle East.[14] Burckhardt's work was admired by his contemporaries, by Rennell especially who felt that Burckhardt 'had contributed very much towards the elucidation of History, Antiquities, Geography, and the State of Societies, in Countries infinitely more interesting to us, than Africa ever can be' – a comment which perhaps says as much about Rennell's continuing interests in Asia and the Near East as it does about Burckhardt's achievements. He is recognised by modern scholars for his detailed commentaries on parts of the world then effectively closed to Europeans. Burckhardt was the first European to encounter Petra in what is today Jordan and Abu Simbel on the Nile.

Yet, Burckhardt failed even to begin the task for which he was appointed. As contemporaries recognised, he left an extensive collection of manuscripts concerning the regions in which he had undertaken his 'preparatory journeys' but 'nothing more than oral information as to those to which he had been particularly sent'.[15] Disguise, pre-exploration toughening-up, linguistic proficiency: all could count for nothing if explorers let themselves be diverted from their brief and died before starting out.

At much the same time as Burckhardt was training for his travels, Henry Salt approached the African Association and recounted his experiences of travel to Abyssinia. Salt, an artist, had travelled there in 1805 with Viscount Valentia, an Irish peer, who had connections with George Canning, Britain's foreign secretary and Lord Wellesley, the Governor-General of India. On learning in December 1808 that Salt was soon to return to Abyssinia and that he had a contact there – Nathaniel Pearce, an Englishman who, following shipwreck in 1798, had been enslaved and had lived there since – Salt (and Pearce too if he could be called upon) was given £500 by the African Association towards 'procuring usefull & curious information, relative to the interior of Africa'.

Specifically, Salt's brief concerned four questions: information about 'The Countries whence the Caravans come that attend the great Mart of Barbary

[Murzuq]'; the 'Chain of Mountains which is said to be not very distant from the country of Darfour' [the supposed Mountains of Kong]; the source of the White Nile; and, lastly – as if the Association was asking Salt directly to test Rennell's theory – 'The course of the River Joliba if any thing concerning it or any large River running from West to East be known . . . or if any information can be obtained of a vast Lake increasing in the Wet & decreasing in the dry Season capable of intempting [sic] and providing a receptacle for the waters of that vast River.' Salt was further instructed to keep a regular journal that he might identify from whom he gathered what facts. In the event, he had no need of it. Salt got no further than what is now Mozambique, Aden, and Cairo (where he met Burckhardt), and he made no use of the previously enslaved Pearce.[16] Yet the word of slaves could be vitally important.

Credible informants?: the slave testimonies of 'Adams', Scott, and Riley

Between 1815 and 1817, three 'shipwreck narratives' concerning enslavement in west Africa and the veracity of slave testimony were published in Britain. Their importance rests in the authenticity and credibility of their authors at a time when Park's second book was being brought together and as expeditions in search of the Niger were being renewed.

The first concerned an 'illiterate African American sailor', as he was later described, found living destitute on the streets of London in 1815. In London, he relayed his tale to Simon Cock, secretary to the Committee of Merchants Trading to Africa, known usually as the African Company of Merchants, the British company which between 1752 and 1821 oversaw British interests in the Gold Coast. He claimed he had been shipwrecked on the Barbary Coast in 1810, taken prisoner, and later sold as a slave to an Arab merchant. He had assumed the name Robert Adams but, in interview, he disclosed that his real name was Benjamin Rose. Adams/Rose had, he told Cock, been taken to Timbuktu, and lived there for four months before escaping and making his way to London via Cadiz.

Adams was interviewed by Samuel Cock, Joseph Banks, and John Barrow. Initially, Adams' account was regarded with scepticism because his story was at odds with what was generally believed, about Timbuktu especially, but Cock, who drew upon his own west African knowledge, was convinced by him. Questions put to him at first interview were subsequently repeated and elicited similar responses. For Cock, the consistency of Adams' answers meant that 'the man's veracity was to be depended upon'. Cock's intention was not just to satisfy himself and others of Adams' veracity as a credible witness to the African interior. It was also to establish his and Adams' credibility as authors. Under Cock's guidance, Adams' fifty-odd-page account became a 200-page book. The publisher John Murray placed the book 'on a footing with the two volumes of Park . . . [and] accordingly printed in a uniform manner with those volumes' – that is, with Park's *Travels* (1799) and his *Journal of a Mission* (1815).[17]

The British periodical press was generally supportive of Adams' account. His description of Timbuktu as less grand than supposed worked in his favour since reviewers saw it as a mark of Adams' authorial honesty. Cock was likewise praised: 'Great pains have been taken to sift the testimony . . . and the proofs of Adams' veracity.' The French and American press and public thought otherwise. To the French map-maker Edme Jomard, whose authority came from having edited the multivolume *Description de l'Egypte*, Adams, being both illiterate and of mixed race, was morally and physically inappropriate to act as an explorer and scientific observer. To them, Adams' race embodied unreason. The publication of an American edition of Adams' narrative in 1817 similarly met with censure and disapproval given his race. The word of a slave was not to be trusted – whether on the death of Park, or in tales of Africa's interior. Any association between truth, knowledge of Africa, and social status was considerably weakened when the testimony in question came from an illiterate black sailor.[18]

At much the same time, a sixteen-year-old Liverpudlian named Alexander Scott was interviewed by the natural scientist Thomas Stewart Traill, and by James Rennell, concerning his own tale of shipwreck and captivity in Saharan Africa. Scott's ship, the *Montezuma*, bound for Brazil, had gone down off the west African coast in November 1810. Scott was taken prisoner, enslaved, marched inland and held in north Africa for nearly six years before, with the aid of 'a friendly Moor', he escaped to Mogador on the Barbary Coast and thence to London. Traill and Rennell quizzed Scott over his claims by showing him pictures of African fauna – which Scott assured them he had seen during his captivity – and by comparing his descriptions with those of Park.

While Scott's descriptions appeared to confirm Park's, Rennell also noted – his emphasis alluding to the different ways in which the information had been gathered – that 'The Lake as described to Mr Park [Lake Dibbie on the Niger, what today is Lake Débo], is much smaller than the one seen by Scott.' Both reports may have been correct. What Rennell did not know is that Lake Débo is seasonal, formed as the Niger floods: different travellers would either have seen or been told of a lake whose dimensions varied at different times of the year. Traill and Rennell were convinced of Scott's claims: as Rennell phrased it, 'The Geographical notices contained in this narrative are scanty, but appear to contain internal evidence of their truth.' Rennell was further assured of Scott's credibility because the young man's claim over where the *Montezuma*

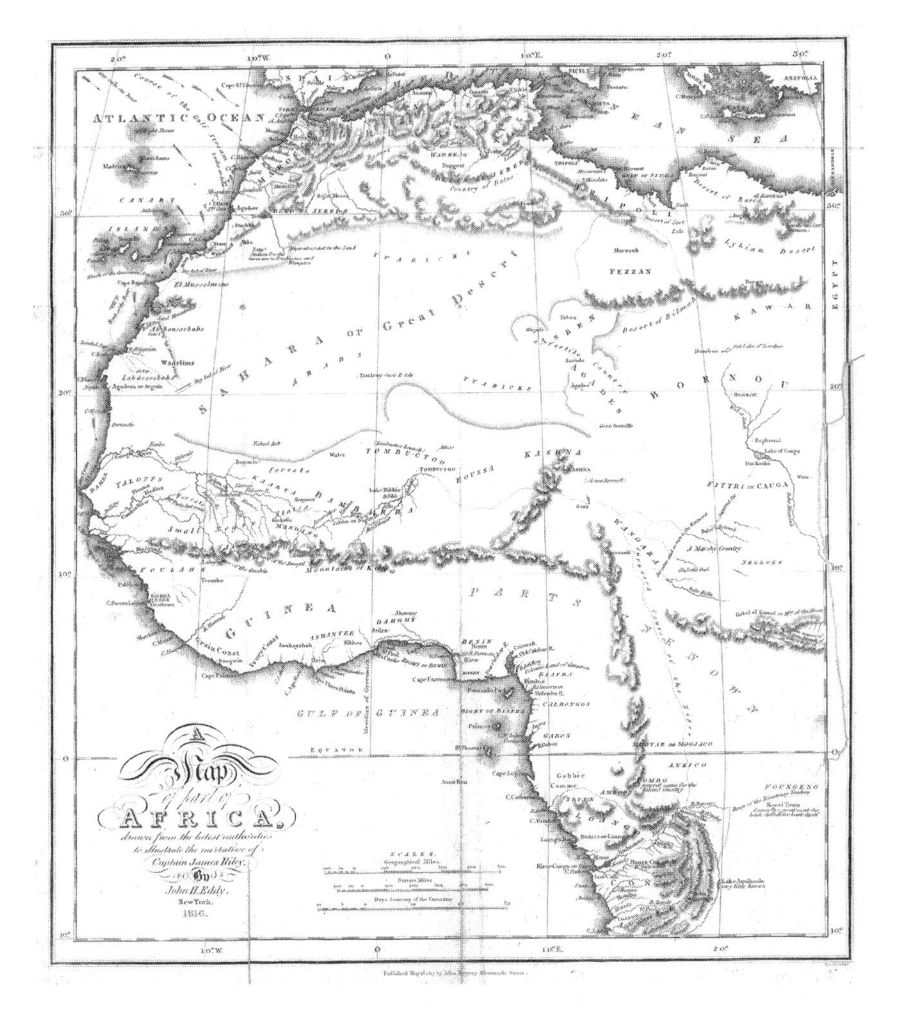

Figure 7.2 In his 1817 map, James Riley shows the 'supposed course of the Niger' running southwards – through a convenient gap in the Mountains of Kong – to join the Congo. Source: J. Riley, *Loss of the American Brig* Commerce (London, 1817). Reproduced by kind permission of the National Library of Scotland.

went aground supported Rennell's findings on the Atlantic's currents off west Africa. Interestingly, neither account – Traill and Rennell's questioning of Scott, Banks's and Cock's of Adams – makes mention of the other.[19]

In his *Loss of the American Brig* Commerce (1817), James Riley, the ship's captain, recounted his shipwreck on the Barbary Coast in 1815. Contemporary interest in the book, which was first published in New York in the same year under a different title, centred on its account of bodily suffering and the inversion of racial hierarchy: a tale of white Americans enslaved by black Africans served to prompt moral reflection on the ignominy of slavery.

The book's geographical importance lies in its account of Timbuktu. Riley had obtained this from Sidi Hamet, the Arab trader who purchased him from his captors. Riley was rescued from enslavement by William Willshire, the British vice-consul at Mogador. There, drawing partly upon his former captor's knowledge, Riley pulled together information respecting Timbuktu and the river Niger. From Hamet's testimony and from other undisclosed sources – Riley refers to them only as 'the best authorities' – Riley had a map prepared by New York state geographer John H. Eddy. This illustrated Riley's belief that the Niger 'discharges its waters with those of the Congo into the Gulf of Guinea'. Riley was not clear on the Niger's exact course (Figure 7.2), but he was insistent that the Congo 'is the outlet of the river Niger'.[20]

Covering as they do events between Park's death and the renewal of expeditionary enquiry, these books are important and not only because each shows the authorship of exploration narratives to be a collective endeavour. We only know the truth of what happened from later published accounts undertaken with the assistance of others, whether largely from memory, as was Riley's, or because both content and putative authors were scrutinised by authoritative figures, the case for Scott and Adams. Authoring words that became authoritative required approving the source of those words, especially when the words were those of slaves.

The timing of these books and what they claimed was similarly important. They appeared as the African Association was losing credibility: the deaths of Ledyard, Houghton, and Hornemann and, in the early 1800s, of Nicholls, Burckhardt, and most significantly, of Park, meant it had largely failed in its aims by the time further expeditions, government-funded and directed by John Barrow of the Admiralty, were underway. These issues – of authorship, credibility, truth, and the collaborative nature of exploration narratives – are exemplified in the book which bears Park's name but to which he contributed only in part.

Making a travel narrative, securing reputations: *Journal of a Mission* (1815)

Mungo Park's *The Journal of a Mission to the Interior of Africa in the Year 1805* was published by London publisher John Murray in March 1815. It bears the subtitle *Together with Other Documents, Official and Private, relating to the Same Mission*, and, finally, *To which is prefixed An Account of the Life of Mr. Park*. As this suggests, the book is a collaborative, multi-part, work.

It was compiled by John Whishaw, a director of the African Institution, which had been established in 1807, following the abolitionists' success in ending the slave trade in Britain, to advance the anti-slavery agenda in west Africa and to redress 'the enormous wrongs which the natives of Africa have suffered in their intercourse with Europe'. Although the Institution lasted only until the late 1820s, its work being effectively replaced from 1823 by the Anti-Slavery Society, it provided an important forum for abolitionism and, through Whishaw, for the promotion of Park's second expedition and his biographical treatment as part of its mission for Africa.[21]

Park's account of his second Niger expedition, which survived because he had handed his notes to Isaaco for safekeeping, makes up only part of the volume: five chapters, in 168 pages, with an accompanying fold-out map (Plate 6). It is unlikely that Park ordered his notes into chapters in the field: we must presume that Whishaw structured them thus as he prepared Park's material. Isaaco's journal, including Amadi Fatouma's report of Park's death, takes up a further fifty pages. The remainder of the book is made up by Park's 1804 memoir to Camden, an account of Park's life, and six appendices, one of which, Appendix IV, itemised the several hypotheses then circulating concerning the Niger as Park had set out for Africa in 1805.[22]

Whishaw made clear the book's twin intentions: that Park's unredacted expedition notes and several of his last letters should stand as an important exploration account (and inform Whishaw's biographical memoir); and that while it might assist the African Institution in its work, profits from its sales should benefit the Park family. While the book's content must be understood in relation to this geographical and humanitarian context, the book was also an exercise in managing reputations. Whishaw had to serve Park well biographically and to establish reputations for them both – Whishaw by being faithful to Park's words, Park for the accuracy of his account, albeit posthumously. As a director, Whishaw had to ensure the book would not do the African Institution a disservice. Banks and Camden had their scientific and political position to consider, which might be damaged by association with an expedition which had ended so badly. Isaaco and Amadi Fatouma were bearers of truth claims over Park's death and so needed to be seen as credible, and John Murray had to protect his standing as a reputable publisher.

Quite when Whishaw began to assemble the several elements that became *Journal of a Mission* is unclear, though it must postdate Maxwell's receipt from Isaaco of Park's notes and his account of Park's death. These materials reached Camden in London in 1813. Things were certainly moving by the end of 1813 following negotiations between Camden on behalf of the government and Whishaw for the African Institution. Whishaw had no motive, he assured Camden, 'except to do justice to Mr Park's memory & furnish correct & useful information to the public'.[23] In January 1814, Whishaw wrote to Murray with copies of elements of the proposed content: Park's letter to Camden of 17 November 1805; 'Mr. Park's Journal' – his notes as passed to Isaaco; Isaaco's journal; and 'several geographical memoranda, which would be of use in communicating a Map of Mr Park's Route; which ought to accompany the intended publication' (Plate 6).

In mentioning the 'geographical memoranda' here, Whishaw was probably alluding to Park's listings of his positional calculations. Park's figures and notes proved no easy basis in this respect: as the printed 'Advertisement' of 1 March 1815 for the book notes, considerable difficulties had been encountered – 'there being occasionally great differences between the latitudes and longitudes of the places according to the astronomical observations, and the distances computed according to the journies'. Despite attempts to reconcile these differences, 'the general result has been, that it was found necessary in adhering to the astronomical observations, to carry Mr. Park's former route in 1796 farther north, and to place it in a higher latitude than that in which it appears in Major Rennell's map annexed to the former volume of Travels.' This later map (Plate 6) is also of interest in showing where Park turned back on his first African journey and in depicting, in rather uncertain terms, the Mountains of Kong on whose existence Rennell's own belief about the course of the Niger depended (Figure 7.3). Fatouma's account was clearly thought reliable on the location of Park's death at least, but it could not be shown on this map since no European then knew exactly where Bussa lay – and Rennell would not have countenanced showing the Niger to run south through the Mountains of Kong.[24]

The Committee of Directors of the African Institution invited prospective publishers to make sealed bids for the purchase of copyright in the book. Murray submitted his in January 1814. On 6 February, he wrote to Whishaw to express thanks at the acceptance of his proposal: 'My offer . . . was made with an honourable feeling towards the noble object to which it was to be appropriated. I have proposed to the best of my calculation all the profit that can in reason be expected from a book, however interesting, that is yet so scanty in quantity, and I content myself with the reputation arising from so popular a volume, & the satisfaction of doing some service at some little professional risk.' Murray paid £1,200 for the copyright, the sums to be paid in three amounts of £400 on the first day of January (that is, before publication), May, and September 1815.

At this stage, any plans to incorporate biographical material on Park rested with Whishaw and the Institution, not with Murray: 'I am authorized to say that, in case you think it advisable to prefix a short biographical

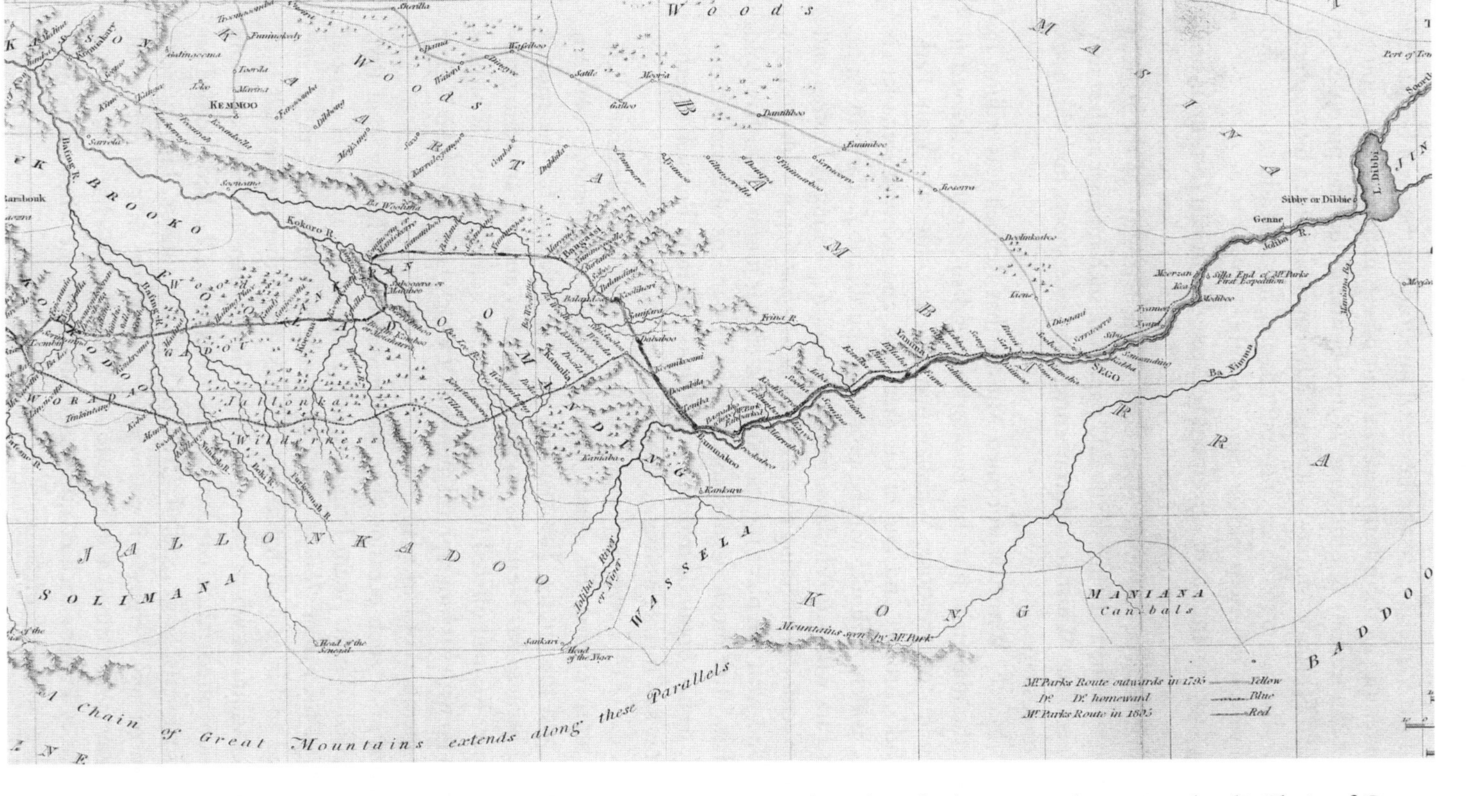

Figure 7.3 In this detail from Park's travels on his second Niger journey (see Plate 6), the map-maker insists that 'A Chain of Great Mountains extends along these Parallels' but does not name or depict it (*cf.* Figure 1.3). From M. Park, *The Journal of a Mission to the Interior of Africa* (London, 1815). Reproduced by kind permission of the National Library of Scotland.

memoir of Mr Parke [sic] to the intended publication, they will be glad to assist in obtaining all the information in their power, both by application to Mr Parke's connections & otherwise, as may be useful for that purpose.' In his letter of 6 February, Murray welcomed Whishaw's plans for a biographical memoir 'with much gratefulness', no doubt hoping it would add substance to what then was an otherwise rather scanty volume.[25]

A formal memorandum of agreement between the African Institution and John Murray was drawn up in March 1814: 'full liberty has been given by the Colonial Secretary of State [Camden] to the Directors of the African Institution to print and publish the said papers, in such manner as they may think proper'. The Whishaw–Murray correspondence shows how the book gradually took shape.

On 26 March, Whishaw informed Murray that Rennell was willing to look over the materials, but that he was busy and so might be delayed in forwarding any observations. Whishaw had that morning met Adam Park, Mungo Park's brother and one of his executors, and received some letters from him 'which, I think, will be proper for publication'. Early in April, Alexander Park wrote to Whishaw from his position as Mungo's brother and principal executor, confirming that the Park family was happy with Murray's proposal and with the plans for the book. Later that month, Whishaw told Murray that, 'after much negociation [sic] and by means of some personal influence', he had obtained material from Banks and from Rennell, together with promises of further assistance from them.

From mid-1814, however, progress on the book slowed. Alexander Park, Mungo's brother, died, his death interrupting the negotiations between Whishaw and the family. The map was proving awkward for the reasons noted and because Mr Neale, its engraver, was slow and not answering correspondence. Rennell offered little given his other commitments. Whishaw was taken up with his biographical memoir for which he met with Sir Walter Scott and members of Park's family.

It was on one such occasion that Scott revealed to Whishaw what Park had confessed to him, Adam Ferguson, and Dugald Stewart in 1804 about the truth of his earlier *Travels*. Park, Scott told Whishaw, 'was rather shy & reserved in his general habits'. He always 'felt rather embarassed [sic] by indirect inquiries which strove to avoid the apparent rudeness of blunt interrogation often much concerning his travels'. But he had acceded to Ferguson's requests, Scott recalled, given the professor's status, 'high interests and advanced age'. Ferguson it was who had spread the map of Africa out before them as they dined at Hallyards and 'made the traveller trace out his whole journey inch by inch & questioned him upon the whole as he went along with characterisk precision'.[26]

In early November, as Whishaw was still deliberating over which of Park's letters were proper to be published, a 'bombshell' arrived from Murray. He was considering pulling out of the whole project. Murray's decision had nothing to do with Whishaw and everything to do with the fact that Murray felt he had been gazumped.

In its November 1814 issue, the *Annals of Philosophy* carried the text of Isaaco's journal in full. In a footnote, the journal's editor, Thomas Thomson, explained that because doubts had been expressed in the press 'over the reality of Mr. Mungo Park's death, we take the opportunity of publishing the only authentic document on the subject which exists, the Journal of Isaaco'. Thomson may well have had *The Times* in mind given its contradictory coverage, but both the *Edinburgh Review* and *Morning Chronicle* had carried a short notice of Isaaco's report on Park's death in September 1812. 'We are not at liberty at present to state the source from which we obtained it', Thomson continued, 'though we have the most complete evidence of its authenticity. It may be sufficient to say that the MS. was sent from the coast of Africa to a scientific gentleman in London, for his private satisfaction.'[27]

Whether the original source was Colonel Maxwell and the gentleman in London was Banks, or Rennell, or someone else entirely, is not known, and probably mattered little to Murray. What concerned Murray was that his intended book about Park's exploration was greatly diminished, his control over the truth of Park's death especially. As he put it to Whishaw, Isaaco's journal 'promised to create an interest nearly as great as Mr Park's work itself. Hence it has always been an object of primary moment to the success of the publication . . . The circumstances as to Mr Park's last scene has been extracted into almost all the newspapers and thus everybody knows what I meant no one to be informed of except from my Book[.] all the sale from this popular curiosity is now gone.' Murray was aggrieved too that while in the *Annals of Philosophy* Isaaco's account took up seventeen pages, he had set aside 'nearly 60 pages & the comparison will be an immediate condemnation of my publication'.[28]

What mattered to Murray, to his finances, and to his reputation, was not the truth of Park's death, but who had the right to publish it and in what form. Despite his misgivings, the matter was agreeably resolved. Murray was reminded that the book was to benefit the Park family, not him – although with Adam Park's agreement he altered the timings of payments – and, as Whishaw informed Banks in recounting the tale of Murray's planned withdrawal, the biographical memoir and additional material would make the book trump anything else.[29]

By the end of November 1814, the project was back on, but things did not run smoothly even then. There was a delay given a shortage of paper. Page proofs had been drawn up by Bulmer, who had printed Park's *Travels*, but Whishaw considered them 'to [sic] loosely and widely printed . . . it would give me great pleasure if some alteration could be made in this respect'. Whishaw and Murray did not see eye to eye over the book's final format. Murray proposed that members of the African Institution might act as subscribers to the work, so helping with production costs.

Whishaw took a different view: 'I have considered what you proposed respecting circular notices to the members of the Association;' he replied ['Association' here meaning the African Institution], 'many of whom are in humble life & ought not to be called upon to be the purchasers of Quartos. Indeed, I have always considered the <u>Octavo</u> as the proper sort of edition to suit the demand of the readers of this work; and the Quarto as fit only for Libraries or to compleat [sic] former Sets.'[30] Murray would win that immediate battle: the book, published as a quarto, made him handsome profits. In June 1816 the Park family's executors agreed that the money arising from the sale of the copyright should be put in trust for Allison Park and her four children, Mungo, Elizabeth, Thomas, and Archibald. The sum of £1,981 was made available to them.[31]

As Park's posthumous book neared completion, Whishaw wrote to Murray in December 1814 with important news. At a recent meeting of the Institution, reported Whishaw, 'we had a communication from Government, intimating that they intend to send another mission into Africa to explore the interior & ascertain the course of the Niger; and desiring that we would furnish any suggestions which in our opinion might be useful. I mention this as a circumstance likely to add to the interest of the present publication.'[32]

Almost a year later, he would write again with further Niger news. There is, he told Murray, 'a sailor now in London, a very illiterate man' who upon interview had given accounts of Africa consistent with those of Mungo Park. His narrative had been taken down 'by a very intelligent man connected with the African Company'. This was the eponymous Robert Adams, interviewed by Simon Cock. If he, Whishaw, could get permission regarding this account, it might be worth Murray adding to the 1816 edition, in octavo format, of Park's *Journal of a Mission*. 'What,' he asked Murray, 'do you think of this?'[33] As we have seen, Murray thought well of the opportunity.

Several individuals in government circles – Banks, Camden, and Rennell – were involved in the book about Park's second expedition, but not to the same extent as they had been in planning it. But the book's existence, and its dimensions and format as a printed work, depended upon others not involved in commissioning Park's second travels. Once published, the book was beyond its authors' control. Readers could make up their own minds over its content, its truth claims, and the role of its participant compilers, dead or alive.

In his commentary in the *Edinburgh Review*, lawyer and statesman Henry Brougham praised both Park and Whishaw: with Park's death, 'the world has lost a great man'; Whishaw's treatment of Park's expedition and life was 'at once instructive and entertaining'.[34] John Barrow lauded the intention to support Park's family, but took issue with Whishaw's treatment of Park, seeing in it a continuing and unwarranted association between Park and slavery, and returning to the question of Bryan Edwards' role in Park's first book. What Whishaw should have done was use his memoir to dismiss these charges altogether.[35]

Barrow was not alone in feeling this. Whishaw received several 'remonstrances' complaining about his treatment of Park. Adam Park wrote to say that while he recognised Whishaw's views, he was 'satisfied as to the integrity and propriety of his Brother's conduct'. Whatever Whishaw may have felt about Adam Park's or Barrow's views about Mungo Park's reputation, slights to his own good name might still be turned to advantage. As he wrote to Murray, 'I hope that whatever may be said against me personally will not be detrimental to the sale of the work – but I believe that such little controversies rather promote the circulation of the book than otherwise.'[36]

Perhaps they did. As Banks, Rennell, and Camden looked to the geographical results of the new expeditions then being planned for west Africa, John Murray recognised the publishing opportunities inherent in them – should the prospective authors and their words return safely.

8

'Collect all possible information'

Searching for Park and the Niger, 1815–c.1825

After 1815, the British appetite for exploration increased for several reasons. War with Napoleonic France ended that year, and restrictions on European travel eased in consequence. As European statesmen re-drew the Continent's political boundaries at the Congress of Vienna, British authorities renewed their geographical ambitions. Their task was helped by the presence of half-pay naval and army officers – men used to discipline, trained in navigation, and with experience of travel – laid off in large numbers at the war's end.

John Barrow, now Second Secretary to the Admiralty and soon to become a leading figure in orchestrating British overseas exploration after 1815 (Plate 8), recognised their exploratory potential: 'To what purpose indeed could a portion of our naval force be, at any time, but more especially in a time of profound peace, more honourably or more usefully employed, than in completing those *minutiae* and details of geographical and hydrographical science, of which the grand outlines have been broadly and boldly sketched by Cook, Vancouver, Flinders, and other of our own countrymen.' Barrow may have had Park and the Niger in mind: these words are from his introduction to James Hingston Tuckey's account of the Niger–Congo expedition.[1]

Barrow had experience of foreign travel to scientific ends, to south Africa and as part of Macartney's failed mission to China. His claims to have done more, made partly to disguise his humble origins as the only son of a Lancashire tanner and in part to secure status in Whitehall, were sometimes questioned. At a dinner party in the presence of Robert Dundas, First Lord of the Admiralty, Barrow's claims to have served on a Greenland whaler were challenged by the Artic explorer John Ross who exposed elements of Barrow's tale since Barrow could not recall the name of the ship, or the captain's name: Barrow was, as Ross put it, 'now completely caught in a Bouncer'.[2] As a sedentary organiser and promoter of others' expeditions, however, the self-made and at times self-promoting Barrow was an influential figure. Barrow's authoritative position and the networks which sustained it reflected shifts in the institutional politics behind exploration: away from Banks and the African Association, and towards the Admiralty and the Colonial Office, and men such as Lord Bathurst, Hanmer Warrington, Britain's consul general in Tripoli between 1814 and 1842, Yusuf Karamanli, the Bashaw, in Tripoli, and Henry Goulburn and Robert Wilmot-Horton, Under-Secretary of State for War and the Colonies between 1812 and 1821, and 1821 to 1828 respectively. These men and others helped establish the routes by which the Niger was approached in the decade after 1815: south across the Sahara, or inland from the west African coast.

Barrow did more than direct the Niger's exploration: he managed – or tried to – what others wrote about it. In this, he worked closely with John Murray. Barrow also reviewed many of these narratives once published, commonly in the *Quarterly Review* which periodical Murray established in 1809. A prolific author and an even more prolific reviewer, Barrow cast imperious eyes over others' work to present a particular interpretation of what exploration should be and to offer his views on the Niger. As Barrow and Murray would come to understand, tales of exploration could be difficult to manage if, like Park, the words returned but the explorer-author did not.[3]

'Two More Disasters': The Niger and the Niger–Congo hypothesis revisited, 1815–*c.*1818[4]

British expeditionary interest in the Niger was renewed on 2 August 1815 in a directive from Goulburn to Barrow, within weeks of the publication of Park's *Journal of a Mission*, and a month or so after Napoleon's defeat at Waterloo. One motive was the still unsolved matter of the Niger's termination. Another was to determine what had happened to Park. Commercial and colonial questions also figured. The war in Europe ended Napoleon's intentions on that continent, but France's ambitions in west Africa remained of concern. So, too, did the continuing slave trade, much of it led by American and Portuguese slavers and facilitated by indigenous leaders for whom human trafficking was a source of income and status.[5]

Two Niger expeditions were planned, one under the control of the Colonial Office, the other under the Admiralty. Although different in their make-up, they were intended as complementary. The Colonial Office expedition was a military affair – its company of volunteers from the Royal African Corps and others was twice the size of Park's 1805 expedition. The smaller Admiralty expedition was more scientific in intent and personnel. In two respects, however, the expeditions were alike. They shared a single purpose – 'discovery of the mouths of the Niger' – although they took different routes to that problem. And both were disasters: delayed marches in intolerable heat, the insistent demands of indigenous authorities, men dying for a geographical goal neither expedition came close to attaining.

The Colonial Office expedition was led by Major John Peddie with Captain Thomas Campbell second in command. Taking his instructions from Lord Bathurst, Colonial Secretary between 1812 and 1827, Peddie was told that the Niger might 'flow into the Nile as has been supposed by some or into the Zair or Congo as has been with a greater appearance of probability asserted by others'. 'In contemplation of the latter hypothesis proving true,' continued Bathurst, a further expedition [the Admiralty's] was to explore the Niger–Congo possibility and, if successful, join up with Peddie's party. Simon Cock advised the Colonial Office how it might proceed. Peddie was to march overland from the coast to the headwaters of the Niger and then follow the river's course to its end: to do, in essence, what Park had set out to do a decade before but with a larger party.

It was not Peddie's expedition for long. He died in January 1817, having made the unwise decision to start towards the Niger further away than was necessary, from the Rio Nuñez, south of the Gambia, and after taking months to come to that decision. As he waited, enduring a fever from which he would not recover, Peddie read Robert Adams' account of his African travels, sent to him by Goulburn. The party did not set out from the coast until February 1817. Campbell was soon too ill to effect command and returned to the coast. The expedition was then led by Captain William Gray and Staff-Surgeon Duncan Dochard. The account of the expedition, which lasted until 1821 but achieved little, was prepared by Gray and Dochard with material from Peddie and Campbell and published in 1825.[6]

The Peddie–Campbell–Gray–Dochard expedition has been described variously: as 'profitless', the narrative 'a rather tedious volume', the whole affair 'a record of elaborate planning, grand misfortune and misunderstanding', a 'forgotten expedition' which was 'lost in place'.[7] Its travails were those of Park's writ large: soldiers ill, dying, and left behind, the constant search for food and water, delays occasioned by weather, and the demands of tribute to travel at all.

Dochard reached the Niger but did not travel down it. Gray had exchanges that could have cast light on Park's death. At one point, he met with Isaaco and travelled with him for several days, even lending him his horse. If the two conversed about Park and the Niger – as, surely, they must – we are not told. Queries about the Niger were met with incredulity as to why explorers were so interested in the river at all and ignorance over where they came from: 'Whenever I spoke of the Niger, or my anxiety to see it, they asked me if there were no rivers in the country (we say) we inhabit; for the general belief is . . . that we live exclusively in ships on the sea.'[8]

Exploration was commonly a matter of encountering and overcoming such moments of mutual incommensurability. However else it has been described, the Peddie–Campbell 'forgotten expedition' was, from its hurried conception to its drawn-out completion and publication, a glorious example of British geopolitical hubris. In this, it is rivalled only by the Niger expeditions of 1832–4 and 1841 – and, perhaps, by the Niger–Congo Admiralty expedition.

The Admiralty expedition set out from Britain in February 1816. Its instructions were principally drawn up by John Barrow, his first major geographical project. They were shaped too by Banks, to whom Barrow turned for advice, chiefly over which scientific men to involve. The expedition was directed to enter the mouth of the Congo, to travel upriver to determine whether the Congo and the Niger were one and the same, and, Barrow stressed, to explore the Congo itself. That only about 200 miles of its course should be known to Europeans was for him 'incompatible with the present advanced state of geographical science'. As Barrow, Banks, and others well knew, geographical science was not in an advanced state where the Niger was concerned.

Initially, the plan was to use what would become that symbol and instrument of nineteenth-century progress and imperial aggrandisement, the shallow-draft steam ship. The idea was Banks's, who considered it necessary to counter the Congo's rapid currents and volume of water. This proved unworkable. Trials on the Thames revealed the engine was too big for the vessel: it consumed vast quantities of wood, increased the draft, and limited the top speed to about five knots – not enough to counter the Congo's currents. On this occasion, steam was not the answer. The expedition's sailing ships, the sloop *Congo* and *Dorothy* the transport, reached the mouth of the Congo in late June 1816.

The expedition was commanded by James Hingston Tuckey. Tuckey had served in the Royal Navy during the Revolutionary Wars before being involved, from 1802, in survey work in New South Wales. It is interesting to speculate that had Park grasped the opportunity of doing likewise, their paths might have crossed there. Like Flinders, Tuckey was held captive by the French. In prison between 1805 and 1813, he there wrote his *Maritime Geography and Statistics*, a history of maritime exploration and trade from a north-west European perspective and published in four volumes in 1815.

Tuckey had command of six officers, a crew of forty-two including fourteen marines, two African translators, and – this the key difference from the Peddie and Campbell expedition – a scientific group whose focus was the natural history of the Congo region. This was made up by Christian Smith, a Danish botanist recently returned from Madeira and the Canaries, David Lockhart, a gardener from Kew, John Cranch from the British Museum whose interest was marine zoology, William Tudor a comparative anatomist, and William Galwey, an Irish friend of Tuckey's.[9]

Perhaps because it was his first venture in directing expeditions, because he was in thrall to Banks, and because, with Banks, he looked back at the instructions given to Houghton and to Park and thought them inadequate, Barrow's instructions to Tuckey and to the scientists were especially long and detailed. He was insistent on the importance of journal writing and on their completeness: 'all your observations and occurrences of every kind, with all their circumstances, however minute, and however familiar they may have been rendered by custom, should be carefully noted down'. Each member of the scientific party was to keep his own journal. Barrow recommended that the journals be kept in triplicate, and that each person carry an abstract of the principal observations, 'in order that, in the event of any accident befalling his journals, he may still preserve the abstract to refresh his memory'. The final directive, a formal requirement, was that the journals were the property of the Admiralty, to be controlled by them and thus by Barrow upon return.[10]

This insistence upon writing in the field, note-taking in situ as an aid to later memory, was being generally encouraged and practised as part of the emergence of more rigorous scientific methods. Barrow's instructions over others' potential authorship additionally reflected his own authoritative position.

Tuckey and his companions had no opportunity for post-travel reflection upon authorship. The Congo proved unnavigable beyond the rapids at Yellala, about 200 miles inland. Attempts to go further by canoe proved difficult. Food was scarce and most in the party were soon ill. Many died of fever within weeks. Instructions on latitude, longitude, and the use of scientific instruments were ignored. The expedition fell apart about 280 miles inland. The survivors returned to the coast after three months. Tuckey and Smith survived only to die on the coast in early October 1817: twenty-one of the fifty-six men who set out failed to return.

Responsibility for Tuckey's posthumous text on the failed Niger–Congo expedition fell to Barrow and to Murray. Papers from the expedition reached the Admiralty on 6 March 1817. Barrow outlined his intentions to Murray in a letter of 8 May 1817, there noting not only his role as editor of Tuckey's journal given his Admiralty responsibilities, but also his hope that profits from the book should go to the families and widows of the expedition's members.

Barrow did more than edit Tuckey's journal. He used a lengthy introduction to justify his own authority, to review Park's views on the Niger–Congo hypothesis and to dismiss Reichard's ideas as 'entitled to very little attention'. He took issue with the account of Park's death: 'No one can tell what the fate of Park may have been, but no one will believe that this enterprising traveller finished his career in the manner related by Isaaco, on the pretended authority of Amadou Fatima [sic]'. Barrow made no mention of the fact that Tuckey did not believe the Niger–Congo hypothesis and had told Barrow so in a letter in August 1815. And, finally, he altered the composition and arrangement of the materials before passing the book to Murray.

One reviewer tendered 'thanks to the editor' for assembling the expedition's materials thus. Another noted that the volume appeared 'greatly amplified by the insertion of his [Barrow's] own reflections and deductions from the account given by the travellers', before concluding that the expedition's members' accounts had been 'properly preserved in a distinct form' following Barrow's interventions.[11] Barrow's plans were at one point hampered by the fact that a pirate edition of Tuckey's expedition seemed to be in the offing, despite his control of the officers' journals: 'rascals have some how or other got access to Tuckey's journal'. The culprit was later revealed to be the purser on the *Congo*, the late Mr Eyre, whose own journal, illicitly kept, was among papers passed to his widow. Because Eyre was not an officer – dismissing Eyre's status as a legitimate source Barrow termed him 'a young man of a corpulent and bloated habit' – Barrow had no formal control over her late husband's journal. But, as he informed Murray, 'I will endeavour to frighten her' over any attempt she might make to bring her late husband's unofficial manuscript into print.[12]

The importance of this to understanding the Niger's exploration rests not in Barrow's frightening a sailor's widow in managing Tuckey's publication, or that reviewers were wise to Barrow's emendations. It lies in the fact that as the Niger's exploration after 1815 became increasingly directed by government institutions, so the officers of those institutions assumed tighter control over who set off and who wrote what about the search for the river and for Park.

Partial insights, further failings, *c.*1818–c.1822

Thomas Bowdich's account of his 1817 mission to the Ashanti kingdom, and Joseph Ritchie and George Lyon's travels in northern Africa between 1818 and 1820 do not figure in most accounts of the Niger's exploration.[13] Their omission is in one sense wholly justified: Bowdich's concerns lay elsewhere; Ritchie and Lyon failed to get further south than Murzuq, Ritchie dying there. In another sense, it is unfortunate. Bowdich's 1819 *Mission to Ashantee* provides evidence relating to Park's death and offers an interpretation of the Niger's course which tries to accommodate the different views concerning the river's termination. Ritchie and Lyon's account is in part a model of how not to conduct exploration: an account from which some later explorers learnt – and others should have.

Thomas Edward Bowdich was an employee of the African Company of Merchants tasked, in 1817, with establishing trading relations between Britain and the Ashanti kingdom (then sometimes Ashantee or Ashante and now Asante), in what is today Ghana. Bowdich's brief included gathering information on the products of the country, the latitude and longitude of its principal places and by whom they were governed, and the physical geography of the territories over which local rulers reigned: in short, a geographical-cum-statistical survey. He was also instructed to enquire into evidence relating to Park, specifically any connections 'between Ashantee and the last places he visited'. The sheer size of the region made both tasks impossible. Bowdich confessed to John Hope Smith, the British governor at Cape Coast Castle, that 'the statistical and scientific desiderata so impressively recommended to my attention, are daily realising beyond my expectations'.[14]

In the face of too large a mission brief, ill companions, and limited resources, Bowdich turned to indigenous Ashanti and to Arab traders to supplement what he learnt first-hand. On 12 July 1817, Bowdich conversed through a translator with one 'Baba', the chief Moor to the court of Saï Tootoo Quamina, King of Ashanti. The next day, Baba sent for another Moor who, unlike many, was unsurprised at Bowdich's appearance because he had seen 'three white men before, at Boussa'. Bowdich 'eagerly enquired after the particulars of the novelty' and they were repeated to him:

> [T]hat some years ago, a vessel with masts, suddenly appeared on the Quolla or Niger near Boussa, with three white men, and some black. The natives encouraged by these strange men, took off provisions for sale, were well paid and received presents besides: it seems the vessel had anchored. The next day, perceiving the vessel going on, the natives hurried after her, (the Moor protested from their anxiety to save her from some sunken rocks, with which the Quolla abounds) but the white men mistaking, and thinking they pursued for a bad purpose, deterred them. The vessel soon after struck, the men jumped into the water and tried to swim, but could not, for the current, and were drowned. He thought some of their clothes were now at Wauwaw, but he did not believe there were any books or papers.[15]

To judge from this account, the locals' intentions were to guide Park and his companions away from danger, not to attack them. No mention is made of arrows or stones being directed at the canoe's occupants. It is possible, of course, that this claim was subject to the same view over the informant's credibility that Whishaw voiced over Amadi Fatouma's belief in the testimony of a slave. Bowdich is silent on this and, we may presume, trusted his informant. He directed one of his companions, William Hutchinson, 'to note any other report on the subject of Mr. Park's death'. Through Hutchinson's efforts, Bowdich was sent an Arabic manuscript on the matter, two translations of which he included in his 1819 *Mission*.

The first was by Abraham Salamé, the Egyptian cultural 'go-between' to Lord Exmouth's 1816 expedition to Algiers and author of an 1819 book about it. The second was by James Grey Jackson. Grey Jackson was for sixteen years British consul in Mogador on the Barbary Coast, what today is Essaouira in Morocco, and author, in 1809 and in 1820, of works on the Moroccan empire, including descriptions of Timbuktu. His 1820 book was undertaken with the assistance of Abdullah Salam Shabeeni who, thirty years before in London, had advised Joseph Banks and Henry Beaufoy, with Simon Lucas translating, about the geography of north Africa.[16]

The translations – of an Arab text provided by Sherif Ibrahim, a resident of Yaud [Yauri] in the Hausa kingdom, where Amadi Fatouma had parted from Park – differ in their wording but not in their substance. Both recount the appearance on the Niger of a vessel 'the like to which in size we never saw before'. In it sat two white men, and – this the only occasion on which female passengers are mentioned in accounts of Park's death – 'one woman, two male slaves, and two maids in the ship' (mention of the maids is given only in Salamé's translation). On encountering the rapids at Bussa, 'the man in the vessel' – presumably Park – killed his wife, threw her into the river, and then the men 'threw themselves into the river, fear seizing them'. While we should dismiss as fanciful the presence of women and the killing of one, this account is otherwise broadly consistent with that of the surviving slave interviewed by Amadi Fatouma. Bowdich despatched the original manuscript to the committee of management in the African Association who, two years later, arranged for its further study.[17]

Following his meeting with Baba and the unnamed 'Houssa [Hausa] moor', Bowdich was handed a chart which he had been at pains to secure for some while. This was a map of the Niger in its middle and lower reaches based on Arab sources. Bowdich lists the territories through which the river ran in this 'Chart No. 1 Course of the Niger or Quolla', with Arabic translations by James Grey Jackson. Listing was all he could do. The original manuscript was 'lost or mislaid' by a gentleman – who is not named – in translating from it, perhaps Salamé, perhaps Hutchinson or Jackson. The existence of further indigenous cartography of the Niger is hinted at in Bowdich's account – and would be established for certain by others within a few years.

In the 'Geography' chapter of his *Mission*, Bowdich scrutinises the work of European authorities on the Niger in the light of what he and Hutchinson were told. His account demonstrates commendable scholarly caution over what d'Anville, Hornemann, Park, Rennell, and others state on the Niger, and considerable circumspection over indigenous claims. Howsoever he was regarded by the African Company of Merchants – he and they later fell out over policy differences regarding Britain's commercial intentions in the Gold Coast – Bowdich was an able natural scientist. He lived in Paris from 1822 before his death in Gambia in January 1824, and there counted palaeontologist Georges Cuvier and naturalist Alexander von Humboldt amongst his associates. As Bowdich was finishing his geographical report, Hutchinson sent him 'a chart drawn by a Jennë Moor . . . confirming all that I had collected in the most satisfactory manner'. It is possible that this was the chart later mislaid, and that Hutchinson was indeed responsible for its loss.

It is possible too that this chart was the basis to the map of the Niger that appears in Bowdich's *Mission* (Figure 8.1). As he remarked, 'My sketch in the map, of course, represents the sketches and descriptions of the natives.' Bowdich mispositioned the river, despite his attention to European authorities and to indigenous testimony. Given their contradictions and the different names by which the Niger was known, and the confusing reports he received in west Africa, this is perhaps unsurprising. In arguing that the Niger divided permanently into several different rivers – Park among others had noted how the Niger divided into two streams after Lake Dibbie before reuniting in one course – Bowdich tried to accommodate the different ideas then held regarding the river's course.

The Niger, he claimed, divided into three streams. One flowed north and eastwards near to Timbuktu to terminate in 'the great lake of Caudee or Chadee' [Chad]. A second, known as the Joliba, flowed northwards. The third and main stream, the Quolla or Quorra, was thought to run south and east before dividing into two. One branch headed eastward and, if one believed the merchants in Ashanti, joined up with the Nile. The other emptied into the Atlantic by several channels, the principal one being the Congo (Figure 8.2).[18]

Reviewing Bowdich's 1819 *Mission*, John Barrow was characteristically forthright: 'Mr. Bowdich seems determined to reconcile himself to every hypothesis that had ever been formed of the course and termination of the Niger. But the way in which he makes his Quolla to perambulate the whole of the African Continent, and literally to quarter it with its divergent branches, some flowing to the east, some to the west, some to the north, and others again to the south, is not only geographically absurd, but physically impossible.' In his review, the geographical commentator Hugh Murray was more

Figure 8.1 Thomas Edward Bowdich echoed Rennell's map title of 1798 in his 1817 map of western Africa but did not echo his style or content (*cf.* Figure 4.1). Source: T. E. Bowdich, *Mission from Cape Coast Castle to Ashantee, with a Statistical Account of that Kingdom, and Geographical Notices of other Parts of the Interior of Africa* (London, 1819). Reproduced by kind permission of the National Library of Scotland.

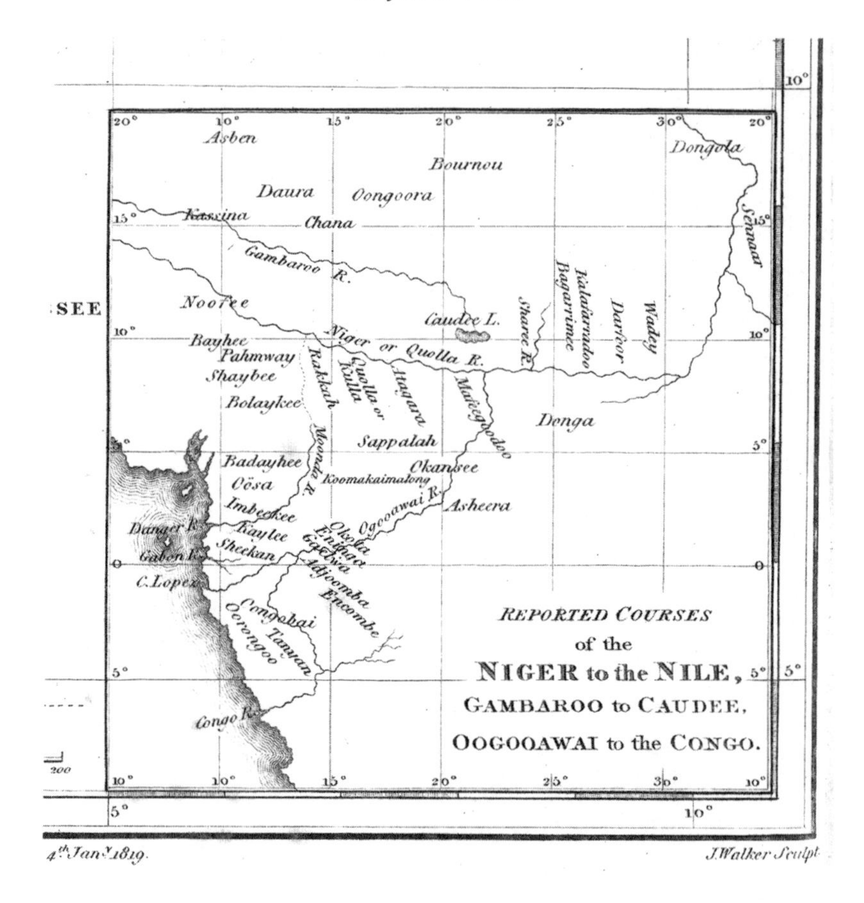

Figure 8.2 In this detail from his larger map (see Figure 8.1) Bowdich has attempted to show the 'reported courses of the Niger to the Nile'. His attempt to show different routes for the Niger was subject to scathing commentary from John Barrow. Source: T. E. Bowdich, *Mission from Cape Coast Castle to Ashantee, with a Statistical Account of that Kingdom, and Geographical Notices of other Parts of the Interior of Africa* (London, 1819). Reproduced by kind permission of the National Library of Scotland.

equable: 'Mr Bowdich's materials afford a fair promise of this great question, – but not exactly in the manner that he himself supposes.' Ending his review, Murray remarked simply 'Much in short remains to be cleared up; but on the whole the probability seems very strong, that this celebrated stream must find its way by more than one channel into the southern Atlantic.'[19] He was right on both counts but no one in Europe then knew it except Reichard, whose views on the Niger the British were not inclined to credit.

Others had also turned their eyes to the Niger. The trans-Sahara expedition of Joseph Ritchie and George Lyon contributed little to an understanding of the Niger – Ritchie's death from fever in Murzuq in November 1819 only added to the death toll. Although the expedition was Barrow's idea, who intended that it supplement what would be secured by Peddie and Tuckey, Ritchie received his instructions from Bathurst. Ritchie, a doctor, and Lyon, a naval officer on half-pay, were accompanied by John Belford, a shipwright from Britain's naval yards in Malta. As with Park, Peddie, and others, optimism about reaching the Niger included the provision to build a boat to navigate it once reached.

The Lyon–Ritchie mission was more diplomatic than exploratory. Bathurst expressed the hope that once Ritchie had established himself as Britain's vice-consul in Fezzan and so established a forward base for facilitating discovery of Africa's interior, he and his companions would then proceed to Timbuktu. Once there, they could 'collect all possible information as to the further course of the Niger', even 'trace the stream of that river with safety to its termination'.[20]

The expedition presented Hanmer Warrington with an opportunity to use his developing connections in the Muslim world for the 'innumerable advantages to geographical science' as he expressed it to Bathurst.[21] But circumstances limited his effective assistance. The expedition was late in departing for Murzuq. They travelled in disguise – Ritchie as Yussuf el Ritchie, Belford as Ali, and Lyon as Said Ben Abdullah Allah – and expressed anxieties about travelling south as part of what was, in effect, a slave-raiding party. Ritchie proved feckless as a leader. He did not engage with the locals and spent much of the £2,000 allocated to the expedition upon instruments, books, cork boards and bottles of arsenic for preserving insects, and 600lb of lead. Lyon only discovered this weeks after Ritchie's death.

The party became ill soon after reaching Murzuq. Lyon and Belford were resourceful even as they were fever stricken, collecting information about Hornemann's death. Lyon scotched a rumour that Park was alive and practising medicine while living under the Sultan's detention in Timbuktu: 'every one agreed that it would be impossible for him, or any other white man, to be confined in the town . . . and there is equal reason to doubt his being in existence, which some have supposed.' From a brief excursion south of Murzuq and from interviewing merchants about the Niger, Lyon was given no single view about its course, different names for the same river, and nothing on Park. 'Thus far,' Lyon observed, 'are we able to trace the Nil' – which, although it was also known as the Joliba, one name for the Niger, he took to be a variant of the Nile – 'and all other accounts are merely conjectural. All agree, however, that by one route or other, these waters join the Great Nile of Egypt, to the southward of Dongola.'[22]

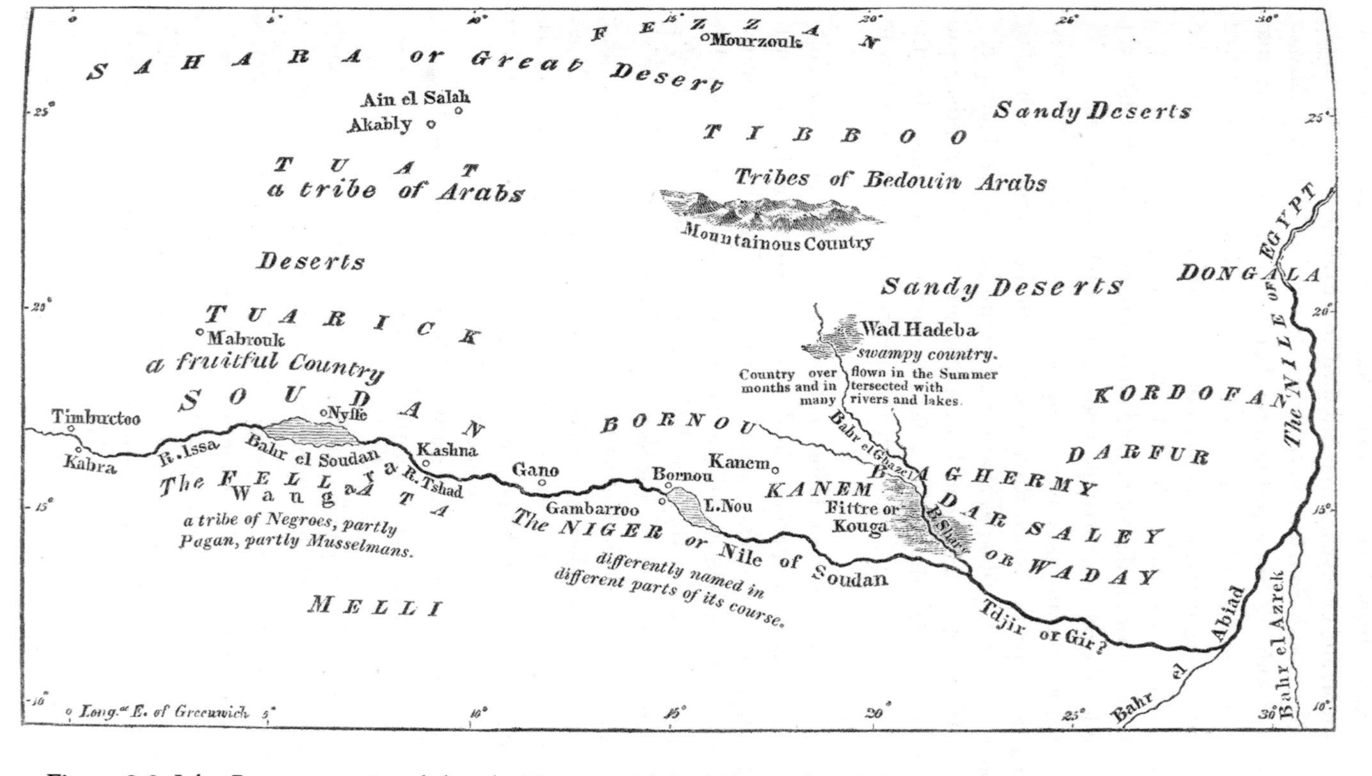

Figure 8.3 John Barrow, convinced that the Niger joined the Nile, produced this map in 1820 in support of his views. Source: *Quarterly Review* 23 (1820), page 237. Reproduced by kind permission of the National Library of Scotland.

Lyon and Belford returned safely to Tripoli in early 1820. Lyon left there in May 1820, taking with him a camel as a gift for George IV. His account of his North African travels has no map of the Niger, but he was convinced that the Niger and Nile were one. Barrow later produced a map to this effect, praising Ritchie and Lyon as he did so, and citing Burckhardt in support of his case (Figure 8.3). This view of the Niger's course, produced in London from his Admiralty office and based on others' words which he could not corroborate, tells us more about Barrow's preconceptions than it does about the truth of Africa's geography.[23]

The Bornu Mission, 1822–1825, and the travails of Alexander Gordon Laing

As Lyon left north Africa – later to achieve fame in Arctic exploration – other men were being lined up to continue the search for the Niger. Barrow and Bathurst were again the principal movers, Hanmer Warrington again the link between Whitehall, the Bashaw in Tripoli, other influential Arabs and African rulers, and the explorers themselves.

The Bornu Mission of 1822–5, so-called for its focus upon Lake Chad and the kingdom of Borno or Bornou, the powerful state between the Niger and the Nile, is one of the best documented and closely studied episodes in the history of European exploration of Africa. For Bovill, the expedition's pre-eminent historian, the crossing of the Sahara and the revelation of indigenous kingdoms south of the Sahara as sophisticated, rich, and with long-distance trade routes, is 'one of the great achievements in African discovery'. It is the more so given the deep personal divisions among the three explorers, its involvement with slave raiding and wars between tribal rulers, and, as Bovill notes, 'in spite of the fact of their having completely failed in their primary duty' – determining the course and termination of the Niger. We should note too that while Murray published the mission's *Narrative of Travels and Discoveries in Northern and Central Africa* in 1826, he did so in the knowledge that it afforded a platform for Barrow's own views on the Niger's course, that Clapperton's involvement was under-represented, and only after omitting the worst of Denham's accusations against his fellow explorers.[24]

The three principals of the Bornu expedition were Walter Oudney, a natural historian and naval surgeon from Edinburgh, Hugh Clapperton, a naval lieutenant from Annan in Dumfriesshire who had served in the Napoleonic Wars and in Canada and who was known to Oudney because they lived on the same street in Edinburgh, and Lieutenant Dixon Denham. Denham, a career soldier and in all accounts of him a most unpleasant man, held ideas on Africa's rivers which supported those of Barrow, despite never having visited Africa.

At Barrow's instigation, Bathurst appointed Denham leader of the mission in March 1821, thus slighting Oudney, the first appointed, and Clapperton, Oudney's appointee. His error in mishandling the appointments was compounded when, without consultation, Bathurst changed Oudney's role to vice-consul – from explorer to political officer – to take effect once the mission had arrived in Bornu. In Tripoli, Hanmer Warrington made things worse by stating that Oudney was in overall charge. Denham disabused both men of this view when he arrived there in November 1821, one month after Oudney and Clapperton, and accompanied by William Hillman, a shipwright from Malta.[25]

The expedition did not arrive in Murzuq until March 1822. Everyone would have known who the explorers were: Clapperton and Oudney had resisted the temptation to travel in disguise being 'determined to travel in our real character as Britons and Christians'.[26] The protection promised by the Bashaw, Yusuf Karamanli, did not materialise. His writ did not in reality extend as far as he had claimed it did when liaising with Warrington and in accepting £5,000 from the British government to provide a military escort. The escort, led by Mustafa el Ahmar, the Bey of the Fezzan, never left Murzuq. Further delays were inevitable. Denham returned to Tripoli to consult with Warrington and the Bashaw. His absence provided Clapperton, Oudney, and Hillman with the opportunity for an excursion to Ghat in the west of the Fezzan under the guidance of a Tuareg, Hatita, who had earlier assisted Lyon. This trip instilled in Clapperton and Oudney an awareness that travel in smaller parties with an appropriate guide and without military accompaniment could be faster moving and more effective.

In Tripoli, Denham and Warrington sought to put pressure upon the Bashaw over his promises to Bathurst. Denham at one point took ship for Marseille, intending to return to London. He claimed that this was to scare the Bashaw into fulfilling his promises: his real reason was to influence Bathurst to advance his, Denham's, own interests. The Bashaw eventually relented – for an additional £2,500 – allowing the mission to proceed south to Bornu, under the guidance of a Fezzan merchant, Abu Bakr Bu Khullum, and with a military escort of 200 men. In total, the caravan was about 300 strong: the explorers' party included a West India merchant seaman, Adolphus Sympkins, a one-time employee of the Bashaw, who spoke Arabic and who now served as Denham's servant: he had sailed so widely in his lifetime that he was known as Columbus.

The mission reached the kingdom of Bornu in December 1822 and Lake Chad by February 1823. They were the first Europeans to see Lake Chad since the Romans. What they saw then was much larger and less distinct than the lake today. Lake Chad is shallow – only a little over five fathoms at its deepest – and over time, its extent has considerably shrunk in consequence of climate change and the drainage demands of intensifying agriculture. Before permission could be granted to explore the lake's western margins to determine if the Niger entered it, and its eastern ones to ascertain if the Niger left the lake, permission had to be sought from the local ruler, Sheikh Mohammed el Kanemi, who was resident in the town of Kukawa. Not for the first time, exploration by Europeans was hindered by warfare among African peoples geared toward providing slaves for the markets of Murzuq and north Africa. Denham became involved with a slave raid led by Bu Khullum, in which the latter died. Denham, lucky to escape with his life, was severely reprimanded by the Colonial Office.

Matters were soon even more unsatisfactory. Clapperton and Oudney became ill. The expedition was delayed and became embroiled in internecine politics. Denham seemed keener on adventuring than exploring, and local rulers continued to restrict the mission's movement. Then things got worse. Denham accused Clapperton of having sexual relations with a male Arab servant. The charge was wholly untrue, and Denham knew it, as he later admitted following an inquiry. Oudney protested about this 'very malicious and wicked falsehood'. Denham did not think to deny it publicly, nor did he ever apologise to Clapperton: he knowingly allowed the rumour to circulate to belittle Clapperton and to strengthen his own position. Important as the Bornu expedition was, its achievements were despite, and not because of, the working relations among the explorers themselves.

By late 1823, the Bornu Mission was in effect two parties. During Denham's involvement in the abortive slave raid led by Bu Khullum, Clapperton and Oudney explored the southern margins of Lake Chad. They recorded how one river, the Shari, flowed into the lake from the south and east, and, on the lake's western margins, how a river called the Komadugu Yobe also emptied into the lake. The Shari they dismissed as the outlet to the Niger because it ran east–west into the lake. The Komadugu Yobe, which ran west to east and so supported Park's words on the direction of the Niger, was thought too small to be the Niger. Although local opinion was insistent that the lake had no outlet, they were unable to investigate the lake's eastern margins and to see things for themselves.

Upon his return from raids, Denham similarly explored Lake Chad and its river systems, in June and July 1824. Like his fellow explorers, he was prevented by warfare from exploring the lake's eastern edge. His attempts were hindered too by the sweeping extent of marshes, a sort of liminal geography of reeds and bulrushes, standing water, and unsafe ground, and by fearfulness over the Biddomah people, lake dwellers with a reputation for attacking all strangers. On one day, Denham spent thirteen hours in the saddle to travel less than six miles, all without catching a clear glimpse of the lake, and returned to camp in the evening with the skin on his face and hands 'falling off' from sunburn and insect bites. His sketch of Lake Chad (Figure 8.4) is in part based on direct encounter (the rivers to the lake's south and west), and in part upon conjecture (the geography of the eastern shore). Unlike Clapperton and Oudney, Denham believed that the Komadugu Yobe *was* large enough to be the Niger.

To deep personal animosities and differences in exploratory temperament was now added contrasting opinions over which river was the Niger and uncertainty as to whether Lake Chad had an eastward-draining outlet.[27]

In December 1823, with Denham involved in further raids, Clapperton and Oudney headed towards Kano in the Hausa kingdom. Clapperton was convinced that the Niger ran to the south of Lake Chad, Oudney that there was a further lake to be discovered that held the truth to the river's course. Both were intrigued by what they had learnt from a Jennë pilgrim in Kukawa in June 1823. Abdel Gassam ben Malaky, en route to Mecca, told the explorers he had heard of Christians before:

> Many years ago, before I was born, white men, Christians, came from Sego to D'jennie [Jennë on the Niger], in a large boat, as big as two of our boats. The natives went out to them in their canoes; they would not have done them any harm, but the Christians were afraid, and fired at them with guns, and killed several in the canoes that went near their boat: they proceeded to Timboctoo, and there the sultan sent to them one of his chiefs, and they held a parley. The Christians complained that the people wanted to rob them. The sultan was kind to them, and gave them supplies. Notwithstanding this, they went off suddenly in the night, which vexed the sultan, as he would have sent people with them, if they had not been afraid of them a little: and he now sent boats after them, to warn them of their danger, as there were many rocks in the belly of the river, all pointed. However, the Christians went on, and would not suffer the sultan's people to come near them, and they all perished.[28]

There are several notable elements in Abdel Gassam's account, recognising that this appears only in translation in Denham's later narrative, that there is no evidence that Park reached Jennë, and that the tale may have been distorted in being an oral account of others' remembered events. Park and his companions are again reported as killing locals, and of vexing local rulers, wittingly or not. There is, again, the suggestion that locals were trying to warn Park of the dangers given the Niger's seasonally low levels, a point supported

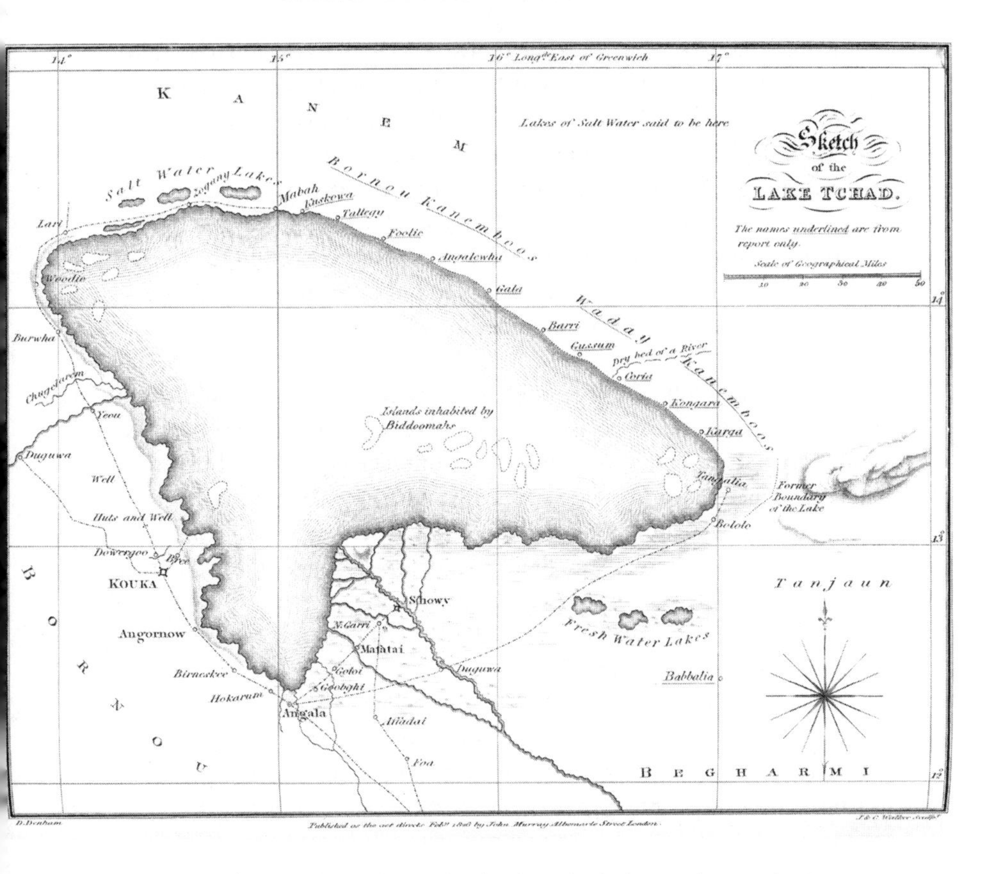

Figure 8.4 Dixon Denham's sketch of Lake Chad, drawn during the Bornu Mission of 1822–5. The lake's eastern shores were not examined at first hand as is apparent from the many underlined names. The 'Dry bed of a river' hints at Denham's presumption that a river did flow eastwards out of the lake. In this, and with others, he was wrong. Source: D. Denham, H. Clapperton, and W. Oudney, *Narrative of Travels in Northern and Central Africa, in the years 1822, 1823, and 1824, by Major Denham, F. R. S., Captain Clapperton, and the late Dr Oudney* (London, 1826). Reproduced by kind permission of the National Library of Scotland.

by the account given to Bowdich. There is nothing about any arrows and stones as the surviving slave had told Amadi Fatouma.

Information of a similar nature was provided by a Mohammed ben Dehmann, whose testimony is included in the Bornu expedition's *Narrative* as 'A Document relating to the Death of Mungo Park':

> Hence, be it known that some Christians came to the town of Youri, in the kingdom of Yaoor, and landed and purchased provisions, as onions and other things; and they sent a present to the King of Yaoor. The said king desired them to wait until he had sent them a messenger, but they were frightened, and went away by the sea (river). They arrived at the town called Bossa, or Boossa, and their ship then rubbed (struck) upon a rock, and all of them perished in the river.
>
> This fact is within our knowledge, and peace be the end.
>
> It is genuine from Mohammed ben Dehmann.

In lines which follow this text, Denham remarked that '[In addition to the above, there is a kind of postscript appended to the document by a different hand; which, being both ungrammatical and scarcely legible, I had some difficulty in translating and giving it a proper meaning. The words, however, are, I think, as follows; though most of them have been made out by conjecture.]' The words then follow: 'And they agreed, or arranged among themselves, and swam in the sea (river), while the men, who were with (pursuing) them, appeared on the coast of the sea (bank of the river), and fell upon them till they went down (sunk) in it.'[29]

This account echoes Abdel Gassam's in several respects: that Park and his companions were fearful of locals' advances, not understanding their motives, and that the intention had been to warn Park, not to attack him. It differs in having the correct place names and in the suggestion that Park's party was assaulted even as they were drowning – 'and fell upon them until they went down'.

These accounts show not just that Park's death was remembered. They show that two decades after its occurrence the event was recounted at different times and in different places to different explorers, and that different versions were circulating in different parts of sub-Saharan west Africa. It is impossible to know now whether the claim that locals were trying to warn Park of the dangers of the Bussa rapids, and not to attack him, was true. Equally, it is impossible to know if that claim – which contradicts the testimony of the surviving slave – was the result of those involved later 'managing' the tale to tell the story as they wished it to be told.

Further insight into the Niger's course, and a little more on Park, was provided to Clapperton alone. Walter Oudney died on 12 January 1824, having had a chest infection since arriving in Murzuq, but sustained his exploration work despite it and the effects of malnutrition and fever. To Clapperton, Oudney's loss 'was severe and afflicting in the extreme'. After a month spent in Kano, Clapperton headed for Sokoto to meet Mohammed Bello, ruler of the Hausa kingdom, a caliphate of several major city-states.

Mohammed Bello was informed, intelligent, and with wide-reaching influence. At their first meeting, he discomfited Clapperton by passing to him books which Denham had lost during his ill-judged raids for El Kanemi. As with other encounters between explorers and indigenous rulers, conversations between Clapperton and Bello did not always run smoothly. Bello was horrified to learn that Clapperton's purpose and the British government's intentions involved the abolition of the slave trade and opening the African interior to foreign trade. To the Hausa and Bornu kingdoms, this would be ruinous. Bello's determination to listen to Clapperton but not to assist in his wider objectives, even to hinder them, may explain his contrasting actions when, on 20 March 1824, the two discussed Park and the Niger. Park, Bello reported, might have passed the rapids had it been the rainy season, but he made no mention of other elements of Park's death save to remark that some books survived, and were now held by his cousin and the Sultan of Yauri.

Of greater interest was what Bello did. Clapperton recounted the incident: 'He [Bello] drew on the sand the course of the river Quarra [the Niger], which he also informed me entered the sea at Fundah' [a little north of where the Niger delta starts to form as it enters the sea]. Such drawing in the soil happened elsewhere when explorers and indigenes first met, the sand acting as a kind of liminal medium between Europeans and non-Europeans as, for want of a shared language, both parties tried to establish their location.[30]

A week later, Bello again 'drew on the sand the course of the Quarra [sic], with the outline of the adjoining countries'. Sand maps are no good if the information they contain has to travel to others elsewhere. Clapperton asked the sultan if one of his learned men would 'make me a chart of the river, on paper'. A month later, Bello presented Clapperton with the paper version (Figure 8.5). In its principal feature – the course of the Niger – Bello's paper map contradicted what he had sketched in the sand weeks earlier. Instead of the Niger flowing into the Atlantic as he had told Clapperton, it was shown flowing eastwards towards the Nile. An accompanying explanatory note read, 'This is the river of Kowarra, which reaches Egypt and which is called the Nile.' What Bello told Clapperton, and what he drew for him, were different things. It is as if Bello had the paper map made as a deliberately deceitful document knowing that disruptive British and foreign trade would follow should the Niger's true course be exposed.

Given his ill health, Clapperton abandoned plans to travel towards Yauri (where he hoped to secure the reported books from Park's second expedition) and returned to Sokoto in late July 1824. He was by now so emaciated that Denham did not recognise him when they were reunited: they had been apart for eight months. Perhaps to hide his own failure in not exploring Lake Chad's eastern margins, Denham chastised Clapperton for not proceeding

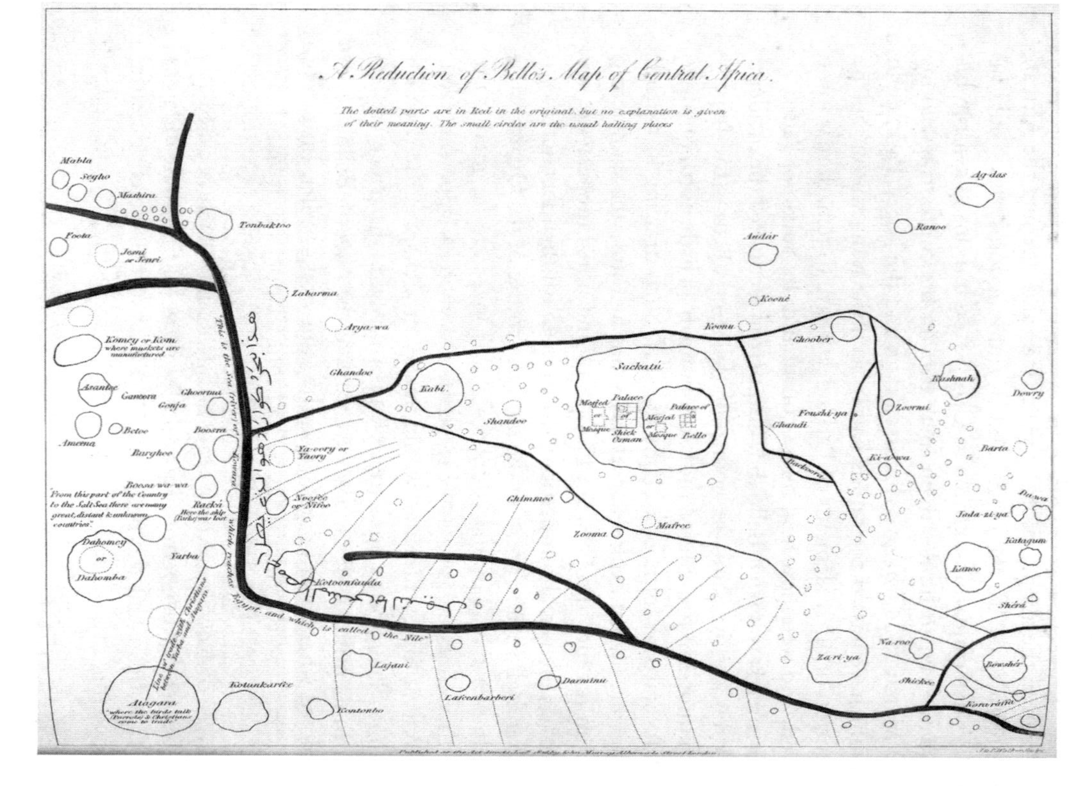

Figure 8.5 Sultan Bello's map of central Africa. Although it distorts the true position and course of the Niger and differs from what Bello told Clapperton and drew in the sand, it is nevertheless an important example of indigenous map-making in this period. Source: D. Denham, H. Clapperton, and W. Oudney, *Narrative of Travels in Northern and Central Africa, in the years 1822, 1823, and 1824, by Major Denham, F. R. S., Captain Clapperton, and the late Dr Oudney* (London, 1826). Reproduced by kind permission of the National Library of Scotland.

further than Sokoto. Both men wanted to achieve more but the only sensible option was to return, to Tripoli and to London.

On their return to London in June 1825, the explorers and their words enjoyed different fates. Barrow controlled the publication of the expedition's narrative, which centred principally upon Denham's exploits. Barrow makes almost no reference to Oudney's work. In public printed comments, Barrow declared himself faithful to Clapperton's journal text as part of the *Narrative*: 'I have carefully abstained from altering a sentiment, or even an expression, and rarely had occasion to add, omit, or change, a single word.' In private exchanges to Murray, he admitted that he had been rather more vigorous: 'dishing and trimming him [Clapperton] as much as I dare'.[31] This may have involved omitting a manuscript map which Clapperton had produced showing his view of the probable course of the Niger, towards the Gulf of Guinea (Figure 8.6).

Barrow praised Clapperton's longitudinal observations in improving the 'best maps of Africa, – all of them bad enough', but made no mention of this map in his prefatory remarks on Clapperton's journal. Nor does he in his later review. Barrow may have excluded it because Clapperton presented a geography for the Niger with which he did not agree.[32] Bello's paper map (Figure 8.5) is included in the *Narrative* no doubt because it supported Barrow's view. As Barrow emphasised in ending his review of the Bornu *Narrative*, the conjunction of eastward-flowing waters issuing from Lake Chad into the Nile 'is not only *possible*, but extremely *probable*'.[33] He was even more insistent when writing to Wilmot-Horton in November 1824: 'I think now there is no doubt that the Niger falls into the Nile.'[34]

Clapperton's reaction to Barrow's redaction of his manuscript map and his dismissal of Clapperton's ideas on the Niger is not known. By the time the *Narrative* was being finalised, he had left again for Africa. By the time Barrow's review appeared, he was dead.

But soon after his return to London and as he was formulating plans to go back to Africa, Clapperton was made aware – to his great displeasure – that another man had designs on the Niger's exploration. Like Oudney, Alexander Gordon Laing was born in Edinburgh. A member of the Royal Africa Corps, his work for them in Sierra Leone, from where he had explored the Niger to within sixty miles of its source, and in the Ashanti kingdom, brought Laing to the attention of the Colonial Office. Laing completed his Sierra Leone book the day before he departed for Tripoli.[35]

Bathurst's plan was that Laing should travel to Timbuktu and explore the Niger from there. Bathurst's compelling interest in the Niger's exploration was marred by his continuing naivete in planning it. Where, earlier, he had mishandled the Bornu Mission, by mid-1825 he and Horton-Wilmot had

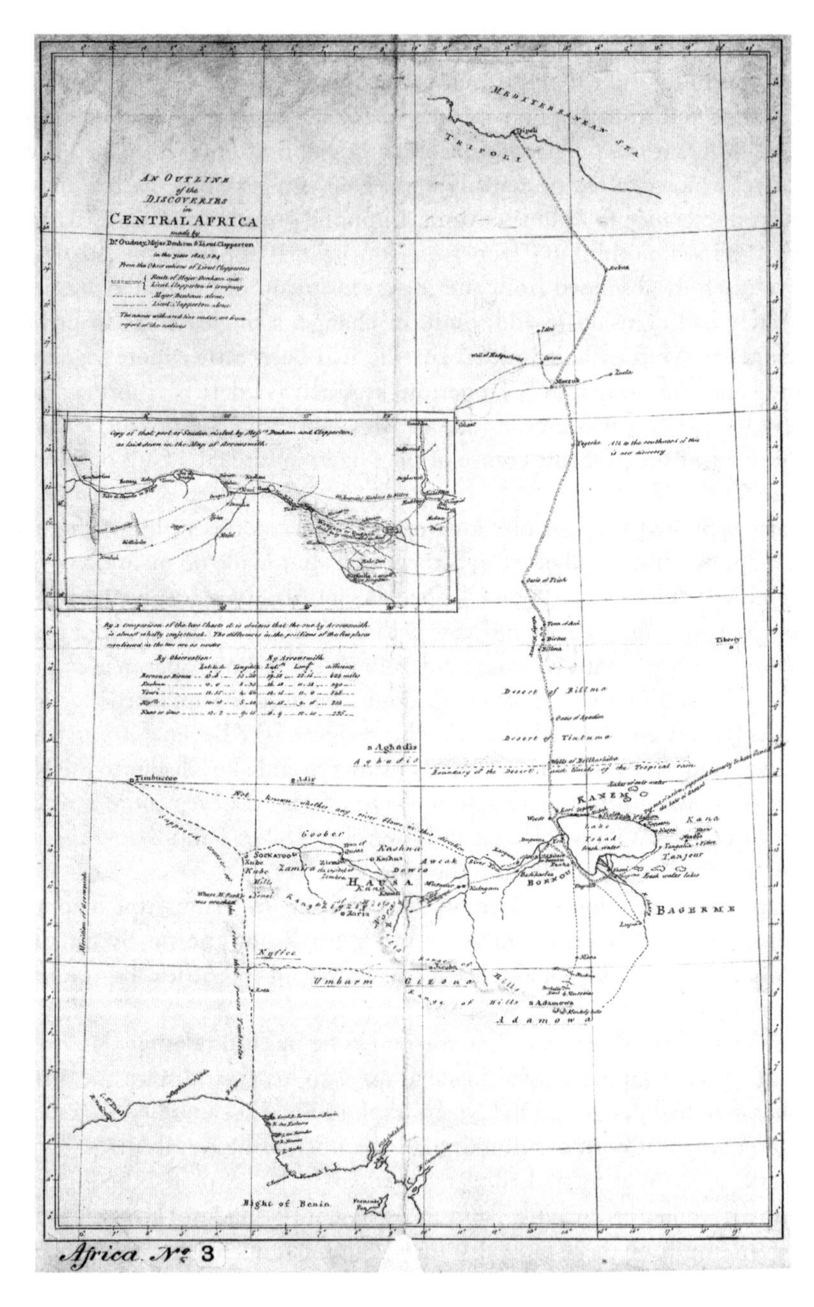

Figure 8.6 Hugh Clapperton was convinced that the Niger flowed south to empty into the Atlantic. His scepticism towards others' ideas – notably those of John Barrow – is evident in his remarks, east of Timbuktu, 'Not known whether any river flows in this direction.' Source: The National Archives, ADM 55/11 (f. 91v).

set in motion plans for Laing to travel south from Tripoli to Timbuktu and for Clapperton to undertake a second African expedition, north from Whydah on the Guinea coast. Both undertakings were thought straightforward. For about eighteen months between March 1825 and September 1826, the Niger's exploration was almost a personal competition between Clapperton and Laing, orchestrated by Bathurst in London and Warrington in Tripoli.

Like several of his predecessors, Laing found it easier to reach Tripoli than to leave it. The Bashaw was again insistent upon payment for his assistance – he eventually secured £2,500 for his involvement in facilitating Laing's travels. Laing was reluctant to depart since he had fallen in love with, and married, Emma Warrington, the consul's second daughter. But orders were orders and leave he did. At In Salah, the principal market of the Sahara's Tuat region, Laing was delayed as he waited for a caravan to be formed. There, to his great surprise, while in conversation with traders about the Niger, he was mistaken for Mungo Park.

Laing recounted the circumstances to a friend, James Bandinel: 'An extremely ridiculous report has gone abroad here that I am no less a personage than the late Mungo Park, the Christian who made war upon the people inhabiting the banks of the Niger, who killed several, and wounded many of the Tuaric [Tuareg].' There was, he continued, 'A Tuaric in this place who received a musket shot in his cheek in a rencontre with Park's vessel, and who is ready to take his oath that I am the one who commanded it.' Laing dismissed the report, but aimed further remarks not at the Tuareg's credibility but at Park's reputation and at the damage his actions had done: 'How imprudent, how unthinking; I may even say, how selfish was it in Park to attempt making discovery in this land, at the expense of the blood of its inhabitants, and to the exclusion of all after communication: how unjustified was such conduct!'[36] Laing's remarks speak to his own self-regard as a putative explorer whose task might be hindered by a predecessor's actions – he was known for his 'consummate conceit' and disliked for it – but this is the first time that Park was criticised by his peers.[37]

Laing was attacked and grievously injured by Tuareg tribesmen in January 1826:

> I have suffered much, but the detail must be reserved till another period . . . in the meantime I shall acquaint you with the number and nature of my wounds, in all amounting to twenty-four, eighteen of which are exceedingly severe. To begin from the top: I have five sabre cuts on the crown of the head and three on the left temple, all fractures from which much bone has come away; one

> on my left cheek which fractured the jaw bone and divided the ear, forming a very unsightly wound; one over the right temple and a dreadful gash on the back of the neck, which slightly scratched the windpipe; a musket ball in the hip, which made its way through my back, slightly grazing the backbone; five sabre cuts on my right arm and hand, three of the fingers broken, the hand cut three-fourths across, and the wrist bones cut through; three cuts on the left arm, the bone of which has been broken but is again uniting; one slight wound on the right leg and two with one dreadful gash on the left, to say nothing of a cut across the fingers of my left hand, now healed up. I am nevertheless doing well.[38]

Miraculously, he survived. Displaying remarkable fortitude for someone whose body, as perhaps no others', bore the stigmata of exploration so evidently, he arrived in Timbuktu in August 1826 and stayed there for about five weeks. Writing to Warrington on 21 September 1826, Laing noted, 'I have been busily employed during my stay, searching the records of the town, which are abundant, and in acquiring information of every kind, nor is it with any common degree of satisfaction that I say, my perseverance has been amply rewarded – I am now convinced that my hypothesis concerning the termination of the Niger is correct.'[39]

Laing's hypothesis was simple – the Niger emptied into the Atlantic into the Gulf of Guinea.[40] If Laing made notes on Park's death, they have not survived. Neither Laing nor the words of his projected journal survived. He was murdered in the Sahara on his return journey in late September 1826. Disputes over his missing journal lasted for years and included unsubstantiated accusations that it had been taken by the French African explorer René Caillié.[41] As Bovill remarks, 'A particularly tragic aspect of Laing's death at the end of a long and terrible journey was that it was to almost no purpose. He had traversed a great part of wholly unexplored Africa for which the geographers craved information, yet he died leaving them hardly any the wiser.'[42]

Much the same might be said, of course, of those others who died in search of the Niger in the decade after 1815 and leading British authorities knew it (Figure 8.7). In February 1826, John Wilson Croker, First Secretary to the Admiralty and Barrow's boss, and Robert Wilson Hay, Permanent Under-Secretary for the Colonies between 1825 and 1828, sat down to discuss what they called 'The Conduct of African Exploration'. Citing the deaths of Houghton, Hornemann, Park, Martyn, Scott, Ritchie, Tuckey, Smith, Burckhardt, and Oudney as proof (Laing's death was not confirmed until 1828), Croker began pithily: 'Nothing can be more indisputable than

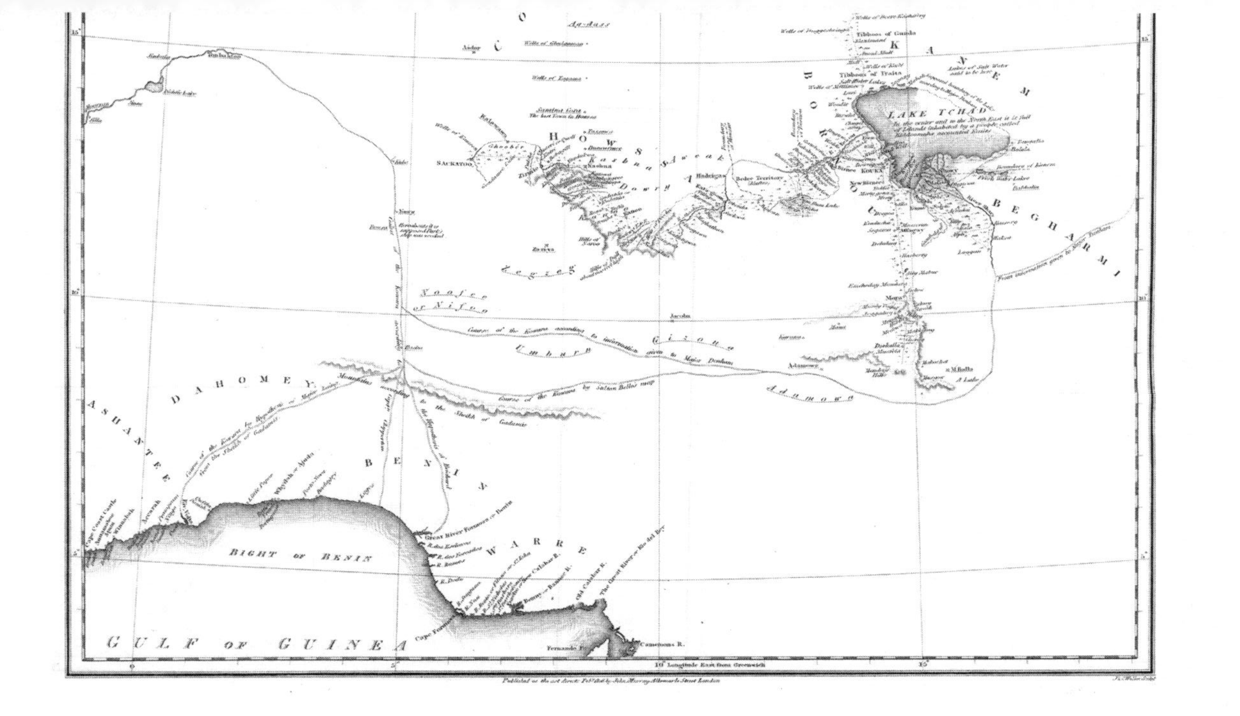

Figure 8.7 The different hypotheses as to where the Niger – here termed the Kowara – ran in its lower reaches are evident in this map of 1826. To the left, Laing's suggested position for the Niger's course; centre of the three, Clapperton's; to the right, that of Bowdich. The course of the river according to Sultan Bello is also shown (*cf.* Figure 8.5) and, north of that, the course of the Kowara according to information given to Denham. Source: D. Denham, H. Clapperton, and W. Oudney, *Narrative of Travels in Northern and Central Africa, in the years 1822, 1823, and 1824, by Major Denham, F. R. S., Captain Clapperton, and the late Dr Oudney* (London, 1826). Reproduced by kind permission of the National Library of Scotland.

that the efforts made by our countrymen for the last twenty years in exploring the interior of Africa have produced very inadequate results and nothing can be more lamentable than the expense of valuable life at which the trifling additions to our knowledge have been purchased.'[43] Others were at work on the Niger's geography, however, without dying in doing it.

9

'*Quorra*, *Quolla*, *Kowarra*, and others similar'

The Niger and critical geography, 1821–c.1829

The exploration of the Niger illustrates different ways of doing geography. In the nineteenth century, a 'critical', 'speculative', or 'armchair geographer,' commonly an independent scholar, often a clergyman or, like Reichard, a civil servant, did geography without leaving home. The object of their enquiries was no different from those men, chiefly military officers, naturalists, and physicians, who set out to know foreign places by personal encounter and with the formal support of government, but their methods and sources were different. Explorers in the field set out to observe geographical phenomena at first hand and, once safely home, there to write about their accomplishments. Critical geographers seldom left their study in turning to others' texts and verbal testimony, and to questions of etymology, philology, and linguistics.

This distinction between types of explorer-geographer was never absolute: James Rennell was both a Classical sedentary critical geographer in his work on Herodotus and, earlier, an East India Company surveyor in Hindustan for thirteen years. Questions of precision and accuracy informed the work of both types, whether expressed through instrumental measurement, or in the credibility of others' words. And until the later nineteenth century, most geographical enquiry of any sort was not done by geographers. Before then, hardly anyone who did geography was professional in the modern sense of that term: salaried, teaching students, practising a discipline.[1]

This context is especially apposite to the Niger's exploration in the three decades following Park's death. At the same time as Lyon, Ritchie, Denham, Clapperton, and Oudney were in the field, elsewhere a clergyman, a retired soldier, a diplomat, an employee of the African Company of Merchants, and a Glaswegian newspaperman and polemicist with interests in the Caribbean each tried to solve the problem of its termination. They turned, variously, to Classical authors, to the rules of grammar and Ptolemaic measurement, to Islamic scholarship and merchants' tales, and to the testimony of slaves. Several were wrong. But two got the Niger's geography right and yet, like Reichard, neither saw the river for themselves.

The Niger traced: the work of John Dudley and Rufane Shaw Donkin

The Rev. John Dudley, vicar in the Leicestershire parish of Humberston and Sileby and sometime tutor at Clare College, Cambridge, is not a name ordinarily associated with the Niger's exploration. He was drawn to the topic following conversations with his local MP, Charles March-Phillipps, on the subject of 'African researches', and from a concern over the high death toll among the river's explorers. Studying the work of ancient authorities was, Dudley argued, a good way to locate the river, far better and less dangerous than sending 'injudicious expeditions'. He emphasised his point by citing Rennell and referring to Tuckey and to Park: 'To ascend the Zaire [Congo] was lately a work of waste to the health, strength, and lives of the adventurers. The attempt to descend the Joliba, or Niger, proved fatal to Park, who in all probability was lost in one of the rapids of that river.'[2]

Dudley's *Dissertation showing the Identity of the Rivers Niger and Nile; chiefly from the Authority of the Ancients* (1821) is a neglected work of critical geography based on a close reading of Classical authors, chiefly Homer, Aeschylus, and Herodotus. While his motive was laudable – a point John Barrow made twice in review – Dudley's argument was diminished by his belief in the literal truth of the story of Jason and the Argonauts.

The ancients proclaimed how Jason and his fellow adventurers had crossed the Sahara after their boat had been stranded in the Bay of Syrtis on the north African coast, the eastern reaches of what is today the Gulf of Sidra. After a twelve-day Saharan journey, the Argonauts reportedly reached a Lake Tritonis in central Africa. Dudley took this to be Wangara. His unshakeable certainty on this was a source of consternation to Barrow since 'neither Burckhardt, nor Ritchie, nor Lyon, could learn any thing decisive' about Wangara – and for good reason: what Dudley took to be a matter of physical geography, a lake, was the name given to a group of nomadic peoples involved in trans-Saharan trade. From Wangara, the Argonauts succeeded in reaching the Nile, aided by the Niger's periodic flooding which assisted their voyage eastwards. Dudley also argued that the dark sediment in the Niger and the Nile was of common origin, a fact supported by his chemical analysis of it – which amounted to citing the leading chemist Humphrey Davy over the common 'earths' found in river-based plants.

Dudley's view that the Niger and the Nile were one and the same is consistent, of course, with one idea concerning the Niger's termination. He was dismissive of others: 'That a mighty river, like the Libyan Niger, should suddenly take such a turn, and work its wide way, as it seems it must, through the *Kong Mountains*, opposing a barrier generally supposed to be insurmountable, can hardly be imagined to be true.'[3] In this, he erred of course but he was not alone. The fact that Dudley's *Dissertation* is substantively in error – Barrow referred to Dudley's 'strange and inconsistent conclusions' – is less interesting than the methods he chose and the fact that it was written at all. As the churchman William Roberts noted in his review, Dudley's argument was let down by its 'reasoning from imperfect data'.[4]

To adopt Barrow's words, modern readers might derive 'considerable amusement, if not instruction', from Dudley's work. We would be wrong to do so – Dudley's *Dissertation* ought not to be casually dismissed. It is evidence that the Niger's exploration in this period was public currency, that explorers' narratives concerning the river were not only read and talked about, including by a clergyman and a parliamentarian in the English Midlands, but also that the Niger problem and the death of explorers was prompting people to think and to write about the river.

Someone who did just that was Lieutenant-General Sir Rufane Shaw Donkin. Following a distinguished military career fighting the French and a three-year posting to India from 1815, Donkin's role in 1820 as acting governor of Britain's Cape Colony marked a shift towards literary and political interests. He was twice a member of Parliament, and, in 1835, became Surveyor-General of the Ordnance. His *Dissertation on the Course and Probable Termination of the Niger* was published by John Murray in 1829.

Donkin's interest in the Niger was prompted, he tells us, by 'the discussions which have taken place of late years concerning that question'. He was of the view that the principal reason behind the failure to locate the river before now was grammar: 'the reason why geographers and travellers had hitherto failed in settling this question was, because they had made a verbal or grammatical error in stating the object of their search to be THE Niger, or rather THE Nile . . . instead of searching for A Niger – or A Nile . . . they have thus confounded a specific appellative with a generic and descriptive one.' Modestly, he aimed 'to reconcile all or most of what has been said of the Niger, from Herodotus and Ptolemy, down to those of Park and Denham . . . and this I hope to do partly by the rectification and proper use of a grammatical particle'.[5] For Donkin, locating the Niger was a matter not of exploration in the field but of grammar and the definite article.

The Niger, wrote Donkin, was not a name known to the indigenous peoples of western and central Africa, but he confessed himself unable to trace the etymology of the several names by which that river was known by them – '*Quorra*, *Quolla*, *Kowarra*, and others similar'. He reasoned from 'what Park,

Clapperton, and Denham say' that these terms were used locally to describe any large river, rather than the Niger alone. Therein lay one reason why the Niger's later course and end had not been determined: a single and definitive geographical descriptor had been employed of several rivers: 'This river has been made to run from west to east; from east to west; into a lake; out of a lake into the Egyptian Nile; into the Atlantic; and to end in a sandy desart [sic]. It is evident that all this cannot be true of any one river.' Perhaps he had Bowdich's odd map in mind as he wrote.

Turning to Ptolemy, Donkin reviewed the Greek's discussion of the 'Ni-Geir' and its west–east flow. He confessed himself 'much disappointed in Ptolemy as a geographer and guide as far as relates to Central Africa'. The source of Donkin's disappointment lay principally in Ptolemy's measurement of the world. Ptolemy, he claimed, had used the westernmost of the Cape Verde islands as the prime meridian in his world mapping, although other commentators had presumed him to have taken Ferro in the Canaries. Donkin on the other hand based his interpretation of Ptolemy's geography of Africa by using Greenwich as the world's 0° baseline as was proposed by Maskelyne in his *Nautical Almanac*. The use of different initial meridians in mapping and in navigation was a common source of geographical confusion at this time.[6] For Donkin, the choice of different prime meridians meant that anyone working with a modern map of Africa had to adjust Ptolemy's names and their longitudinal location to fit modern metrology and toponomy. Only then was a concordance between ancient and modern geography possible. Unsurprisingly, and following Donkin's realignment, Ptolemy's view that the Niger ran into a lake in central Africa affirmed his own thoughts on the matter (Figure 9.1).

Donkin still faced a key problem. Ptolemy was unclear on the course of the river east from Lake Chad. Denham and Clapperton had been unable to see if a river flowed eastwards out of the lake. Yet a river must run there, Donkin reasoned, even if it had not been found (Figure 9.2). His solution was simple and rather imaginative: the river in question, the Niger, Ptolemy's Ni-Geir and the Nile of Bornou, flowed beneath, not across, the Sahara. He argued this not by turning to those ancient authorities who had claimed similarly – Pausanius and Pliny the Elder according to Dudley – but by reference to the siliceous quality of the Sahara's sands which allowed water to flow beneath and across them, rather than to percolate through it and disappear. Simply, and certainly inventively, Donkin added a further variant to the several ideas then circulating concerning the Niger's termination: 'I need hardly add, that the sea to which I look as the receptacle of the Nile of Bornou is the Mediterranean . . . the gulph of Sidra, the ancient Syrtis, as the point at which the Nile of Bornou enters the sea.'[7]

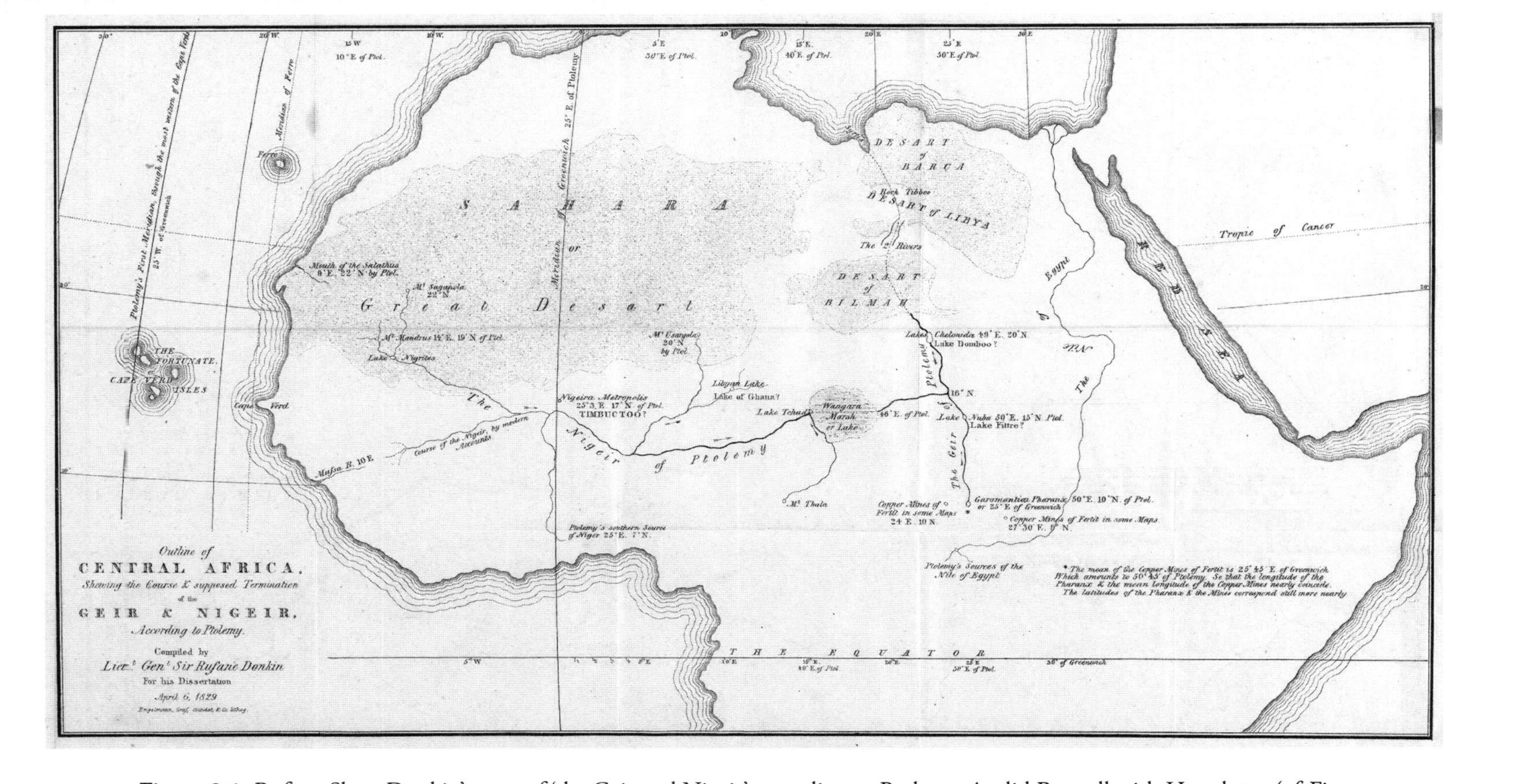

Figure 9.1 Rufane Shaw Donkin's map of 'the Geir and Nigeir' according to Ptolemy. As did Rennell with Herodotus (*cf.* Figure 2.3), Donkin turned to Ptolemy as a key Classical source about the Niger. Source: R. S. Donkin, *Dissertation on the Course and Probable Termination of the Niger* (London, 1829). Reproduced by kind permission of the National Library of Scotland.

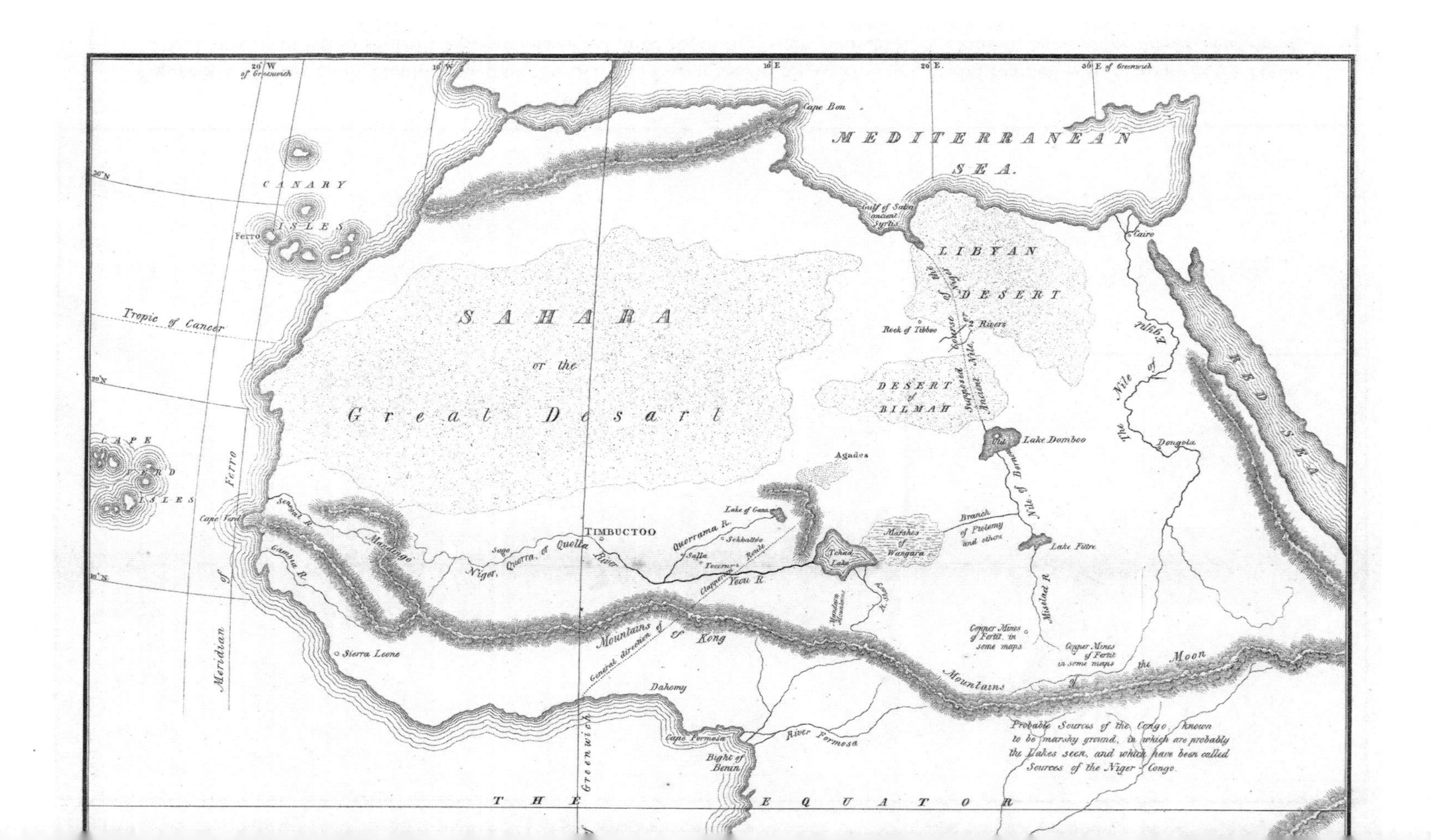

MEDITERRANEAN SEA.
RED SEA
Cape Bon
Gulf of Sidra ancient Syrtis
Cairo
LIBYAN DESERT
Rock of Tibboo
2 Rivers
Suggested Course of the Niger Ancient Nile
DESERT of BILMAH
The Nile of Egypt
Dongola
Lake Domboo
Nile of Bornou
Lake Fittre
Misselad R.
Branch of Ptolemy and others
Marshes of Wangara
Tchad Lake
Agades
Lake of Gana
Schkalts
Querrama R.
Salla
Tocrrers
Clapperton's Route
Yeou R.
TIMBUCTOO
Sego
Niger, Quorra, or Quolla River
Mandingo
Senegal R.
Gambia R.
Cape Verd
Sierra Leone
Mountains of Kong
General direction of
Dahomy
Cape Formosa
Bight of Benin
River Formosa
Mountains of the Moon
Copper Mines of Fertit in some maps
Probable Sources of the Congo, known to be marshy ground, in which are probably the Lakes seen, and which have been called Sources of the Niger Congo.
SAHARA or the Great Desart
CANARY ISLES
Ferro
CAPE VERD ISLES
Tropic of Cancer
Meridian of Ferro
Greenwich
THE EQUATOR
20 W of Greenwich
10 W
10 E
20 E.
30 E of Greenwich
30 N
20 N
10 N

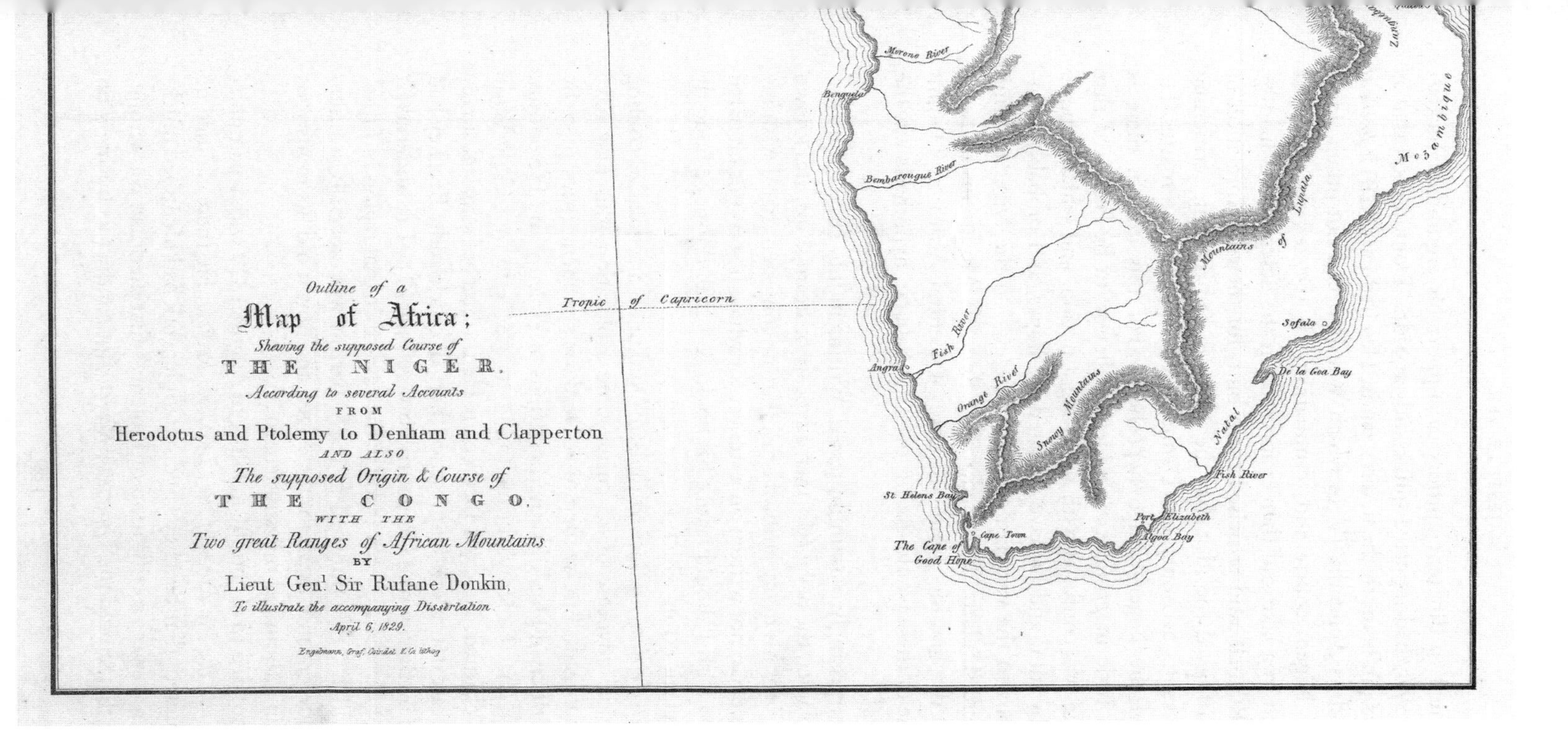

Figure 9.2 Donkin's map of Africa to illustrate his Niger thesis shows two different prime meridians – necessary to his adjustment of Ptolemy's work (*cf.* Figure 9.1) – and depicts his presumed course for the Niger into and out of Lake Chad en route on its subterranean course to the Mediterranean. Source: R. S. Donkin, *Dissertation on the Course and Probable Termination of the Niger* (London, 1829). Reproduced by kind permission of the National Library of Scotland.

Donkin turned to different authors to support his argument. Tuckey's *Maritime Geography* had reported quicksands at Sidra. Donkin took that as proof that rivers ran beneath the desert's sands. He cited Park's 1815 *Journal of a Mission* and its Appendix IV, in which Whishaw laid out others' ideas concerning the Niger's course and termination. He rejected the idea that the Niger ended in a lake and then evaporated in the desert, views held by d'Anville and Rennell, and criticised them both for their interpretation of Ptolemy. He dismissed the idea that the Niger joined the Nile, criticising James Grey Jackson's accounts and Denham's credulity on the matter. He was scornful of the Niger–Congo hypothesis: 'For this notion, there is not, I think, a tittle of *prima facie* evidence.' It was 'to be lamented that poor Park's head was confused with so wrong a notion, and which appears to have stuck by him, and to have actuated his hopes, and probably influenced his actions to the very last'. Reichard's views on the Niger entering the Atlantic were similarly dismissed, citing Rennell's Mountains of Kong as a barrier – 'the granitic mountain base of Central Africa' as Donkin termed it. None of these would do: his own sub-Saharan hypothesis was the correct solution.[8]

Barrow, unsurprisingly, was unimpressed. Writing in July 1829 to an associate, Charles Napier, the naval reformer, advocate of steam ships, and known as 'Mad Charlie' for his many enthusiasms, Barrow confessed that 'I wish my friend Sir Rufane had not written on a subject for which he was not prepared . . . He seems to wish I should notice his book, & I shall do so, & honestly point out to him his errors which are not few.'[9]

In the *Quarterly Review*, Barrow praised Donkin as 'a scholar' – coming from the principal advocate of adventuresome exploration, this may have been a measured insult. But he then took Donkin's argument apart, piece by piece. Barrow criticised Donkin's interpretations of Ptolemy and Herodotus, and pointed out that Ptolemy's prime meridian was Ferro, not Cape Verde as Donkin had claimed. He cited Lieutenant Henry Beechey's coastal survey of Sidra which cast doubt on the presence there of quicksands: Beechey had been supported by Barrow and Goulburn from February 1821 to undertake a coastal survey of Cyrenaica, what is today the eastern coastal region of Libya, the results of which were published in 1828. Barrow was especially scathing of Donkin's 'underground' thesis for the Niger, and how it hid 'its monstrous tail of more than a thousand miles in length'.[10]

Donkin responded to his public mauling in August 1829 with a pamphlet entitled *A Letter to the Publisher of the Quarterly Review*. He there claimed he had been misquoted, that Beechey's hydrographic work had identified quicksands, that the reviewer's rectification of Ptolemy's first meridian was wrong, and, pointedly, that should the *Quarterly Review* not conduct its business 'on the principles of truth, and fair showing, rather than on those of trickery, false colouring, misquotation, and unfair suppression?'[11]

Barrow did not respond, at least not in public. In September, he again wrote to Napier on the matter: 'I perceive that a review of Sir Rufane's book, which he pressed me to write, has called down his ire and thrown him into a transcendent fit of the bile – of which I mean to cure him – or, at all events, prescribe for him. The real grievance is, I damaged his Greek, which is almost as bad as his Geography.'[12]

Critical geographers in the field: William Hutton and Joseph Dupuis

To call William Hutton and Joseph Dupuis 'critical geographers' is both appropriate and not: appropriate because both studied the Niger without seeing the river's full course, deriving their knowledge from others' words; inappropriate in that both knew west Africa at first hand, principally the Ashanti kingdom where Bowdich had worked.

Hutton and Dupuis knew one another, and both were involved in others' Niger experiences. Hutton, like Bowdich a colonial administrator on behalf of the African Company of Merchants, travelled with Dupuis in 1819, assisting Dupuis in his role as British consul at Coomassie (today Kumasi) near the Cape Coast, the capital of the Ashanti kingdom. In his earlier position as vice-consul at Mogador, Dupuis had helped free Robert Adams, whose slave testimony and evidence on the Niger he strongly supported. On his journey to Africa, Hutton met Peddie at Gorée in July 1816. Although Hutton was employed by the African Company of Merchants, Peddie there appointed him his secretary with a £300 annual salary, a further £200 to fit himself out, and the promise of £2,000 to come from Lord Bathurst 'for risking my life for the public good, should our exertions to discover the course and termination of the Niger be successful'.[13] Peddie later cancelled the arrangement and denied he had ever promised such a sum. Hutton's return to work for the African Company of Merchants almost certainly saved his life.

Dupuis's *Journal of a Residence in Ashantee* (1824) is principally a record of the political negotiations between British officials and Ashanti rulers following recent conflict, and after Bowdich had been in the region and Laing had been in Sierra Leone. Dupuis there discusses the 'Geography of West Africa', based chiefly on Muslim verbal testimonies and documents, several of which he lists in an appendix.

Given contemporary prejudices, aired as they had been over Fatouma's and Isaaco's evidence, Dupuis' dependence upon Muslim sources raised questions of credibility: what was told could be questioned if the teller was not the 'right sort' – as Amadi Fatouma and John Whishaw recognised. Dupuis understood it too. While he could not guarantee the veracity of what was told him – no Niger explorer-author could, without adequate corroboration or by seeing the feature in question – Dupuis was at pains to stress the credibility and moral standing of his Islamic informants: 'I depend with as much reliance upon the information I obtained through their means, as I should upon the testimony of any class of honourable men in Europe. And', he asked, 'why not?': these men were indigenous authorities, of high social status in their local communities, and with first-hand knowledge of the region. It was true, he stressed, echoing Park, that 'Every thing short of ocular demonstration, it cannot be denied, is defective, or at least liable to error,' but that was why the word of authoritative others had to be taken seriously.[14]

Dupuis brought together what he was told in a map (Figure 9.3). Difficulties in putting his map together included having to equate statements over time distance and camel travel with linear distance in British statute miles, and locals' ignorance of latitude and longitude such that places could not be accurately located. But he was confident that it improved upon Bowdich's geography which, Dupuis concluded, was 'a fabrication of his own' – a view Barrow shared given his condemnation of Bowdich's 'absurd' map (*cf.* Figures 8.1 and 8.2). The geography of west Africa that Dupuis's map showed was 'carefully compiled from a series of manuscripts, and from information purely native, whose authors were in frequent habits of traversing the African Continent from south to north, (or vice versa).' Dupuis suggests that the Niger and the Nile were one and the same, but he distanced himself from any controversy that might result:

> The reader is naturally at liberty to establish his own reflections upon the material before him, while the author of the treatise . . . will simply claim the privilege of asserting, that it could not concern him intimately or remotely, if it even should be shewn by possibility or demonstration, that the Niger, the Khora [sic], the Ghulby, &c, pour their waters into the western or eastern ocean, into the northern African sea, (the Mediterranean) or into the swamps and lakes which have been supposed to exist between the equator and the tropic of Cancer, and to which the erroneous name of Wangara seems to have been given.[15]

Dudley and Donkin are unusual in their views on the Niger. Dupuis is unique among those looking for the river both in being in Africa as a critical geographer and in professing not to care which and whose hypothesis on the river's termination was correct. In contrast, William Hutton, Dupuis's

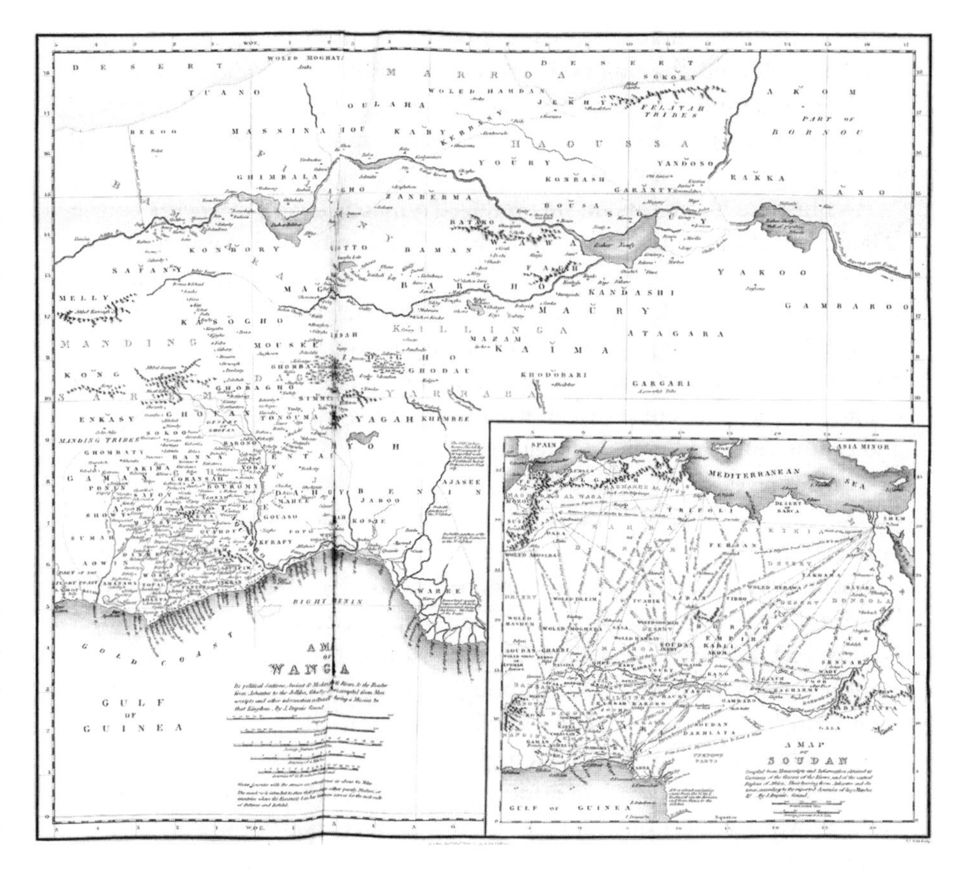

Figure 9.3 Joseph Depuis's map of 1824 shows a developing knowledge of settlements in the west African interior and of the trans-Saharan trade routes (inset), but no clear understanding of the course of the Niger. Source: J. Depuis, *Journal of a Residence in Ashantee* (London, 1829). Reproduced by kind permission of the National Library of Scotland.

one-time travelling companion, was convinced about the Niger's termination and knew that his view mattered.

The Niger, Hutton unequivocally states, 'terminates in the Bights of Benin and Biafra'. Hutton's dependence upon indigenous testimony is apparent throughout his 1821 *Voyage to Africa*. So is the fact that however Hutton went about gathering his information, he did so with a degree of deception. The African Company's instructions for his diplomatic mission cautioned him that while geographical facts were essential, he was to be discreet in gathering them: 'you will carefully treasure up every thing you can learn, without shewing you attach any importance to the obtaining of it'. Hutton generally followed this injunction with the result that while his view on the Niger's termination is clearly stated, how he came to think it is not: he writes only of 'various enquiries and observations during my residence in Africa'.[16]

To illustrate his argument, he produced a map showing the river's course (Figure 9.4). His positioning of the Niger is inaccurate, but he was correct in his hypothesis – as was Reichard nearly twenty years earlier and Clapperton in 1822, albeit with less conviction (*cf.* Figures 5.1 and 8.6). That Hutton got the Niger's geography right but was silent over how is frustrating. The reference to 'Best Authorities' in the title to his map is no guide: this was a common conceit designed to assure readers that the work was both up to date and produced by individuals of repute.

Hutton reveals only a little of the context behind his map and his view on the Niger's geography. The death of Henry Nicholls was particularly to be lamented, Hutton remarked, since he had been best placed to explore upstream from the river's delta but died before doing so. Hutton had outlined his ideas on the Niger in a letter to Henry Mackenzie, president of the Royal Society of Edinburgh, who, Hutton tells us, had read it to the fellows 'in April last', that is, April 1820. Although no record of this letter has been found, nor of the exact occasion at which Hutton's thoughts on the Niger were laid before this learned Edinburgh audience, this is additional evidence to show that the Niger work of critical geographers, like the texts of explorer-authors, was a subject of wider public discourse. As Hutton was at pains to point out, he had arrived at his view and written to Mackenzie about it at much the same time that others were coming to the same conclusion: 'Among all the hypotheses which have been submitted to the public, that which has been lately published by Mr M'Queen, carries with it the greatest probability of being correct.' 'I have read it,' Hutton continued, 'with increased satisfaction, from the circumstances of that gentleman's sentiments being so much in accordance with my own.'[17] Someone else, it appears, had solved the Niger problem – and recently.

Slave testimony and critical map-making: the work of James MacQueen

James MacQueen is the epitome of a nineteenth-century critical geographer. Born in Crawford in Lanarkshire in 1778, he published on topics ranging from Africa's geography to Russia's soils and agrarian economy, improvements to Britain's trans-Atlantic and national postal systems, and Britain's West Indian colonies. He contributed articles on trade statistics and politics

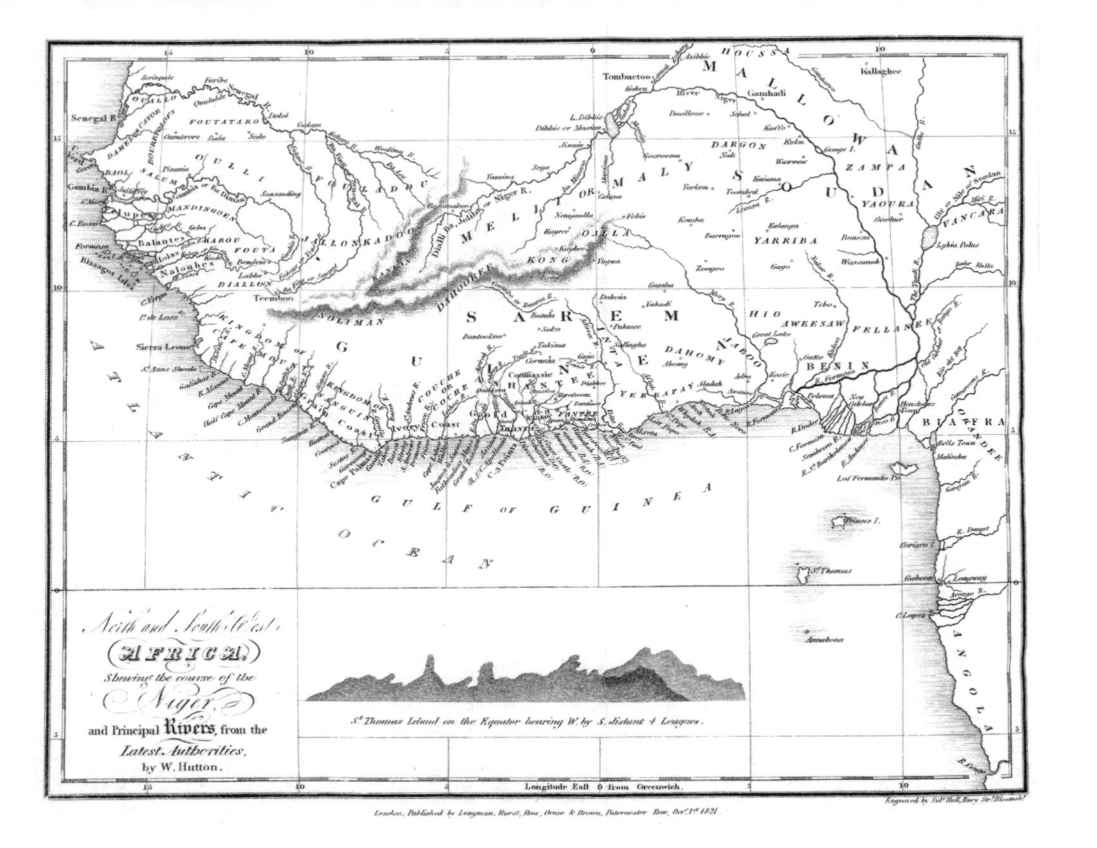

Figure 9.4 William Hutton's 1821 map of the course of the Niger. His depiction of its course is not topographically accurate, but Hutton was correct in his view that the river emptied into the Atlantic. Source: W. Hutton, *A Voyage to Africa* (London, 1821). Reproduced by kind permission of the National Library of Scotland.

to the periodical press, notably to *Blackwood's Edinburgh Magazine*. By 1796, he was employed as an overseer on the Westerhall estate in Grenada, later becoming the estate's chief overseer and manager. He was, from 1821, editor and part-proprietor of the *Glasgow Courier*, which newspaper he used to promote what he called the 'West India interest' – the 'rights' of plantation owners and their attitudes towards the Caribbean and west African economies in a period of heightened anti-slavery sentiment. He never visited Africa.

In 1821, he drew together his thoughts on Africa's geography, and its connections with the Caribbean, in his *A Geographical and Commercial View of Northern Central Africa: containing a Particular Account of the Course and Termination of the great River Niger in the Atlantic Ocean*. MacQueen there makes clear what prompted his interests and how he proceeded on the Niger problem:

> When Mr. Park returned from his first journey I was resident in the island of Grenada (West Indies). There I had Mandingo Negroes under my charge, who were well acquainted with the Joliba. They knew the name perfectly from hearing me pronounce it in reading Mr. Park's book. I also knew a Housa Negro, who said he rowed Mr. Park across the Niger. These things naturally attracted my attention; and being fond of geographical subjects, I endeavoured to collect all the accounts which I could concerning the country on the Upper Niger, as well as from Negroes as from gentlemen of my acquaintance, who had obtained their information from similar sources. Though it was scarcely possible to reduce these, standing by themselves, into regular order, yet, connected with other accounts, they became satisfactory, and formed the commencement of my labours and collections on this subject . . . Numerous authorities regarding the interesting portion of Africa have been examined with much care, and the most striking facts elicited from their pages. This investigation in the geographical department, has led to the conclusion which is now submitted to the world.[18]

MacQueen, it would appear, had been resident in Grenada as Park arrived in Antigua en route to London and, later, had a copy of Park's 1799 *Travels* to hand on his Westerhall estate. Why MacQueen should have chosen 'Mr Park's book' to read to his plantation workers he does not say. Perhaps in reading from it, he showed his listeners Rennell's maps or used Park's list of Mandingo vocabulary to address them. Being spoken to in their own language, even to hear a few words of it, would certainly have caught their attention. It is possible, of course, that among his listeners were those slaves, or their descendants, who had been shipped west on the *Charlestown*, the American slaver which Park boarded in October 1797 and with whose captives Park had spoken and perhaps tended on his voyage to the Caribbean.

MacQueen is silent on the further conversations which, presumably, followed his reading aloud and quite how and when he came to know the Hausa man who claimed to have rowed Park across the Niger. He drew from others' accounts including that of George Robertson, a Liverpool-based west African trader, who argued in his 1819 *Notes on Africa* that the Niger emptied into the Gulf of Guinea, his conclusion also being derived from indigenous testimony, including African merchants, and his understanding of the coast and its river deltas following numerous trading voyages.[19] MacQueen knew Bowdich's 1819 account, a copy of which was given to him by Robert Jameson, professor of natural history at the University of Edinburgh, as MacQueen was finalising his 1821 book.

While it is not clear how and when MacQueen otherwise accessed the many sources to which he refers in his *Geographical and Commercial View of Northern Central Africa*, it is nonetheless a vital work of critical geography which involved the textual scrutiny of ancient and contemporary geographical writing alongside the assessment of 'captive knowledge' – 'exploration-at-distance' was time-consuming, a process of scholarly accumulation, textual exegesis, and patient critique.[20]

The conclusion which MacQueen advanced was that the Niger entered the Atlantic in the Bight of Benin. To illustrate this argument, he included in his book 'A Map of Africa North of the parallel of 7° South Latitude' (Figure 9.5). The map, which is dated 6 June 1820, was engraved by the Edinburgh firm of Lizars. It was included in a report which MacQueen sent in late June 1820 to the Admiralty, the Colonial Office, and to the Board of Trade – and so would have come to the attention of Barrow, Bathurst, and Goulburn among others, and may have had an even earlier expression.[21]

MacQueen's depiction of Africa's geography was in general inaccurate, the position and representation of Africa's physiography and topography especially so, including the Mountains of Kong (here conveniently divided to facilitate the Niger's passage). Much like Reichard in 1803 and Hutton in 1821 (*cf.* Figures 5.1 and 9.4), the Niger is shown swinging southwards through 'a rich soil & level country' towards the Atlantic.

Accuracy was not MacQueen's end purpose. MacQueen's 1821 book and his 1820 map and report were motivated by commercial interests in Africa, of which the Niger's course and termination was but part. In a letter of 9 December 1820 to his publisher, William Blackwood, in which he referred to his 'Skeleton Map of Africa which will cost little & will I conceive do your

Magazine credit,' MacQueen admitted that his 'chief object with regard to Africa is publicity' regarding plans for Britain's (and his own) commercial expansion in west Africa. MacQueen's position as 'West Indian apologist and West African expert' as he has been neatly described, his broadly pro-slavery arguments, and his view that Britain should concentrate its colonial interests around the island of Fernando Po and not in Sierra Leone, meant that his views on west Africa's geography and its commerce were both themselves consistent and at odds with official British policy.[22]

MacQueen's 1820 map is of considerable importance, however, with respect to positioning the Niger and tracing its course. Its purpose, as he stressed, was not topographical accuracy but to illustrate an argument: 'Perfect accuracy on these subjects is at present unattainable, nor is it here pretended to. The delineation of the general features of the country was all I had in view, and this I flatter myself has been done with sufficient accuracy to establish all the leading points which were contemplated.'[23]

To judge from correspondence with Blackwood, MacQueen continued to revise his map, later in 1820 and throughout 1821 and 1822. He did so by drawing from others' narrative accounts including that of Denham, Clapperton, and Oudney. He had a further version of his map ready by early April 1823, informing Blackwood that 'It is so striking when coupled with the facts related a glance must convince the most credulous,' but noting, too, that corrections were necessary. It may be that these corrections, combined with MacQueen's other writing at this time, delayed the map. MacQueen's 1823 article on the Niger was mainly concerned with responding to Barrow's review of his 1821 *Geographical and Commercial View* whose arguments Barrow had rejected, Barrow placing his faith in the expedition of Denham, Clapperton, and Oudney which had just left Britain.[24]

In early May 1826, MacQueen was again in touch with Blackwood advising him how he, and Lizars the engravers, should alter his earlier map:

> I return the map which I have received from Msrs Lizars – it will do nicely when finished.
>
> Tell him to attend particularly to my alterations to look only at the lines I have seen and at the true Courses of the Rivers – some parts of these laid down he will have to carry out some alterations in their stead. Some additions are also made and which when carefully inserted will be sufficient to make the dullest reader comprehend the article in the magazine.
>
> Tell Msrs Lizars to keep my old map by him and also a copy of Denham's Journal by which means he will be able to spell out the names properly should he not be able to make out my hand

> correctly – I want him to be particular on this point and also in their true position.
>
> Also say to him to continue the double dotted line up to Timbuctoo in the line which the Quarterly [the *Quarterly Review*] marks as the course of the Niger. It passes through a mass of names I have laid down but Msrs Lizars will easily trace it.[25]

The results of MacQueen's revisions and these instructions are apparent in a further article and map on African geography and the Niger published in *Blackwood's Edinburgh Magazine* in June 1826. While this was in part a review of Denham, Clapperton, and Oudney's *Narrative of Travels and Discoveries in Northern Central Africa* (1826), it was also a none-too-thinly veiled critical commentary upon Barrow's review of Denham and his colleagues' account and of Barrow's support for the Niger–Nile theory.[26]

MacQueen's 1826 map, 'Africa North of the Equator', is like its 1820 predecessor of interest both for his depiction of the Niger's actual course as he believed it, and for the way he shows, in a faintly stippled line, the 'Course of the Niger, according to the Quarterly Review', that is, according to Barrow (Figure 9.6). In contrast to his 1820 map (Figure 9.5), MacQueen's spurious physiography has gone, excepting his vague delineation of 'The Mountains of the Moon' to the south and east. To judge from a letter to Blackwood in November 1826, MacQueen was planning to make yet further revisions to this 1826 map in order that, through maps, he could bring his theories on the Niger and his commercial plans to the attention of a wider public:

> I return you one copy of the Map with the very Trifling additions which I would make. It makes the fact of the termination of the Niger in the Sea self evident . . . What I would like is that you would get Lizars to engrave the additions & turn me off 12 copies to be put in the copies of the magazine which I mean to send to official and influential people. It is impossible with such a map before them that they could not see the sense in my arguments and information. If however you cannot do this without much trouble I must content myself with the former maps and as soon as I know this I will send you the names of the persons to whom I want the Magazine sent.[27]

It is not clear if MacQueen did despatch copies of *Blackwood's Edinburgh Magazine* with this map to official and influential people as he intimated he would: no list survives of prospective names. But he stressed to Blackwood the importance of the later map in order that, among other people, Members

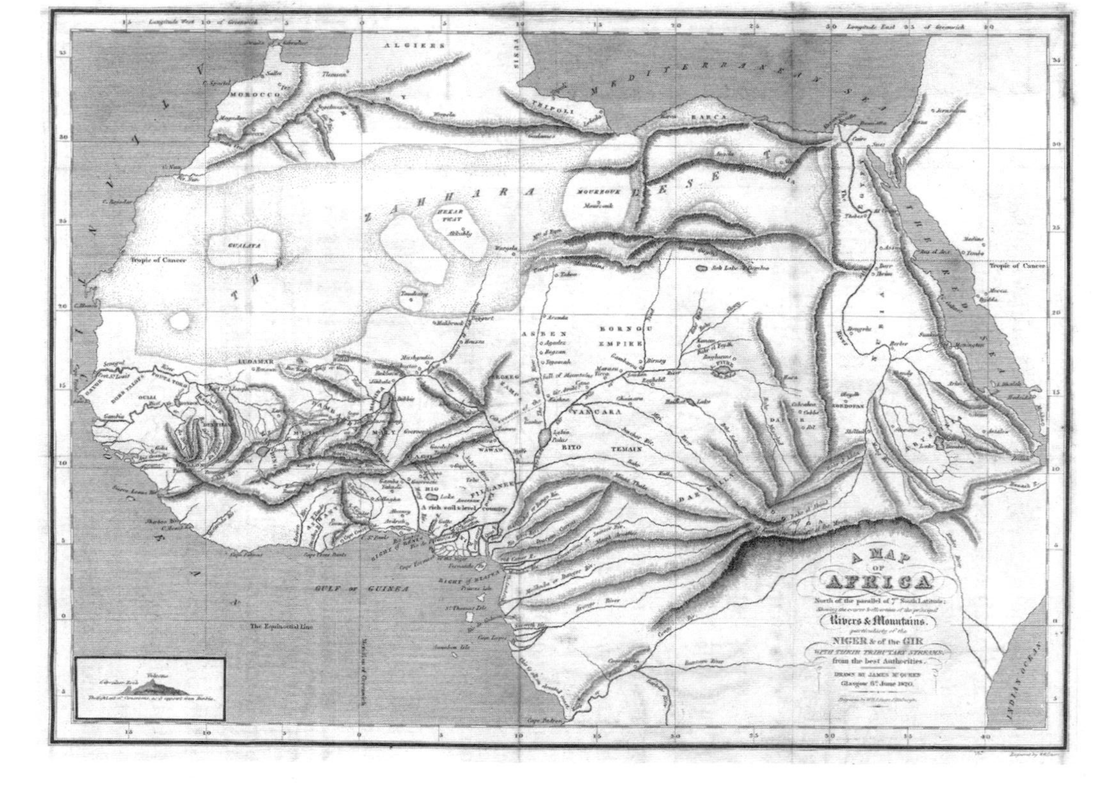

Figure 9.5 James MacQueen's 1820 map of Africa is not topographically accurate – a fact he acknowledged – but shows the course of the Niger as it enters the Atlantic. MacQueen's correct but 'armchair geography' was fiercely criticised by John Barrow. Source: J. MacQueen, *A Geographical and Commercial View of Northern Central Africa* (London, 1821). Reproduced by kind permission of the National Library of Scotland.

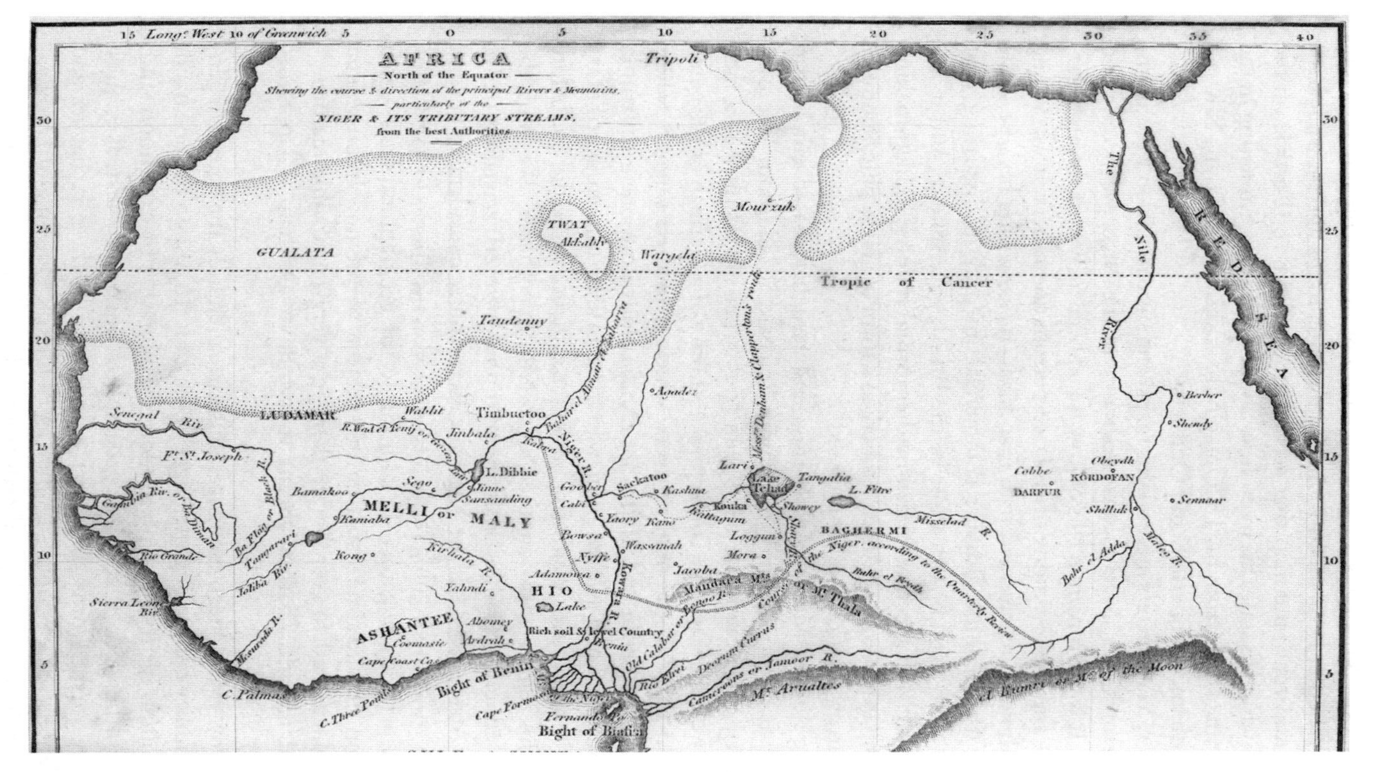

Figure 9.6 James MacQueen's 1826 map of Africa. MacQueen shows the Niger, here termed both the 'Niger' and the 'Kowara', and, in order to discredit the views of John Barrow, he also depicts the 'Course of the Niger according to the Quarterly Review' flowing eastwards to join the Nile. Source: J. MacQueen, *Blackwood's Edinburgh Magazine* 19 (1826). Reproduced by kind permission of the National Library of Scotland.

of Parliament should understand his argument: 'you have no idea how careless members of Parliament are if they have not everything placed before their eyes'. Whether he planned to make more alterations and on what basis is similarly uncertain: the map that accompanies a later 1831 article on the Niger, in which he re-states his hypothesis about the Niger terminating in the Bight of Benin, differs from his 1826 map only in showing the trans-Saharan route of Denham, Clapperton, and Oudney.[28]

To Barrow, MacQueen's method, his maps, and his conclusion were anathema. In Barrow's eyes, MacQueen's geography of the Niger was wrong and multiply so. It was wrong in terms of how it had been arrived at through textual criticism and the testimony of enslaved persons and others. It was wrong in terms of how he represented it in map form. It was wrong because while MacQueen had in mind Britain's commercial interests in discussing west Africa, they were at odds with those of the British government. And it was wrong because MacQueen, who had neither formal training in map-making nor first-hand experience of Africa, embodied a form of geographical enquiry to which Barrow was implacably opposed.

In taking aim at MacQueen, Barrow endorsed field-based empiricism and used it as an opportunity to dismiss Reichard in addition: 'We cannot, however, agree, with Mr McQueen [sic], when he assumes as a fact, (grounded, we believe, on the conjectural authority of a German of the name of Reichard,) that "all the mighty rivers, which send their sluggish waters into the bights of Benin and Biafra, are ramifications from one great trunk, the Niger, supplied and swelled in its western course by numerous tributary streams".' He continued, dismissing all other evidence as he did so, before reiterating his earlier view on the Niger–Nile hypothesis: 'We must be excused for adhering to the consistent testimony of every travelling native of northern Africa, as to its [the Niger's] eastern course considerably beyond Bornou, which, without going farther, renders all conjecture as to its Atlantic termination perfectly nugatory. In addition to the geographical facts which we have from time to time stated on this subject, we are now in possession of others which tend to corroborate them in a very remarkable manner.'[29] By his final remarks, Barrow meant the work of Denham, Clapperton, and Oudney.

The Barrow–MacQueen exchanges illustrate splendidly the nature of geographical enquiry concerning the Niger in the early 1820s: on the one hand, the metropolitan institutional promotion of exploration by expedition, and, on the other, the work of armchair geographers working by textual enquiry and trust in others' words. Yet closer examination of these exchanges and other evidence in this period – Dudley and Donkin, Hutton and Dupuis – complicates this twin typology of geographical knowledge.

These differences were epistemological – that is, they concerned the methods chosen to address the Niger's unknown termination. They were also political in terms of the culture and personnel of exploration. They were to a degree also personal. The fact that Barrow described MacQueen as being a 'humble copyist' and that his map had been 'constructed from materials collected in his closet' was a deliberate slight on two grounds: that MacQueen had copied from Reichard, and that, in being done in the closet rather than in the study or the cabinet (the more usual spaces associated with critical geographers), MacQueen's work evoked meanings of effeminancy and introspection, qualities far removed from Barrow's judgement as to the proper forms of geographical discovery: 'There is nothing so easy as to fill up the vacant spaces of maps with points and lines according to some favourite hypothesis; but to fix with precision the exact spot that the point ought to occupy, to give the flowing line of river or mountain its proper direction, require not only personal presence and actual and minute observation, but for the most part great patience and perseverance, much bodily fatigue and danger, and but too frequently loss of health and life itself. This has been peculiarly the case with regard to African geography.'[30]

The Niger was also a matter of public interest – more so probably than the facts of Park's death alone – in part because published narratives upon the topic were produced by Murray as expensive quarto editions. It is noteworthy that none of the works of Robertson (1819), Dudley (1821), Hutton (1821), and Dupuis (1824) were published by Murray: their utilitarian commercial-political focus, and in the case of Dudley niche interest, would not have been to Murray's taste.

Public interest in the Niger as a matter of critical geography was shaped too by a growing and differentiated periodical culture: between Murray's *Quarterly Review*, pro-Tory in its sympathies and which conveyed the government's perspective while appearing to be independent of it, the Whig-leaning *Edinburgh Review*, and the more strongly anti-establishment and so popular *Blackwood's Edinburgh Magazine*. What one might learn of the Niger could depend on who you read and in which periodical.[31]

The same point may be made of the Niger in geography books for schools and in texts and atlases for the interested public. Most geography school texts in this period presented systematic descriptions of continents and countries, with Africa the least well understood and descriptions of the Niger contradictory if the river was mentioned at all. In this respect, books of geography had not changed from those written and compiled in Park's time. To keep production costs and purchase price low, few schoolbooks were illustrated. In those that were, the maps were not at a scale sufficient to show the Niger in detail. In his *A System of Geography, Ancient and Modern*, for example, published in six volumes between 1810 and 1814, James Playfair, churchman, historian-geographer and, for nearly twenty years, principal of the University of St Andrews, simply noted of the Niger that 'Various opinions have been entertained concerning the rise and termination of this river.' While the 'enterprising and intrepid' Mr Park was credited for the most up-to-date information on the river, and Rennell was cited, Playfair offered no definitive conclusion – for the very good reason that there was none to give. In his *A New General Atlas: Ancient and Modern* (1822), Playfair included a map showing the Niger's geography, one that seems to owe much to Rennell's 1798 map of Africa in showing the Niger running in an almost straight course across Africa (Plate 9).

Park's life and work featured similarly in other educational contexts. In Sarah Atkins Wilson's *Real Stories: Taken from the Narratives of Various Travellers*, a geographical primer published in 1827, Park's experiences are the subject of two chapters: his captivity among the Moors, and his escape from them. The narrative, a combination of geographical instruction, historical information, and moral message, is structured through conversations between a mother and her two children. 'Mamma,' asks Bevan the son, 'I wish you would tell us something about the celebrated traveller, Mungo Park. Did he not go into Africa, in hopes of discovering the source of the Niger?' The son is puzzled that the croaking of frogs should give Park 'such peculiar joy'. The daughter, Caroline, understands at once and chastises her brother that he does not: it is a signal that water is near, 'or at least some damp, marshy place, was at hand', and with that, a chance for Park's survival.

While not accurate in its facts – Park's concern on his first journey was with the direction of the Niger, not its origin – his travels are here being used to impart a sense of achievement in the face of danger and to extol the moral qualities of the adventuresome traveller. From this female author, the lesson was also to highlight both the kindness and compassion shown to Park by African women and the latent intellectual capacities of the female sex.[32]

This is to make the point that as for other books and periodicals at the time there was a geography of reading regarding the Niger, and that Park's travels, especially his first journey, featured as an instructive guide in different geographical-cum-educational contexts. The river and the solution to the Niger problem might appear different at school and in home learning, in schoolbooks and atlases, from those quarto narratives of exploration and journals read silently in the drawing rooms of the well-to-do – or when Park's *Travels* were read aloud on a Grenada plantation, by a mother instructing her children, or when, after reading the account of the Bornu Mission, a churchman talked about the Niger with his MP. In different forms Park, and news about Park, was public currency.[33]

Park was influential, or thought to be, even when dead. In 1822, various individuals including Thomas Anderson in Selkirk and Charles Napier, Barrow's correspondent, presented a memorial to Lord Bathurst on behalf of Archibald Park, Mungo's third and youngest son. Archibald's brother Mungo, Park's eldest, was then in India as an assistant surgeon to the East India Company (but would die of cholera there in January 1823) and Thomas, Park's second son, was on his second voyage on a whaling ship in the south Atlantic. Archibald, then nearly eighteen, was without a patron, and the petitioners hoped 'that the celebrity & misfortunes of his Father may plead in his favor'. In urging Bathurst to appoint Archibald Park to a cadetship in the East India Company, they stressed the dead father's qualities over the living son's: 'the son of a man whose talents, had he lived, would have found ample means of providing for his Family – whose genius, intelligence & adventurous spirit led Government to select him as the fittest person for an Enterprize [sic] which they considered was called for to promote the Interests of the Country & of Science, and whose melancholy fate has created a deep interest throughout the world.'[34] Bathurst did help. In December 1824, Adam Park wrote to Thomas Anderson to say that he had heard from Archibald 'who seems to like Calcutta very much and all his duties'.[35]

Barrow's trenchant criticism of MacQueen is a particularly academic expression of this wide public interest in the Niger and in Mungo Park. Their quarrel was geographical, political, and personal. Barrow had his own and others' interests to protect. He placed his faith in the results of Denham and Clapperton's Bornu Mission because the results, partial though they were, 'confirmed' his own views about the Niger flowing into the Nile. More specifically, the results of that expedition were, essentially, those of Denham alone. At the same time, Barrow may be criticised, and not simply because he was wrong about the actual course of the Niger and did not yet know it.

Barrow redacted Clapperton's notes despite claiming not to have done so and he may have omitted a manuscript map as he prepared Denham's notes for publication. While Barrow and MacQueen agreed on one point – both thought Sultan Bello's geographical knowledge poor and his paper map fanciful (Figure 8.5) – Barrow chose to overlook the fact that, for MacQueen, confirmation of the solution to the Niger problem would come only from observation and encounter in the field: 'Till a European, however, ascertain, all these points from ocular demonstration, and with scientific precision, the public mind will not be satisfied on these important topics. Doubt and uncertainty can, however, remain but a very little time longer. Captain Clapperton, with three companions, have [sic] proceeded on a second journey of discovery, and whatever be the result, they have at last taken the proper road to accomplish their object.'[36]

By the time MacQueen wrote this, Hugh Clapperton had again departed for Africa. His objective was to confirm what on his earlier expedition with Oudney and Denham he had thought to be the case and had sketched in outline. Intrepid as he was, the fact remains that by the early 1820s, the remaining Niger problem had been solved by others.

10

'The greatest geographical discovery'

The Niger problem solved, 1825–1832

Even as he arrived back in England in June 1825, Hugh Clapperton was determined to return to Sokoto – he had promised Mohammed Bello that he would. Quite apart from having left matters incomplete following his earlier expedition with Denham and Oudney, it was important that he should. The Niger's exploration was by now crucial to Britain's commercial and political interests in west Africa and to extending and enforcing its abolitionist agenda. Convinced that the Niger ran into the Atlantic in the Bight of Benin, Clapperton was anxious that the discovery of the river's later course and end should be his, and not Alexander Gordon Laing's.

The Niger problem would finally be solved in 1830, but not by Clapperton or by Laing. By canoeing downstream from Bussa where Park drowned and where Clapperton secured further evidence about the explorer's death, two Cornish brothers, Richard and John Lander, provided proof of the river's course and its termination in the Atlantic. The Landers' was a triumph for a certain sort of geography: empirical confirmation of others' speculative geography, not least the solution-by-critical-reasoning of Reichard, Hutton, and MacQueen. The Landers' achievement was recognised by the Royal Geographical Society, which was established in London in the same year as the Landers' success, and made further public by the Landers' book in 1832. For Europeans, the 1830 solution to the Niger problem marked an end to 2,000 years of geographical speculation.

For Barrow, however, the solution was an embarrassment: the Landers were the sons of a Cornish innkeeper, not Admiralty officers, and his Niger–Nile belief, to which he held forcefully in the face of mounting evidence to the contrary, was shown to be wrong. For MacQueen, the Landers' solution was an opportunity for personal vindication.

Clapperton's second journey – the Niger mapped and Park's death investigated

Clapperton's plans for a second expedition were drawn up within weeks of his return from the first. Anxious about promises to Bello and that Laing might steal his thunder, Clapperton was concerned too about the timing of his return: Park's second expedition was reminder enough of the perils of travel in the rainy season.

The primary objective of Clapperton's second journey was 'to establish a friendly intercourse' between Britain and Mohammed Bello in Sokoto with a view to ending the slave trade in the region. Clapperton was instructed to make Bello aware of 'the very great advantages he will derive by putting a total stop to the sale of slaves to Christian merchants . . . and by preventing other powers of Africa from marching Koffilas [coffles] of slaves through his dominions'. The advantages to him in doing so were to be made clear: once the road was open to the coast, 'he will receive whatever articles of merchandise he may require at a much cheaper rate than he now pays for those which are brought across the long desert'.

The Niger's exploration was important but Bathurst's phrasing deemed it so more in relation to Britain's commercial needs and abolitionist ends than as a long-standing geographical problem:

> If this river, contrary to ancient and modern testimonies, should, according to the information which you received at Sockatoo, be found to bend its course southward and fall into the Bight of Benin, instead of continuing to flow to the eastward, as has hitherto been supposed, and if it should be found navigable through the Sultan's territories, or any part thereof, such a discovery may prove of the utmost importance in facilitating the objects of the present Mission and our future intercourse with that Sovereign.[1]

Barrow involved himself in preparing Bathurst's instructions. He wrote to Hay in the Colonial Office on 25 July 1825 – 'I have done with the enclosed [the instructions] & I think you may put them forward.' Barrow would have recognised that Clapperton's record of achievement on the Bornu Mission meant his ideas could not easily be dismissed, though they did not agree with Barrow's.[2]

Clapperton was to be accompanied by Captain Robert Pearce of the Royal Navy, Thomas Dickson, a doctor, and Robert Morison, a naval surgeon and naturalist. Each had separate directives from the Colonial Office. As a naval man Pearce was probably Barrow's choice: his specific brief was to confirm

Barrow's views about the Niger's course, to renew contact with El Kanemi, and make good on Denham's failure to explore the eastern shores of Lake Chad. Dickson and Morison were to remain in west Africa: Dickson the medical practitioner that Bello had earlier asked for, Morison to accompany Pearce as his medical officer.

The expedition party was completed by four 'domestics' as the ship's log termed them: the much-travelled Columbus from the Bornu Mission; a Hausa man named William Pascoe (sometimes Pasco) who had accompanied Giovanni Battista Belzoni, the Italian explorer-archaeologist of north Africa (and one-time circus strongman) who died of dysentery on the Guinea coast in December 1823; a Patrick Murray of whom nothing is known; and a twenty-one-year-old Cornishman named Richard Lander. Young though he was, Lander had experience of travel in Europe and to the West Indies, where he had contracted yellow fever, and he knew Clapperton. This was the party that boarded HMS *Brazen* on 28 August 1825, and which arrived off the kingdom of Whydah on the Slave Coast in November to join the British anti-slavery patrol. Whydah, like Cape Coast Castle and Badagri, was at the centre of the slave trade: the latter two settlements based around 'slave castles' where the slaves were held before being shipped to the Americas, the former a kingdom under Portuguese and then French rule which transhipped over one million slaves before the trade ended in the 1860s.

Things started to go wrong from the start. The expedition was four months behind the schedule which Clapperton had outlined to Bello with the result that there were no guides to meet them as had been promised. Under the guidance of John Houtson, an English merchant resident at Whydah, the expedition moved along the coast to Badagri, an easier point of departure for the interior. Columbus died soon after: an earlier suggestion that Columbus should precede the others was never acted upon.[3] Dickson accepted the offer of a Portuguese slave trader to conduct him to Katunga, the Yoruba capital, with the plan to rejoin his companions at Sokoto. Although one report suggested that Dickson did reach Yauri, where Amadi Fatouma had taken leave of Park in 1806, nothing further was ever heard of him. Pearce and Morison died of malaria within three weeks of the expedition's December departure. By early January 1826, only Houtson, Clapperton, Lander, and Pascoe remained alive.

The party arrived at Katunga, the residence of the King of Yoruba (or the Alafin as the king was called), on 23 January 1826. They were only about forty miles from the Niger. Further onward travel north towards Kano and Sokoto was not immediately possible. The Yoruba kingdom was at war with the Fulani and the intended route through Nupe was closed in consequence. On 6 March, after six weeks in Katunga in which they were well treated but delayed yet further from meeting Bello, Clapperton and Lander were allowed to proceed to Yauri. Houtson returned alone to the coast and died there in April.

Back in London, Barrow railed at the misfortunes and the rising expense of the expedition now underway – 'It is exceedingly provoking that these African travellers will not adhere to their instructions, and that so little reliance is to be placed on their discretion.' Barrow was especially anxious about the cost of the instruments which Clapperton had purchased – four watches and three chronometers – observing to Hay that he had told Clapperton 'over and over again, that they were to consider themselves mere pioneers, that we did not want the longitude of places to the last second or within a handful of seconds'.[4] He even proposed recalling the expedition and paying its members only half of their annual salary, the remainder to be used to pay off the debts they had incurred 'outside the limits of their instructions'. One month later, Barrow admitted to Hay that he was grieved to learn from Captain Willes of HMS *Brazen* of the deaths of Pearce and Morison but confessed himself heartened 'that Clapperton was going on well, had crossed the Kong Mountains' and was planning to meet up with Dickson.[5]

Clapperton appears not to have realised that the route to Yauri took them to Bussa where Park had died twenty years earlier. Once this was made clear to him, he was keen to proceed: 'I must go out of the way to visit the Sultan of Boussa [sic]', he observed. 'I would not on any account miss the opportunity of going to Boussa, being most anxious to see the spot where Park and Martin [sic] died, and perhaps get their papers.'

Clapperton's initial uncertainty in this respect is surprising given his earlier travels in the region and his previously expressed interests in retrieving what he could of Park's second expedition. It is surprising too because unlike earlier Niger explorers in the field who if they produced a map of their travels did so after the event, Clapperton had a map of west Africa to hand at the start of his second expedition. We know this from a remark in 1826 by James MacQueen. En route to Whydah, HMS *Brazen* had anchored off Freetown in Sierra Leone in late October 1825. MacQueen recounts the circumstances: 'The editor of the Sierra Leone Gazette states that Captain Clapperton there exhibited to him a map of the interior parts of Central Africa, in which the river Niger was laid down as flowing southward from Nyffe till it entered the Atlantic in the Bight of Benin. That such a map was constructed before he left London we are credibly informed.'[6] MacQueen does not disclose when the map was drawn or by whom.

Clapperton is similarly silent on the map in his journal and field notebooks.[7] It is unlikely that it was one or other of MacQueen's Niger maps (*cf.* Figures 9.5 and 9.6). While there is no evidence to prove that Clapperton read MacQueen's articles on the Niger in the month or so he was in England between his west African travels, it is highly likely that he was aware of MacQueen's claims. It is probable that the map referred to is the map Clapperton had prepared following his first expedition (Figure 8.6) and that it did not feature in the narrative of that expedition, not because Barrow deliberately excluded it in preparing the work for publication – though that remains a possibility given Barrow's admission of his emendations of Clapperton's work – but because Clapperton had retained it from the first expedition and had it in his possession as he left again for Africa.

Clapperton is remarkably terse in his field notebook, what he called his 'remarks book', with the result that his published journal offers rather clipped prose throughout. When, on 31 March 1826, he arrived at Bussa and for the first time encountered the Niger he noted simply, 'At 3.30 arrived at a branch of the Quorra.' Bussa was important in providing a base for determining the later course of the Niger, and as the place of Park's death.

Moving north from the coast, Clapperton asked the locals about Park and was told 'various stories at different places on the road'. At the town of Wawa, they were told that Park and Martyn had survived the initial grounding of their canoe but drowned in trying to reach the riverbank. Crowds had watched, recounted Clapperton. What he was told regarding Park's last moments differs from others' accounts:

> the white men did not shoot at them as I had heard; that the natives were too much frightened either to shoot at them or to assist them; that there were found a great many things in the boat, books and riches, which the sultan of Boussa has got; that beef cut in slices and salted was in great plenty in the boat; that the people of Boussa who had eaten of it all died, because it was human flesh, and that they knew we white men eat human flesh.

This last – that Europeans bought slaves only to eat them – was a commonly held view in west Africa.[8]

In Bussa, Clapperton enquired of the local chief about 'the white men who were lost in this river near this place twenty years ago'. His question was met with unease. The chief was only a child when the event occurred, and Clapperton was told not to go to the site where they were lost – 'it was a very bad place'. Clapperton assured the chief that he wanted only to retrieve any remaining notes and papers of Park's: 'if he would give me the books and papers it would be the greatest favour he could possibly confer on me'. He was told that nothing remained with the chief – 'every thing of that kind had gone into the hands of the learned men' – but should any prove still to exist he would procure them and pass them to Clapperton. Clapperton then asked to be allowed to enquire of the old people in the town about the particulars of the affair, reasoning that 'some of them must have seen it'. But the chief 'appeared very uneasy, gave me no answer, and I did not press him further'.[9]

Clapperton sent Shereff Mohammed, a servant hired at Tabra on the journey north, to retrieve Park's books from the chief at Yauri. Upon meeting the chief, Mohammed was told that the chief would only give him them 'whenever I [Clapperton] went to Youri [sic] myself for them, not until then'. He was additionally informed that the books in the chief's possession 'were given him by the Imam of Boussa; that they were lying on the top of the goods in the boat when she was taken; that not a soul was left alive belonging to the boat; that the bodies of two black men were found in the boat chained together; that the white men jumped overboard'. On the evening of 17 June 1826, Clapperton talked with a man who said he was an eyewitness to Park's death. What he was told adds yet further to what we know of the event.

Park and his companions had sailed into regional conflict, their canoe appearing 'just at the time that the Fellatas first rose in arms and were ravaging Goober and Zamfra'. The Sultan of Bussa, on learning that the persons in the boat were white men and that the boat was unlike anything seen on the river before, 'attacked and killed them, not doubting that they were the advance guard of the Fellata army then ravaging Soudan, under the command of Malem Danfodio, the father of the present Bello'. There were only four men in the boat, 'two white men and two blacks'. One of the white men, Clapperton was told, 'was a tall man with long hair', and that 'they fought for three days before they were all killed'.

This is the first account that refers to Park and Martyn fighting off an attack after their canoe had capsized. 'Great treasure' was found in the boat, and 'the people had all died who eat of the meat that was found in her'. Clapperton ended his account with a remark about his informant's credibility: 'This account I believe to be the most correct of all that I have yet got; and was told without my putting any questions, or showing any eagerness for him to go on with his story.'[10]

Clapperton sketched the rapids where Park's expedition had come to grief, picturing himself as an observer in his own scene (Figure 10.1), and he drew a plan view of Bussa and the course of the Niger with the location of the rapids where Park and the others drowned (Figure 10.2). Clapperton's drawings are an important visual record: depicted cartographically since the early eighteenth century, the Niger was for the first time appearing pictorially. His sketch is the first depiction by a European of the rapids at Bussa, and his plan of the town is a record of a settlement lost, like the rapids themselves, to the development of the Lake Kainji dam scheme in 1968.

Keen though he was to solicit further information, and to head downstream to confirm his views on the river's course, he did neither. His instructions were to renew contact with Bello and so Clapperton, Lander, Pascoe, and their servants proceeded north.

On arrival in Kano in July 1826 the party found themselves in a war zone. Conflict between the kingdoms of Sokoto and Bornu had been precipitated by incursions by El Kanemi, Denham's old comrade in arms. By now, Clapperton was very ill: his notebook entries for late July read simply 'ill all day', 'a little better', or just 'ill'. He had not properly recovered from his first expedition before returning to Africa. But, ignoring his poor health, he was determined to reach Sokoto despite the fighting and the difficulties of travel and, given Pearce's death, to undertake the exploration of the eastern shores of Lake Chad. On Clapperton's instructions, Lander remained in Kano as Clapperton set out alone for Sokoto in August 1826. He arrived there on 20 October, having been met on the way by the *gidado*, Bello's viceroy. On the way, he had his pocket journal, notebook, and his writing materials stolen. This occasioned a hiatus in his writing for some weeks. In December 1826, to his great surprise, Clapperton was joined in Sokoto by Lander and Pascoe who had been instructed by Bello to make the journey north.

Clapperton's reunion with Bello was not the pleasant meeting of old acquaintances that it should have been. Bello was unhappy at Clapperton's meeting with the King of Bornu, with which kingdom Bello was at war. Furthermore, Bello was suspicious about a letter for El Kanemi from the British government which Clapperton carried in a sealed metal container. He even suspected Clapperton of gun-running for his opponents. The explorers' travel was, as a result, restricted for over two months. Plans for British–west Africa commerce and the cessation of the slave trade hung in the balance before Bello relented and allowed them to return southwards, a decision which suited Clapperton who wanted to follow the Niger south from Bussa.

But by mid-March 1827, after nearly seven months of travel, Clapperton was dying. He tried various treatments including, like Park, dosing with calomel. On one occasion, he blistered his side with heated metal cups to relieve the pain of a severely swollen spleen (it did not work). His last written entry was on 11 March 1827, and by early April, Clapperton knew that the end was near. Lander was instructed to 'Take care of my journals and papers after my death' and to return to England by heading northwards, across the Sahara to the Fezzan and from there to Tripoli and the assistance of Hanmer Warrington. Warrington had been informed by the Colonial Office that he should continue to facilitate trade with Bornou and to be alert to any news of Clapperton.[11] On 13 April 1827, Clapperton died in Lander's arms, the two men by then close friends rather than master and servant.

Figure 10.1 Hugh Clapperton's sketch of 'the Banks of the Quorra or Niger' at Bussa, drawn on 2 April 1826. This is the earliest known representation of the site of Mungo Park's death. Source: The National Archives, ADM 55/11 (f. 92).

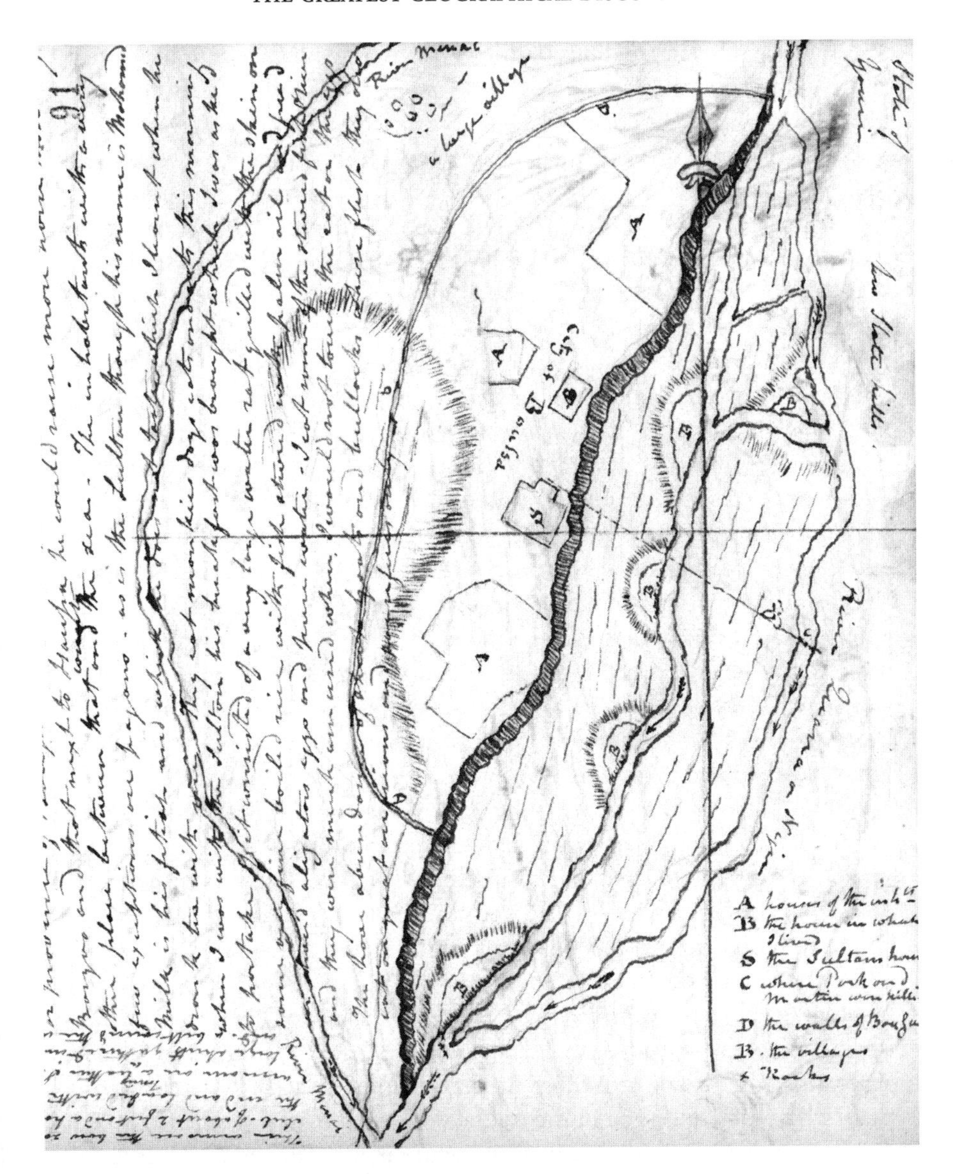

Figure 10.2 Hugh Clapperton's sketch plan of the town of Bussa on the Niger drawn on 1 April 1826. The small 'c' in the river, to the right centre of the sketch, marks 'where Park and Martin were killed'. Source: The National Archives, ADM 55/11 (f. 92).

Lander had his instructions but now, entrusted with Clapperton's journals and making his own observations as he proceeded, he was effectively alone, ill, and without the authority vested in Clapperton. He was, however, accompanied by Pascoe. As Lander makes clear in his own journal, he owed his safe return to a combination of Pascoe's assistance, the high regard with which the late Clapperton was held, and to the decision to head south for the coast rather than north across the Sahara. Not following orders probably saved his life.

As the two travelled south, Lander in turn picked up further information regarding Park. On the evening of 3 September 1827, a local imam came to his tent: 'You are not, Christian, the first white man I have ever seen. I knew three of your countrymen very well. They arrived at Youri at the fast of the Rhamadan [April]'. This corroborates the claims of Amadi Fatouma who thought that Park had perished four months after his letters from Sansanding in November 1805. Landers' informant had even met Park and Martyn: 'I went with two of them three times to the sultan [of Youri]' and witnessed gifts being given, including a 'handsome gun, a cutlass, a large piece of scarlet cloth, a great quantity of beads, several knives, and a looking-glass'.

The 'person that appeared to be the head of the party' – we can take this to be Park – was described as 'a very tall and powerful man, with long arms and large hands, on which he wore leather gloves reaching above the elbows. Wore a white straw hat, long coat, full white trousers, and red leather boots. Had black hair and eyes, with a bushy beard and mustachios of the same colour.' The Sultan of Youri advised the explorers to travel the remainder of the way on land 'as the passage by water was rendered dangerous by numerous sunken rocks in the Niger, and a cruel race of people inhabiting the towns on its banks'. This advice was ignored, the explorers stating 'that they were bound to proceed down the Niger to the Salt Water'.

Upon learning of the explorers' death, the Sultan of Youri was reportedly much affected but, continued the imam, 'it was out of his power to punish the people who had driven them into the water'. Almost immediately afterwards, the sultan and most of the inhabitants at Bussa, 'particularly those who were concerned in the transaction', died from a pestilence, which was interpreted as a 'judgement of the white man's God' and so everything remaining from the canoe was burnt. Lander was told that the saying 'Do not hurt a Christian, for if you do, you will die like the people of Boussa' was now common in the African interior. With that, his informant left, Lander thanking him for the information 'thus voluntarily given'.[12]

Lander and Pascoe arrived back in Badagri in late November 1827, where Lander faced the hostility of Portuguese slave traders before taking ship for England from Whydah early in 1828, Clapperton's papers and his own journal in hand.

Most of what we know of Clapperton's second expedition comes from the published version of his and Lander's notes. Clapperton's 1829 *Journal of a Second Expedition* combines Lander's journal, as the book's full title indicates, and has an account of Clapperton's life by his uncle Samuel, a captain of Marines at Chatham. Barrow wrote the introduction and prepared the book for the press. Clapperton's journal, Barrow noted, 'is written throughout in the most loose and careless manner . . . Much, therefore, has been left out, in sending it to the press, but nothing whatever is omitted, that could be considered of the least importance.' Somewhat uncharitably, Barrow also observed that Clapperton 'has not contributed much to general science, but, by his frequent observations for the latitude and longitude of places, he has made a most valuable addition to the geography of Northern Africa'. The journal did include, Barrow admitted, 'two objects of peculiar interest, which are still open for further investigation', namely, the course and termination of the Niger, and the recovery of papers belonging to Park. Barrow was satisfied that 'The exact spot on which he [Park] perished, and the manner of his death, are now ascertained with precision'.[13]

For Barrow, 'it is pretty clear, that the particulars of the death of Mungo Park', and the spot where the fatal event happened, 'are not very different from what was originally reported by Amadoo Fatima [sic] and has since been repeated in various parts of the continent'. Thanks to Abraham Salamé, who translated other Arab language accounts of Park's death, Barrow included further such documents in his introduction. Clearly, as Clapperton and Lander had been in Africa, Barrow had spent time in London gathering further evidence over Park's demise, and from Arab sources.

One, from the Sharif of Bokhary, offers further evidence that Park and the others had not met their end at once but had gone down fighting: 'As their ship or vessel was proceeding down the river, it came to a narrow place or creek, into which they pushed it, and remained there three days'. The people of Bussa – 'Boossy' as the Sharif has it – observing them there, 'assembled, and went and fought them for three days.' As the fighting escalated, the Christians 'began to take up their goods and throw them in the river' until, realising the situation was desperate, they threw themselves in, leaving their two slaves in the canoe. What is important in this version was not any denial over Park's and Martyn's drowning but the absolution of Bussa locals from their murder: 'that the hands of the people of Boossy did not reach so far as to kill them'.[14] Barrow may have been confident about the place of Park's death, but as noted different versions were current about the particulars.

Barrow continued to verge on the dismissive regarding Clapperton's role in solving the Niger question. 'With regard to its termination,' he wrote, 'the reports continue to be contradictory, and the question is still open to

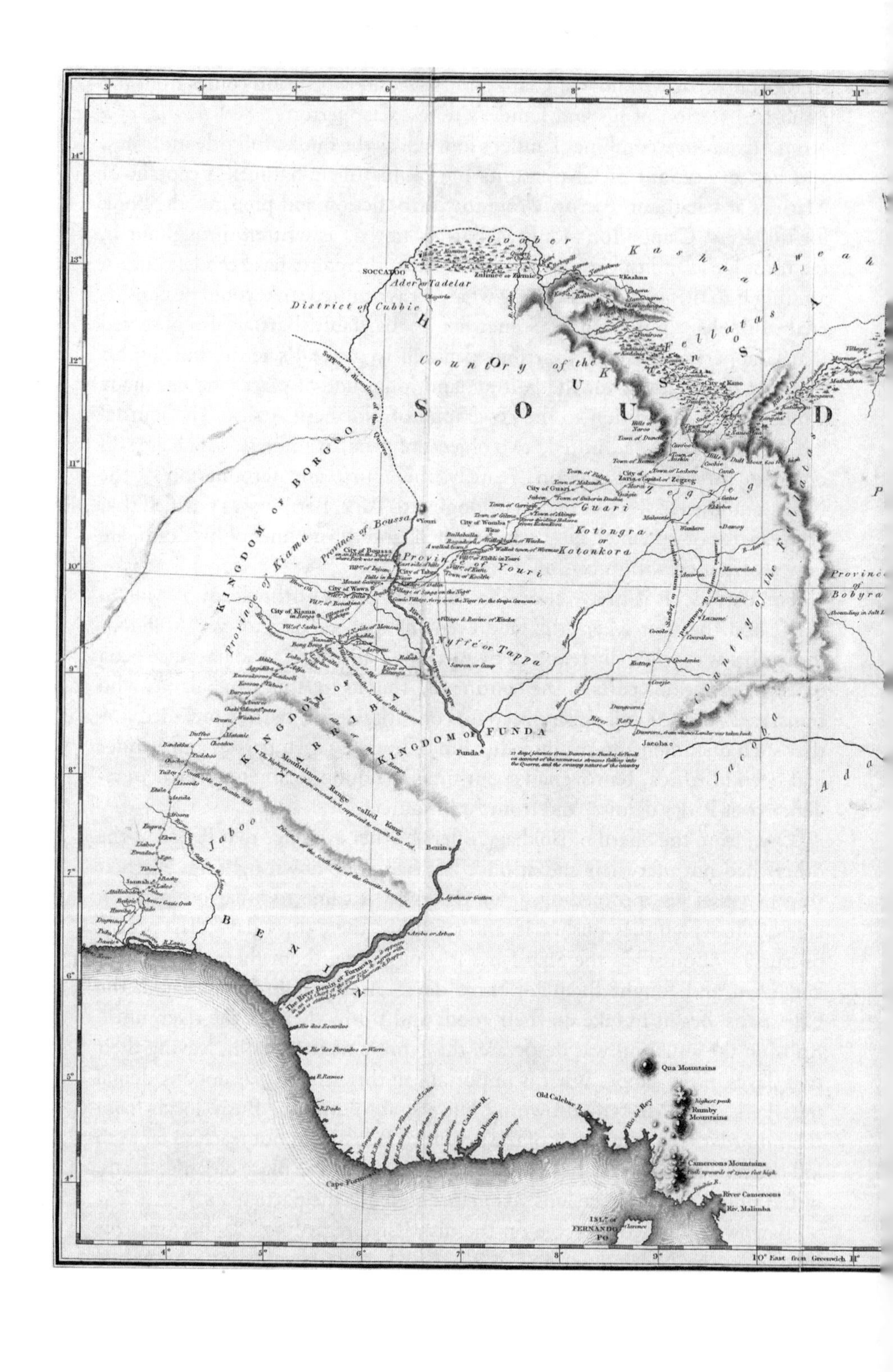

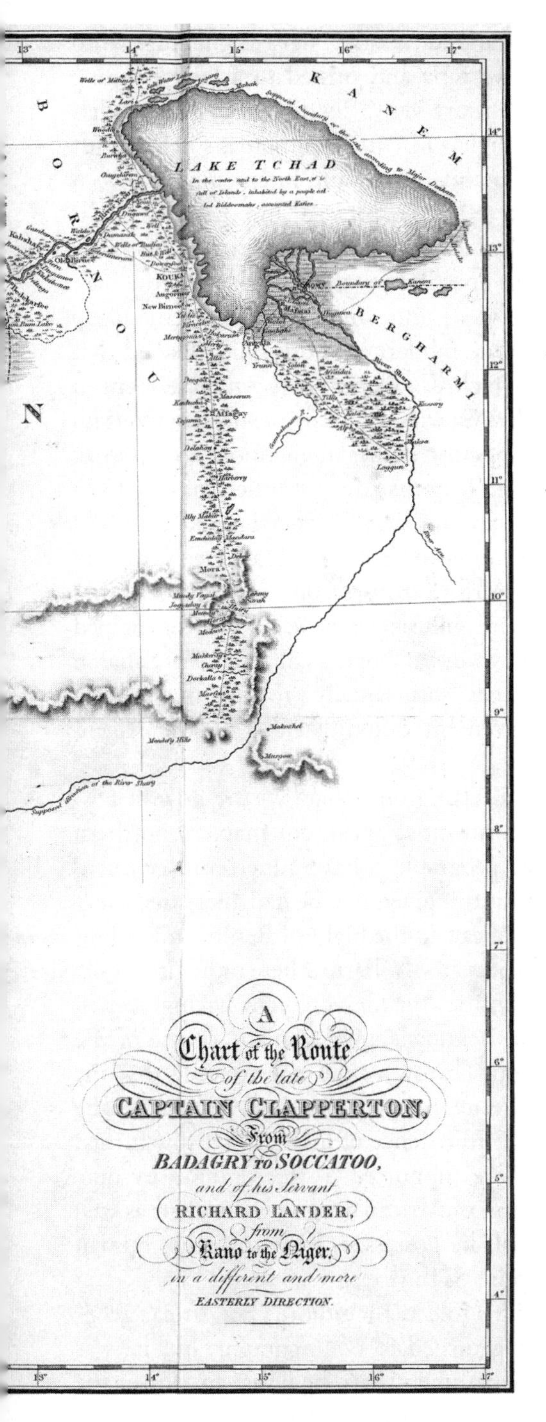

Figure 10.3 Hugh Clapperton's 1829 map of his route with Richard Lander shows the 'River Quorra or supposed Niger' flowing south towards Funda and then a gap before the course of the river Benin or Formosa is shown: an almost complete course for the Niger – but not quite. Source: H. Clapperton, *Journal of a Second Expedition into the Interior of Africa from the Bight of Benin to Socatoo, to which is added The Journal of Richard Lander from Kano to the Coast, partly by a more eastern route* (London, 1829). Reproduced by kind permission of the National Library of Scotland.

conjecture.' Even as he grudgingly admitted that the Niger might empty into the Atlantic, Barrow took a defensive tone and turned to a long-standing explanation in support of his still-contrary view: 'Its direction, as far as has now been ascertained, points to the Bight of Benin; but there is still a considerable distance, and a deep range of granite mountains intervening, between the point to which with any certainty it has been traced, and the sea-coast.'[15]

Neither in his introduction nor in his lengthy discussion of Clapperton's 1829 *Journal* in the *Quarterly Review* does Barrow directly mention the map which illustrates Clapperton and Lander's route and which shows the likely course of the Niger (Figure 10.3). One modern authority has claimed that this map was based in part on one which Clapperton drew while resident in Katunga and partly on material in the *Narrative* of his first expedition, but no evidence is provided for this and Clapperton makes no mention of map work during his residence in Katunga. It is likely that the published map of 1829 was based on Clapperton's earlier map and that Clapperton added to and amended it before and during his second expedition. As MacQueen's remark makes clear, Clapperton arrived in Africa on his second expedition with a map in hand. The 1829 map certainly illustrates how Clapperton helped close the gap between what was known of the Niger's course in sub-Saharan regions and the river's delta on the coast. It is equally probable that Barrow chose not to discuss it in detail given his differing views on the Niger's later course.[16]

In the *Quarterly Review*, Barrow declared only that 'We are now in possession of authentic materials to reform those gratuitous maps of northern Africa which are a reproach to the geography of the nineteenth century.' Clapperton was afforded due credit for this given that he had 'measured every degree of latitude from the Mediterranean to the bight of Benin, and of longitude from the lake Tsad [Chad] to Soccatoo' – Barrow here rather ironically publicly praising Clapperton's propensities for measurement having decried them in private to Hay. But despite acknowledging that the course of the Niger seemed to be in that direction, Barrow would not concede that the Niger entered the Bight of Benin: 'We are inclined, therefore, to consider the Quorra to empty itself into the Tsad.' In defence of his view, Barrow turned to an established Niger figure: 'we are supported in this opinion by one, who has done more for the elucidation of African geography, ancient as well as modern, from the slender materials he possessed, than any other human being . . . Major Rennell . . . now in his 87th year'.[17]

Rennell had long since conceded his role as an authority on Africa's geography to Barrow, his interests having turned to oceanography, but he was still in touch with Niger news. In September 1826, he wrote to Hay at the Colonial Office to say that 'the <u>easterly</u> course of the Niger' – the underlining for emphasis appearing in his letter – had been known in 1764 to Mr George Barnes, the then British governor of Senegal, but because the question of the Niger was at that time of little interest, 'he did not think it worth making known'. Rennell had only recently been told this in a conversation with Barnes's son.[18]

News about the Niger circulated between government officials and, elderly though he now was, Rennell was still seen as an authority. In a letter of March 1827 to William Martin Leake concerning news from Hanmer Warrington in Tripoli, Robert Hay told Leake that he would pass the news on to Rennell: 'I shall not neglect to transmit to him for perusal any communications which we may receive from our Travellers in Africa, as I know that by so doing, I shall at once fulfil the wishes of Lord Bathurst and gratify my own by affording amusement to our distinguished Geographer.'[19]

Closing his review of Clapperton's *Journal*, Barrow remarked that 'It is by means of single travellers that we shall eventually be able to settle the geography of northern Africa'.[20] He would be proved nearly right on the number involved even as the two Cornishmen concerned would prove him wrong on the Niger. James Rennell would never know that he too was wrong on the Niger. He died on 29 March 1830, one week before Richard and John Lander stepped ashore at Badagri.

The triumph of the Landers

Like Clapperton before him, Richard Lander was keen for a second expedition to west Africa to complete unfinished tasks, and, like Clapperton, he did so as an explorer-author. As plans were made for departure, Lander was keen to see his own journal into print, to correct what had appeared under his name in Clapperton's *Journal*, his comments there having been 'drawn up in haste', and to advance his own standing. He approached John Murray with his proposal, and Murray, inevitably, turned to Barrow.

Barrow thought Richard Lander's contribution to Clapperton's *Journal* to be that 'of a very intelligent young man'. His attitude to Lander's own manuscript was quite the opposite however. 'It appears to me,' Barrow confided to Murray, 'that Mr Lander's "Wanderings in Africa" which I have attentively read, would scarcely justify the experiment of substantive publication.' Lander was, in Barrow's view, guilty of 'deficiencies of style' and 'sins of egotism'. The scientific material offered was 'deplorably meagre'. Barrow suggested that to be intelligible the proposed book would need the map from Clapperton's 1829 narrative (Figure 10.3), but even that would not save it: 'his "Book" I am afraid, would have but little chance of ultimate consideration or immediate popularity'. Steered thus, Murray declined Lander's text. It was published, in 1830, as a two-volume work by the London firm of Colburn and Bentley, and received generally positive reviews for its engaging prose and Lander's loyalty to the dying Clapperton.[21] By the time the reviews appeared, Richard Lander, explorer-author was in west Africa with his brother John, and was reunited there with Pascoe.

The Landers were encouraged to recover such of Park's papers as were believed to be with the Sultan of Yauri but the main object of their mission was clear: 'to ascertain the course and termination of the Niger'. They departed Portsmouth on 9 January 1830 and from Badagri, where Richard had narrowly escaped death two years earlier, on 31 March 1830. As was the case with Clapperton and others before him, white men were a curiosity, white men whose purpose in travelling lay in determining the course of a river doubly so. Travel was hampered by the constant attention paid to them and by demands for tribute which drained their funds and stock of gifts, but Richard Lander's familiarity with the route and several local rulers made the journey inland easier.

Making their way to Bussa they were there promised a canoe – and given a stark warning. The king could give the travellers no protection 'from insult and danger beyond his own territories' and it would be necessary to solicit the goodwill of other rulers on the banks of the Niger. Recognising that 'we shall be beset with dangers from the shore', the explorers were nevertheless expectant: 'we are in high spirits, and humbly hope that, by attending to the necessary precautions, we shall be able to overcome them'.

On 18 June 1830, the Landers viewed the Niger near where Clapperton had sketched it almost two years earlier. The site prompted a moment of sombre reflection: 'The rock on which we sat overlooks the spot where Mr. Park and his associates met their unhappy fate; we could not help meditating on that circumstance, and on the number of valuable lives which have been sacrificed in attempting to explore this river, and secretly implored the Almighty that we might be the humble means of setting at rest for ever the great question of its course and termination.'

In conversation the next day, the Landers were told that the king had in his possession a tobe (a length of cotton or other cloth, worn as an outer garment, from the Arabic *tawb*), 'which belonged to a white man who came from the north many years ago'. The Landers asked to see it and it was gifted to them. Contrary to their expectation, the tobe turned out 'to be made of rich crimson damask, and very heavy from the immense quantity of gold embroidery with which it was covered'. Because the king had purchased it at much the same time as Park's death, and because they had never heard of any white man to have come as far south as Bussa, the Landers took it to be 'part of the spoil obtained from the canoe of that ill-fated traveller'. They were at a loss to say whether Park wore the tobe himself, which they took to be 'scarcely probable on account of its weight', or if it was intended as a gift.

No other account mentions this tobe. There is nothing in Park's writings about it and it is not mentioned by others as being in the canoe at his death. It can reasonably be discounted as belonging to Park – whether it was brought or bought by him or by Hornemann, another white man who came from the north many years ago, we shall never know. The Landers later made good use of it to secure their safe passage.

Like Clapperton before them, the Landers tried to obtain any surviving books and papers belonging to Park but were disappointed when they were brought only Park's book of logarithms, into which other papers had been folded: observations on the height of the water in the Gambia river, a tailor's bill to Alexander Anderson, and a dinner invitation to Park, dated 9 November 1804, from a Mr and Mrs Watson of the Strand (Plate 10). Because the book had talismanic value to the locals, the Landers did not purchase it. After 'much trouble and anxiety, and some little personal sacrifices likewise', the Landers ended their attempts to secure Park's papers: 'all our hopes of obtaining Mr. Park's journal or papers, in this city [Bussa], are entirely defeated'.[22]

The Landers departed Bussa on 24 June 1830, headed first upstream to Yauri. Rocks and sandbanks prevented easy travel, and their canoes were several times unpacked or even lifted from the river and transported on the banks. At Yauri, repeated requests to the sultan for sight of Park's papers were denied before the travellers realised that he had never had them and had claimed to only so he could barter for Western goods: 'It is a satisfaction at least for us to know that the long-sought for papers are at present no where in existence.'

The brothers were held in Yauri for weeks. They did not depart downstream from Bussa until 20 September and then managed only a few miles before discovering their canoes to be overloaded and leaky. Repairs were made but progress was hindered by the curiosity and hospitality of locals, suspicion as to the travellers' purpose, and conflicting advice over the best course to take. Richard Lander headed inland to Wawa where he was handed a hymn book belonging to Alexander Anderson and a letter to Park from Lady Dalkeith.

Whenever they landed as they went downstream, they were delayed, often for days. Upon leaving the town of Lever in early October, they made good progress, the Niger 'a noble river of magnificent appearance' with banks forty feet or more in height (Figure 10.4). The river varied in its course and size as they travelled: 'Sometimes it seemed to be two or three miles across, and at others double that width.' The current carried them forwards, at times with great force, but sandbanks and islands also presented obstacles. In places, the Niger's waters had covered the fields and the houses on the river's edge – even the extensive rice plantations: 'the tops of these plants only were visible, and no cultivated land anywhere appeared'. Elsewhere as they progressed downstream, villages stood clear of the waters, their fields fertile and the crops abundant. Progress was important, but stopping was a necessity. Every landing required that they meet the local chief or village head man, spend a day or two paying tribute and, often, haggle over hiring a new crew as their canoeists returned home after rowing a stretch of the river.

On 6 October, their progress was hindered further by having to pay their respects to the 'King of the Dark Water', as one chief was known. Richard Lander put on an old naval coat which 'I had with me for state occasions'. John was dressed 'in as grotesque and gaudy a manner as our resources would permit'. The party stood beneath the Union flag which, 'unfurled and waving in the wind . . . made our hearts glow with pride and enthusiasm'.[23] Despite the Landers' canoes being outnumbered by the twenty of the King of the Dark Water, this rather odd ceremony might be taken as the first symbolic appropriation of the Niger by the British.

In the town of Rabba, Richard Lander was displeased to be reacquainted with an Arab servant he and Clapperton had employed on their first expedition. Sent back for tent poles which he had forgotten, this servant had decamped with Pearce's sword and a sack of cowries. Lander's mood quickly improved upon being promised a new, larger canoe and a local to act as guide and interpreter as far as the sea. It was as swiftly deflated however as the chief in the town of Zagozhi let it be known that the goods he had received were inadequate and that the party could not proceed. In their published account, the Landers note how 'At this moment we thought of Mr. Park's tobe, which was given to us by the king of Boosà' [sic] and that this might 'prove an acceptable present for the covetous prince'. They 'deeply lamented the necessity to which we were reduced on parting with this curiosity, but it was inevitable'.

In his manuscript journal, John Lander provides a fuller picture: 'The tobe of Mr Park is extremely showy, and heavy with the amazing quantity of figured gold lace, and spangles of gold, by which it is not only embroidered but liberally covered.' Reckoning that it was impenetrable to dart or spear, 'its value to a Falatah Chief who is almost perpetually engaged in war or predatory excursions can scarcely be estimated'. Reluctantly, the tobe was handed over. Progress was now possible but as they came to leave on 14 October, they confessed themselves worried: 'our goods are very nearly exhausted; and so far from being in a condition to make further presents, our means will scarcely be adequate to procure the bare necessaries of life'.[24] It is a sentiment, and a circumstance, uncomfortably reminiscent of Park's.

Much as Park had been jostled and stared at in Benowm, the Landers were the object of fascination and disbelief in nearly every village and settlement they landed at, their white skin and clothing provoking laughter and genuine fear. In one village, on market day, Richard Lander recounted the scene: 'If I happened to stand still even for a moment, the people pressed by thousands to get close to me; and if I attempted to go on, they tumbled one over another to get out of my way, overturned standings and calabashes, threw down their owners, and scattered their property about in all directions.' Children hid behind their mothers. Women in particular seemed fascinated by the first white man they had ever seen: 'My appearance seemed to interest them amazingly, for they tittered and wished me well, and turned about to titter again.'[25] In the town of Egga, the Landers were forced to blockade themselves in a hut, so great was the crowd of disbelieving onlookers: 'The people stand gazing at us with visible emotions of amazement and terror.'

Figure 10.4 Banks of the Quorra. The sketch conveys something of the size of the Niger in its lower reaches before it reaches its delta. Source: R. and J. Lander, *Journal of an Expedition to Explore the Course and Termination of the Niger; with a Narrative of a Voyage down that River to its Termination* (London, 1832). Reproduced by kind permission of the National Library of Scotland.

A few miles south of Egga, they spotted a seagull – 'a most gratifying sight to us. It reminded us forcibly of the object we had in view.' But they and the seagull were still 330 miles from the sea. Mostly, the Niger took them south-east or south, but on Monday 25 October 1830, the river changed direction to south-south-west and ran between immensely high hills (Figure 10.4). They were at this point the first Europeans to encounter the junction of the river Benue (known locally as the Tsadda) with the Niger.

As they headed further downstream, the Landers became increasingly aware of travelling through a kind of 'contact zone'. The signs of European trade were more and more evident: in clothing with the appearance of Manchester cottons in the town of Damuggoo, in language with the smatterings of English spoken by locals and, less pleasingly, by the signs of Spanish and Portuguese slavers.

On 5 November, near the market town of Kirree, they observed several canoes flying the Union flag with many crewed by Africans dressed in European clothing. Richard Lander confessed himself 'quite overjoyed by the sight of these people, more particularly so when I saw our flag and European apparel among them, and congratulated myself that they were from the sea-coast'.

Appearances proved deceptive. In an instant, Lander recounted, 'all my fond anticipations vanished'. The explorers' canoes were boarded and unloaded 'with astonishing rapidity', their content placed in their assailants' boats. Richard Lander was stripped of his jacket and shoes. He and Pascoe fought off their attackers as best they could, clouting one with 'one of our iron-wood paddles that sent him reeling backwards, and we saw him no more'.

Similarly assaulted in the other canoe, and like his brother fooled by the Union flag and the trappings of European civilisation, John Lander narrowly escaped drowning. The Landers pursued their attackers into the market at Kirree. There, after some discussion, the locals decided to return some of the stolen belongings: in acting as they had the robbers, who were not locals, had dishonoured Kirree and not for the first time; Moslem merchants did not want their trade interrupted by pagan incursions; townsfolk dare not be seen to side against their king who had been friendly to the explorers. But while the Landers retrieved some of their belongings – a box of books, John Lander's journal, a carpet bag with a few clothes – their guns, one of which had been Park's, were lost, as was their compass. Distressingly, so was Richard Lander's journal, except for a notebook detailing their travels since Rabba. The Landers were held captive. On 8 October, they were taken downstream to the town of Eboe where the Obi, chief of the Ibo, pronounced upon their fate. The Landers were to be ransomed to the value of twenty slaves. Their African servants, Pascoe and his wives included, were to be sold into slavery.

Salvation, and with it the solution to the Niger problem, came from an unlikely source. In Eboe, a merchant by the name of King Boy from the town of Brass arranged with the Obi that the Landers should be ransomed off for money and a further fifteen slaves to the captain of an English merchant ship, the *Thomas*, then anchored to the south. King Boy would receive an additional payment and a cask of rum for helping make the exchange. Directed by King Boy, Richard Lander set out from Brass – one of the most 'wretched, filthy, and contemptible places in this world of ours' as he later put it – leaving his brother as hostage.

As he headed south, his passage on the Niger on 8 November was hampered by a fog so dense 'that no object, however large, could be distinguished at a greater distance than a few yards'. The canoes had to be lashed together to keep Lander prisoner and in sight. As the fog cleared, Lander was presented not just with the Niger but with other smaller tributaries and water in all directions: 'We now found ourselves on an immense body of water, like a lake, having gone a little out of the road, and at the mouth of a very considerable river, flowing to the westward, it being an important branch of the Niger; another branch also ran from hence to the south-east, while our course was in a south-westerly direction on the main body; the whole forming, in fact, *three* rivers of considerable magnitude.'[26] As the Niger seemed close to being traced to its final end, so it seemed less distinct.

Just after seven o'clock on the morning of 18 November 1830, Richard Lander arrived at the river Nun, or the First Brass river as the main branch of the Niger was known locally. Lander admitted that he knew nothing of it or of its peoples – though they had 'some acquaintance with our countrymen, and some slight knowledge of our language'. His senses told him that the sea was near: 'We were evidently near the sea, because the water was perfectly salt, and we scented also the cool and bracing sea-breeze, with feelings of satisfaction and rapture.' A few minutes later, he sighted two merchant vessels lying at anchor before them. Lander was struck dumb: 'The emotions of delight which the sight of them occasioned are quite beyond my powers of description.'[27] The remaining Niger problem was at last solved in the field: from Bussa, where Park drowned and where Clapperton had sketched and the Landers had mourned his passing, the Niger flowed south to the sea (Figure 10.5).

The Landers' situation however remained difficult. One of the two ships moored was Spanish. Boarding this first, Richard was well treated but left hurriedly on realising that the crew was fever stricken. Expecting better on the English *Thomas*, he was astonished and humiliated when its captain refused

Figure 10.5 The Niger's end mapped. In the title of this map illustrating their achievements, the Landers associate the course of the river – the Quorra, the Joliba or Niger as it is variously named – with Mungo Park. Source: R. and J. Lander, *Journal of an Expedition to Explore the Course and Termination of the Niger; with a Narrative of a Voyage down that River to its Termination* (London, 1832). Reproduced by kind permission of the National Library of Scotland.

to accept the deal brokered by King Boy or to assist the explorers despite being shown Lander's instructions from Bathurst. He was persuaded only after repeated entreaties and then not from any sense of duty or morality: the *Thomas*'s captain needed the Landers, Pascoe, and their African servants to help crew the ship since four of his crew had just died. Those remaining lay ill in their hammocks. A week later, John Lander was reunited with Richard. King Boy later received his goods in full from the British government.

The Landers sailed first for the British colony on Fernando Po before, on 20 January 1831, they departed for Rio de Janeiro. From there, they left for England on 20 March 1831 on HMS *William Harris*, arriving in Portsmouth on 9 June. On 10 June 1831, Richard Lander arrived in London to report his discovery to Lord Goderich, Bathurst's successor as Colonial Secretary.

Bovill describes 18 November 1830 as 'one of the great days in the stirring annals of geographical discovery'. The Niger problem, 'around which bitter controversy had raged ever since Mungo Park's discovery of the Niger in 1796, had . . . at last been solved'.[28] His claim requires qualification. The Landers' solution and 'discovery', crucial as it was to European geographical commentators and political institutions, may be interpreted, to use Park's words, as proof by 'ocular demonstration'. It was empirical confirmation of others' speculative reasoning regarding the second part of the Niger problem. There is in that sense a parallel with Park's earlier confirmation of the direction of the Niger's flow in solving the first part of the Niger problem: 21 July 1796 was an equally great day in the annals of geographical exploration.

In both cases, 'discovery' for the Europeans involved was a revelatory moment whose significance was established by seeing the feature in question for themselves and, later, in being believed by others in what they wrote about it. For the moment and the means of revelation to be regarded as credible, and for the discoverer to be thought trustworthy, the moment of discovery and the processes of discovering – what, together, we have come to call exploration – needed to be put into print. Only then could others make up their own minds, about claim and claimant both. For the Landers, that meant their notes would be passed to the man who had helped direct their Niger travels but who had reasons to disbelieve their stated outcome and his own reputation to manage.

In the event, John Barrow was not given the task of preparing the Landers' account for publication. There is no doubt that Barrow would have read the Landers' notes given his Admiralty position, but John Murray gave the job of putting them to order and of writing an introduction to the Admiralty hydrographer, Lieutenant John Becher. It is possible that Barrow delegated the job to Becher, but there is no evidence for this, and Murray had final say over who prepared his narratives for publication. Murray was probably wary of Barrow's judgement given that he had been so condemnatory of Lander's 1830 'Wanderings in Africa' and given Barrow's treatment of Clapperton's narratives. Another interpretation is also possible: that Barrow refused to undertake the work because to have done so would have meant formal association with a solution he had long opposed – and more than opposed, dismissed, in the case of Reichard, MacQueen, Hutton, and Clapperton.

Becher's job was complicated by the fact that Richard Lander had lost his journal, 'with the exception of a note-book with remarks', and John his memorandum, note and sketchbooks, with a small part of his journal, even after elements of his narrative were returned to him in Kirree. Becher's role, that of blending partial field notes into a consistent uniform text – the partiality of the text a possible further reason why Barrow did not undertake the job – proved too difficult for him. He was as one modern authority puts it, an 'infuriating editor' whose search for clarity was 'the cause of [textual] confusion'.[29]

The scientifically minded Becher seems to have shared Barrow's pejorative assessment of the observational capacities of the Lander brothers, especially on the map which illustrated their accomplishment (Figure 10.5):

> In conclusion, a word or two may be said respecting the map which has been constructed from the journals. The accomplished surveyor will look in vain along the list of the articles, with which the travellers were supplied, for the instruments of his calling; and the man of science to inform his opinion of it need only be told, that a common compass was all they [the Landers] possessed to benefit geography, beyond the observation of their senses. Even this trifling though important assistance was lost at Kirree, below which place the sun became their only guide. Too much faith must not therefore be reposed in the various serpentine courses of the river on the map, as it is neither warranted by the resources nor the ability of the travellers. The map, in its most favourable point of view, can be considered only as a sketch of the river, authenticated by personal observation, which will serve to assist future travellers, from whose superior attainments something nearer approaching to geographical precision may be expected. Even under these circumstances, the present travellers will always derive ample satisfaction in reflecting that they have served as pioneers of African discovery.[30]

Despite thanking Becher for his 'friendly aid and valuable assistance' in the 'Address to the Public' which prefaced their book, the Landers had reason to feel aggrieved at his view of their work. John Lander wrote indignantly to John Murray over Becher's treatment of him: 'I cannot help regretting exceedingly that because accident has thrown me into a humble sphere of life, my veracity is questioned & my promises treated with indifference & contempt. I am very much afraid Mr Beecher [sic] had not behaved to me with the candour & Sincerity of a Gentleman.'[31] Whatever difficulties Becher had with the book, and with John Lander's view of Becher's editing, John Murray astutely recognised the public's interest in the Niger's solution. Instead of his usual expensive quarto works, he published the Landers' account in duodecimo format in the firm's Family Library Series at 15s for the three-volume set.

While the Landers felt slighted and Barrow was embarrassed, James MacQueen was vindicated. In June 1831, as news of the Landers' achievements became public, he wrote to William Blackwood on the matter: 'Every effort will be made to rob me of the prize of the discovery by an invidious and unnecessary Ross and by Lander and his friends' (the reference is to Sir James Clark Ross, a close associate of Barrow in the Royal Geographical Society). MacQueen felt his own work had been overlooked, and not just by Barrow and the Landers:

> My name merely descended the River the course of which I pointed out to them & I also know that it was from my book and map that Clapperton formed his opinion about the outlet of the River being in the Delta of Benin when he undertook his last journey in which Lander as a servant accompanied him. I know this from two persons who saw a map he made from mine and according to them before he left England. Lander has made few discoveries except the mere fact of having seen with their eyes the descent of the Niger in its lower course through the delta of Benin into the Sea.[32]

Reviewing the Landers' book in the *Quarterly Review* in 1832, Barrow took aim at those whose 'hypothetical or speculative' geography had been proven correct: Reichard whose happy conjecture 'was grounded on false data, he had not a single fact to guide him', and MacQueen, a 'humble copyist, with an equal poverty of facts'.[33] The true discoverer was Richard Lander and he had been rewarded by the Royal Geographical Society with its first Founder's Medal and the sum of fifty guineas (Figure 10.6). As the result of the Landers' travels, observed Barrow, 'Two questions may be put – is this Quorra in reality the continuation of Park's Joliba? And is the Joliba or the Quorra the Niger?' He proceeded to answer his own queries: 'To the first we reply, without hesitation, YES; but, to the second, if by Niger is meant the river so named in the works of ancient geographers and historians, we say, decidedly NO.'

Figure 10.6 Portrait of Richard Lander. Source: R. and J. Lander, *Journal of an Expedition to Explore the Course and Termination of the Niger; with a Narrative of a Voyage down that River to its Termination* (London, 1832). Reproduced by kind permission of the National Library of Scotland.

This statement of geographical casuistry is not only a veiled reference to his own belief, now proven false, that the Niger flowed into the Nile. In arguing that the Niger in its variant names such as Quorra and Joliba could not have been known by those names to ancient authorities, Barrow's comment is also Donkin-like in seeming to argue for *a* Niger, different Nigers, rather than *the* Niger. Barrow even referred to his earlier review of Donkin's *Dissertation* in making this point. Because this single river was known by different names and from his insinuation that the Niger proper was only what flowed eastwards before it ran south, he proposed that the name of the river he had worked so tirelessly to get others to explore should not now appear on any maps at all: 'we trust that the word Niger, so vaguely employed by the ancients, will be expunged from the map of Africa'.

This statement is simultaneously ironic, given his earlier dismissal of Donkin's work, and pathetic, Barrow having been exposed as wrong-headed but without the courtesy to admit it. Barrow makes no mention of his Niger work in his autobiography, in which he emphasises the support he gave to Arctic exploration – and he was not one to shy away from self-promotion. A leading figure in facilitating the Niger's exploration, John

Barrow was humbled, his authoritative status reduced, by the solution to the Niger problem.[34]

MacQueen had no time for Barrow and his unflattering remarks, nor for those who overlooked his own earlier work: 'the writer in the Quarterly Review knows very well that the map of Northern Africa, constructed by me, and the researches made to shew [sic] the course and the termination of the Niger in the Atlantic, was not made, nor undertaken, for the purpose of seeking applause, or medals, or rewards from Government, or from any other quarter, but made to establish clearly an important geographical fact.'[35]

Direct in print, MacQueen was even more direct in correspondence to William Blackwood. Barrow, he wrote, 'must feel some of his obstinancy & error – but he ought to do more – he ought to feel sorrow, regret and shame for sacrificing so many valuable lives, so much money and so much to this country in searching for the course and termination of the Niger'.[36]

11

'The river is not correctly laid down'

The Niger navigated and revealed, 1832–1857

With the Niger problem solved, west Africa was presumed to be open for business as never before: 'With the Landers' triumph, the exploration of the Niger ended its heroic period and became profitable.'[1] Commercial intentions were to the fore, combined with renewed attempts to abolish the slave trade and to solicit the support of indigenous rulers in that objective.

Yet while the Niger problem was solved in terms of the river's lower course and termination, it was not wholly solved geographically. MacQueen and Barrow continued to bicker over the right way to conduct exploration and who was right over the Niger. MacQueen chided Barrow for the 'partiality and intolerable arrogance' of his discredited views over the river's course. Barrow's tone and manner, argued MacQueen, was unlikely to encourage others to become involved in the newly established Royal Geographical Society: 'it is not by conduct like this that the society will encourage geographical research or collect useful geographical information; nor is it by giving publicity to articles so erroneous, yet written in such a contemptuous, domineering style, that the Quarterly Review is to maintain or to spread its name and fame for a superiority over its brother periodicals.'[2] The Society's vice president, William Martin Leake, considered the Niger problem still one of Classical philology and contemporary toponomy – whether Pliny, Ptolemy, Herodotus, and others knew the river's variant names of Quorra, Joliba, and so on – but he regarded the Landers' achievement as 'the greatest geographical discovery that has been made since that of New Holland [Australia]'.[3]

The Niger's course and termination may have been determined, but the river's characteristics, and those of its tributaries, had not. How deep was the Niger? Large-scale commerce required vessels larger than canoes: but the Landers had several times encountered sandbanks. Park and his companions had drowned in rapids. Was the river navigable only in the rainy season when the volume of flow would reduce the chances of grounding but when mortality was likely to be greater? How deep into the interior did the navigable river go? Where was safe to establish a secure anchorage and store goods for shipment?

Becher's comments on their map had offended the Landers, but he was right: 'the various serpentine courses of the river on the map' were not an accurate guide to the Niger's actual geography. Future travellers, especially those with profit in mind, needed something approaching to geographical precision. Further in-the-field work was both inevitable and indispensable: as Leake phrased it, 'nothing short of the ocular enquiries of educated men is sufficient to procure the requisite facts'. In the quarter of a century that followed the Landers' journey, the Niger afforded access to traders, missionaries, and abolitionists. For them and for explorer-geographers it became again the focus of ocular enquiry – and at considerable cost.

Conquered?: the 1832–1834 Niger expedition

The first mission to the Niger after the Landers was a private and commercial affair under the direction of MacGregor Laird, a Greenock-born Liverpool trader and the son of William Laird, founder of the Birkenhead shipbuilding firm. The Admiralty and the Colonial Office remained interested, and in Lieutenant William Allen the Admiralty had its own observer on the expedition. But they lacked funds, Barrow was chastened by circumstances, and in 1832 the government was concerned more with managing cholera outbreaks in Britain's towns and cities and in extending suffrage through the Great Reform Act than it was in supporting another Niger expedition so soon after the Landers'.

MacGregor Laird's plan was to use steamships. In 1815, Banks's faith in steamships for the prospective Niger–Congo expedition had foundered on the vessel's draft and the engine's inadequacy. Laird's paddle steamers were more powerful and had a shallower draft. Even so, the government could not commit to their use for the Niger's exploration alone: as Robert Hay explained to Barrow in August 1831, 'If any object connected with the suppression of the Slave Trade can be obtained by so doing, there may be no difficulty in fitting out a vessel for that quarter, but I suspect that for the purpose of exploration alone, the Government will not consent to authorize the expenses.'[4]

MacGregor Laird's instruments were not just technical. In 1831, he established the African Inland Commercial Company, its name a clear indication of his concerns and ambitions. That he did so before the Landers' book was published indicates his awareness of the opportunities signalled to by the Landers and reflects heightened public interest given news of the Landers' work and exchanges in the periodical press between Barrow and MacQueen. His third and principal instrument was Richard Lander.

Laird's initial plan was for two ships. Lander disagreed, seeing the plan as not ambitious enough. He proposed three: two steamships and a supply vessel. Lander liked steamships, describing the steam engine as 'the grandest invention of the human mind'.[5] Laird appointed Lander captain of the *Alburkah*, a steam-powered paddle wheeler – it became the first iron ship to make an ocean voyage. Laird travelled on the larger wooden paddle steamer *Quorra* under Captain Harries, who had experience of west Africa from serving on the Africa Squadron blockading slave ships. Together with the brig *Columbine*, the storeship, the MacGregor Laird expedition left Milford Haven in south Wales on 25 July 1832. Liverpool, then suffering a cholera outbreak, could not be used as their port of departure. After putting in at Sierra Leone to supplement the crew by taking on the aptly named Kroomen (sometimes Kru), members of a local people much employed by European nations for their seamanship and their proficiency at 'wooding' (chopping and storing wood), it arrived at the Niger in October 1832. There they several times met King Boy. On one occasion he appeared in full Highland dress, a gift from MacGregor Laird's father. King Boy, naturally enough, was staggered to see Lander again.

However, the dying soon began. On the *Alburkah* first Harries died, then Curling the engineer, then the boatswain, and, as the two steamships headed north leaving the *Columbine* at the river's mouth, crewmen on the *Quorra*, the chief mate, the sailmaker, the chief engineer. MacGregor Laird and William Allen were soon incapacitated and watched as the seasoned Lander, seemingly unaffected by the fever ravaging the crews, administered to them and to others. Lander was attentive but untrained as a physician. His treatment was little more than bleeding and blistering and kind words. Laird's remedies in the face of fever were even less effective: his schemes to 'keep up the excitement of the crew in every possible way' included tripling the rum and brandy allowance, keeping them constantly busy lest fever catch them at rest, and playing music on board.

By the end of November 1832, the *Quorra* had lost thirteen men, the *Alburkah* ten. The health of the surviving crews made carrying out their duties difficult. Between 13 November and 5 December, Laird was so ill that he could not write his journal: he later completed it from memory. Trade with the locals – when it happened at all – was similarly hindered. In an age that believed in miasmatic theory, that the air itself was the cause of ill health, it was as if the environment itself warped both body and mind: 'In those accursed swamps, one is oppressed, not only bodily but mentally, with an indescribable feeling of heaviness, languor, nausea, and disgust, which

Figure 11.1 The steamship *Quorra* aground on a Niger sandbank. While the Niger's course and width was becoming understood by the early 1830s, its change of depth according to the seasons was not. This made exploration difficult despite having the latest technology. Source: M. Laird and R. A. K. Oldfield, *Narrative of an Expedition into the Interior of Africa, by the River Niger, in the steam-vessels Quorra and Alburkah in 1832, 1833 and 1834* (London, 1837). Reproduced by kind permission of the National Library of Scotland.

requires a considerable effort to shake off.' By December, they reached the confluence of the Benue and the Niger which Lander had last seen with his brother three years earlier and where Laird built his warehouse. They had spent over two months on the river, lost many European crew to fever, and amassed 'not half a ton of ivory'.[6]

Matters then got worse when the *Quorra* grounded on a sandbank. Despite numerous attempts to refloat it, including the removal of its armaments and cargo to the riverbank, it remained stuck fast for five months (Figure 11.1). Laird read – 'Shakspeare [sic] and Scott were my favourite authors'. His companion, Dr Briggs, learned Arabic, and the two played chess. Lying under awnings on deck to protect them from the sun and to capture what breeze there was, everybody sweltered.

Laird's reading included *Blackwood's Edinburgh Magazine*. Perusing James MacQueen's arguments there over the Niger in his letter to Lord Goderich,

Laird could scarcely believe that MacQueen had not seen the river for himself: 'His opinion has been verified by experience; and discovery has so fully proved the accuracy of his descriptions, that it is difficult to imagine that he must have been anywhere than on the spot when he penned that memorable letter, which will ever remain an evidence of his sound judgement and geographical skill.' In a footnote in his later *Narrative*, Laird observed, 'It may truly be said that the honour of the theory belonged to Reichard; but his opinion was founded on the extent of alluvial deposits alone; it was strengthened and confirmed by the additional evidence collected with so much zeal and ability by Mr. McQueen [sic].'[7]

Leaving Laird and Briggs behind, Lander headed upriver on the *Alburkah* as far as Rabba where, in October 1830, he and his brother had handed over Park's tobe. Laird was soon without Briggs' companionship. His words on Briggs' passing echo Mungo Park's upon the death of Alexander Anderson and Richard Lander's upon Hugh Clapperton: 'I cannot describe the feeling of anguish and desolation that came over me.' Laird set off to navigate the Benue in search of commercial opportunities, despite being so ill he could barely walk. Laird's experiences in the village of Fundah mirror those of Park in Benowm: the locals never having seen a white man, he was the object of curiosity and contempt. Trade was impossible, and he was threatened with being poisoned. Laird managed to send a messenger back to the *Quorra* who returned with some Congreve rockets. The launch and explosion of these had the desired effect: Laird returned to the *Quorra* in early 1833 to be reunited with Lander.

By then the expedition was finished as a commercial enterprise, and in July 1833, it split into two. Resigned to his failure, Laird headed south on the *Quorra*. He recuperated for two months on Fernando Po, leaving the *Quorra* in Clarence Bay, and returned to Britain on the *Columbine*, arriving in Liverpool on New Year's Day 1834.

MacGregor Laird was nearly broken by his encounter with the Niger. His health never recovered, and he incurred considerable financial loss and the bankruptcy of his African Inland Commercial Company. Yet he remained convinced about the region's commercial potential and the importance of its geographical exploration and in 1849 founded the African Steamship Company, the first regular steamship service between Africa and Britain. Laird later wrote to Roderick Impey Murchison, president of the Royal Geographical Society, to express his 'earnest desire that this steam communication may be made available for geographical purposes' and offering free passage to Africa for anyone travelling there under the auspices of the Society.[8]

MacGregor Laird helped establish Britain's imperial interests on the lower Niger. This interpretation is especially valid if his pioneering venture is read in relation to later events, and no distinction is made between the commercial and the geographical results of the expedition. MacGregor Laird took a particular view regarding his expedition's success in his *Narrative of an Expedition*, written jointly with Oldfield: 'As far as proving the navigability of the Niger, and the ease and facility with which that mighty stream may be used for the purposes of commerce; it was successful in no ordinary degree, considering the novelty of the undertaking, the complicated nature of a steam-vessel, and the excessive mortality of the crews.'[9]

It is more accurate to distinguish the effects of the MacGregor Laird Niger expedition between its commercial and its geographical consequences. Commercially, it was defeated by a mix of colonial hubris, over-reliance upon an untested technology, poor medical knowledge, and, in the many problems arising from uncertain navigability, by the river itself.

It was not the only Niger expedition to come to grief at that time. In London in December 1831, Leake and Lander had offered their guidance to a Mr Coulthurst and a Mr Tyrwhitt who were planning, at their own expense, to sail to the mouth of the Niger and explore eastwards from there. Overcome by sunstroke on the voyage, Tyrwhitt turned for home immediately upon arrival in the Gambia. Coulthurst changed his plans and determined 'to penetrate, in the steps of Mungo Park, from the Gambia to the Djoliba, and thence to Timbuktoo and Funda'. He never did. Refused passage to travel, he returned to the coast and died there of fever on 15 April 1832. His death was duly reported to the Royal Geographical Society as a further Niger failure.[10]

In April 1840, unaware that a larger expedition was being planned, Captain John Becroft (sometimes Beecroft) commanded the steamship *Ethiope*, owned by Robert Jamieson of Liverpool, up the Niger. Among his crew was Mina, who years before had been interpreter to Clapperton and Lander. Becroft's purpose was clear: 'if an intercourse with the interior of Central Africa of any extent or practical utility is ever to be established by the navigation of the Quorra, or Niger, by Europeans, some new body of approach to the main body of that river must be found by which the pestiferous swamps of its Delta will be avoided.' Most ships entered the Niger delta via the river Nun, down which the Landers had come in following the Niger to the sea, but high surf and sandbars at its mouth made access difficult. Becroft thought the Benin river, also known as the Formosa – a central feature in Reichard's arguments over the termination of the Niger – might provide easier ingress. After eleven days sailing, their progress was hindered by carpets of aquatic plants which choked the river, and he returned to the coast.

Using other points of entry, the *Ethiope* got as far as Eboe by 20 May, but no further for two months: the Niger's volume of flow was insufficient and the ship frequently grounded. As the river's flow increased with the rainfall upriver, the *Ethiope*'s steam engine proved no match for the current.[11] Encountering the Niger was one thing. Navigating it was quite another and required knowledge of the river's levels and the volume and rate of flow according to the seasons.

Laird's expedition was more successful geographically. It led to a further map of the Niger, prepared by William Allen (Figure 11.2), its style different from the Landers' (*cf.* Figure 10.5). Given Allen's naval background, this was probably much more what Becher had in mind. As the vessels moved up and down the Niger, or were stuck fast like the *Ethiope* before them, the officers calculated longitude and latitude. On board the *Alburkah*, Captain Hill found he was also correcting others' observations. Hill reported that 'the river is not correctly laid down in Mr Landers book chart nor is he so far in the interior as he reported to me'. While Hill admitted that Lander's map may have been incorrect because Lander had lost his journal and instruments on his earlier journey, his report is evidence of continuing attempts to improve geographical understanding of the Niger.[12]

William Allen was one of only nine survivors from a total British complement of forty-eight men. In 1837 and in 1838, he addressed the Royal Geographical Society on the Niger and its tributaries. Ill though he was, part of his time on the *Quorra* had been spent, like Hill, taking astronomical observations and undertaking a survey of the Niger's course. Preparing his findings in London, Allen left Clapperton's map and longitudinal observations untouched, but reckoned Lander's earlier itinerary and calculations 'very erroneously laid down'. Allen's argument was that the river Chadda (a tributary of the Niger known as the Tsadda, Shadda, or the Benue), flowed out of Lake Chad. He approached the topic 'with great timidity and caution' – in deference to Denham who had twice been thwarted in looking for an outlet from the lake, and to Rennell 'for whose sagacity I have the highest veneration'.

Allen reflected too on the distinction between in-the-field workers and armchair geographers. Once the former, he was now the latter: 'For myself, if I succeed in proving my proposition, I shall consider that I have made a greater discovery in Africa within my closet, than while personally labouring on the spot.'[13] He was wrong: Lake Chad has no outlet to the Niger. He and Oldfield also debated and disagreed over the rivers of the Niger delta.[14]

They may not have agreed – the entire history of the Niger's exploration between 1788 and 1830 might reasonably be described as one of geographical disagreement – but, little by little, the Niger was being revealed to European eyes: in books, reviews, maps, papers to learned societies, even via Africans themselves. At Newcastle in the summer of 1838 men of science gathered for the annual meeting of the British Association for the Advancement of

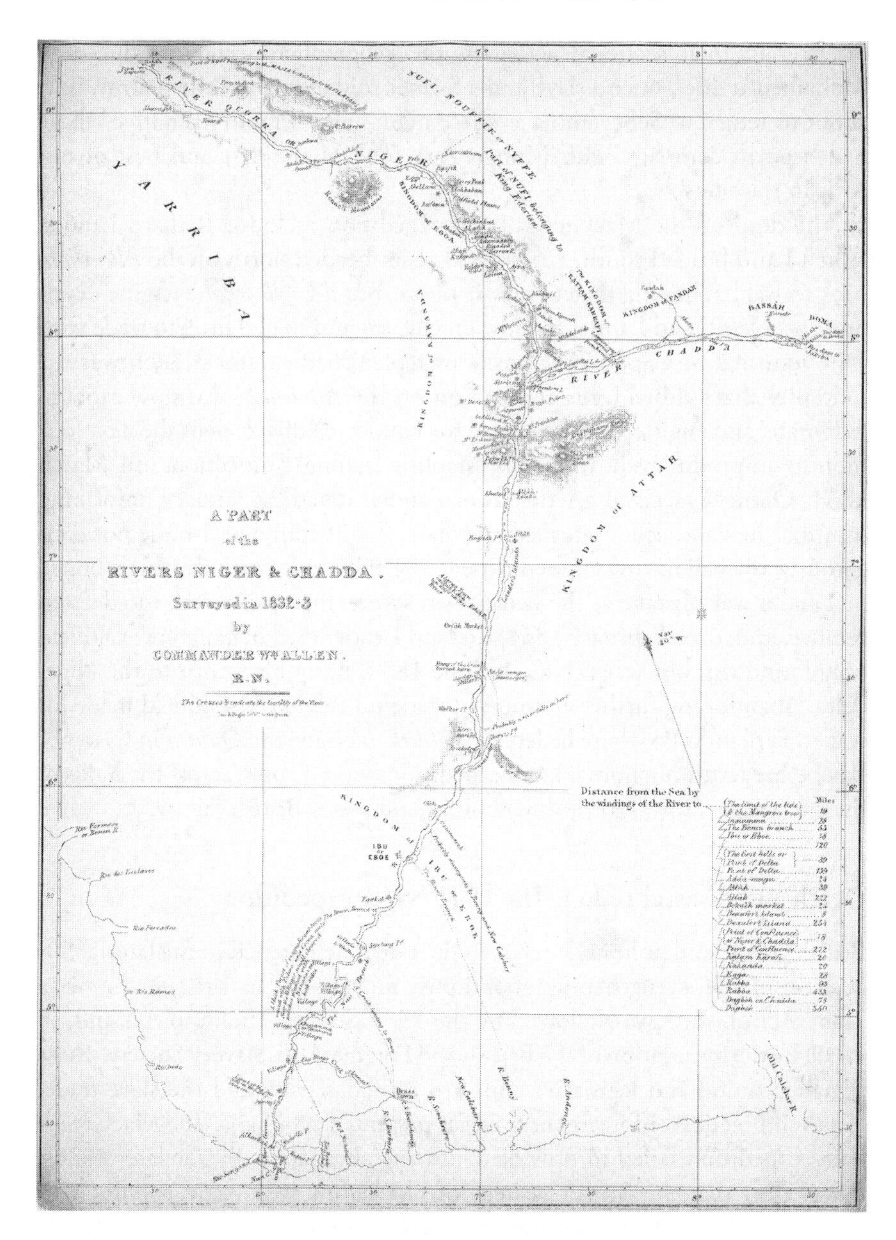

Figure 11.2 William Allen's map of the lower Niger and of the river Chadda (the Benue). On the west bank of the Nun (one of the outlets of the Niger's delta), Allen describes the 'black forest of significant trees' mixed with palms and coconut trees and with clearings for bananas (*cf.* Figure 11.4). Source: W. Allen, *Picturesque Views on the River Niger: sketched during Lander's last visit in 1832–33*. Reproduced by kind permission of the National Library of Scotland.

Science. In their sectional activities, the geographers were introduced to Mohamed-u Siseï, once a slave and a former soldier in the British army, now about to return to Senegambia as a free man – and 'who in his native village had been in company with Mungo Park, one of the first and best of our African travellers'.[15]

The dead on the MacGregor Laird expedition included Richard Lander. When Laird headed south, Lander had again headed north on the *Alburkah*, back to Rabba, hoping to get as far as Bussa. But the *Alburkah*'s engine developed problems, and Lander ran low on cowries and so had little to trade with – he returned to Cape Coast Castle to replenish their stock. Such was the mortality that Oldfield, his companion on the *Alburkah*, was now captain, first mate, and engineer. As he waited for Lander, Oldfield spent the next four months unproductively trying to establish trading connections. In March 1834, Oldfield received a letter from Lander, dated 22 January, informing him that he had been ambushed and shot: 'I am wounded, I hope not dangerously, the ball having entered close to the anus and struck the thigh-bone.'

Lander was mistaken. The wound was severe, the musket ball too deep to remove, and, on 6 February 1834, Richard Lander died of gangrene. Oldfield only found this out weeks later, in June 1834, upon his return to the Niger delta. Abandoning further attempts at trade on the Niger, Oldfield made his way to Fernando Po where he left the *Alburkah* beside the *Quorra* in Clarence Bay before returning home. Later expeditions would come across the hulks of the two steam ships – rotting symbols of ambitious British failure.

Death and disaster redux: The 1841 Niger expedition

British public and political interest in the Niger heightened in the later 1830s because of the strengthening abolitionist movement. In Britain, the Slave Trade Act of 1807 was followed by the 1833 Slavery Abolition Act and, in 1839, by the foundation of the British and Foreign Anti-Slavery Society. These signalled moral and legislative concerns to end slavery and the slave trade, but their practical enforcement overseas presented problems. The MacGregor Laird expedition failed to provide a humane alternative. By the later 1830s, it was clear that the British strategy of blockading west African ports with the navy's African Squadron was expensive and not always effective: slavers' vessels, Portuguese, Brazilian, American, still eluded the navy's patrols.

In his 1840 *The African Slave Trade and its Remedy*, the leading abolitionist Sir Thomas Fowell Buxton proposed a different strategy: educate and civilise Africans, teach them the love of God and the use of the plough. Strictly, his policy was not new. Clapperton, the Landers, and Laird had each tried to persuade local rulers away from trading in slaves and towards trade in British goods in exchange for ivory, palm oil, indigo, and other African produce, but without sustained success. But Buxton's plans were lent impetus by Britain's abolition of the slave trade, the failure of MacGregor Laird's scheme, and the ineffectiveness of the African Squadron. To give his plans institutional expression and to secure wider social support, Buxton established the Society for the Extinction of the Slave Trade and the Civilisation of Africa. Its president was Prince Albert, the prince consort: its members included numerous MPs and others of influence. For it to work, Buxton's plan depended upon a further expedition to the Niger.

Buxton had not been to Africa and never would, but he drew upon the works of those who had. Park's *Travels*, the work of Denham and Clapperton, the Landers, and Laird and Oldfield were read and referred to. Their accounts were treated both as statements of human accomplishment despite the authors' failure or death and as moral lessons from which others should learn, the British government especially.

The Niger, Buxton stressed, was the key to west African commerce. The safe navigation of the river and the establishment of mission stations and trading settlements along its course were the means to end 'the two great scourges of Africa, superstition and an External Slave Trade'. In order that Africans should have 'new methods of earning wealth by honest industry', Buxton's plan was three-fold: demonstrate to local inhabitants the nature and the advantages of legitimate trade; have local chiefs sign treaties committing them to end slavery within their territories; and establish a model farm at the confluence of the Niger and the Chadda/Benue rivers – where Laird had tried to establish his warehouse for goods – where Africans could be shown the twin advantages of Christianity and settled agriculture.[16]

For advice on the Niger and the implementation of his plan, Buxton turned to a recognised authority but an unlikely associate – James MacQueen. MacQueen's work probably came to Buxton's attention through MacGregor Laird and through Buxton's secretary, Andrew Johnston, who had read his 1821 *Geographical and Commercial View*.

MacQueen's work influenced the proposed 1841 Niger expedition in several ways: by advancing Fernando Po as the base for Britain's west African interests, not Sierra Leone; in maps which MacQueen drew of west Africa; in offering advice on the African interior – even over the vessels best suited to navigating the Niger and which rivers on the delta afforded access to the Niger – and in soliciting information on Buxton's behalf from the Royal Geographical Society. MacQueen's involvement in Buxton's scheme also involved a degree of self-interest – to extend to the proposed settlement of west Africa those rules of colonial authority for a plantation economy that MacQueen had advanced in the Caribbean.

Buxton was prepared to involve MacQueen in his plans because of MacQueen's reputation: 'he knows more about Africa than the whole Admiralty'. A prominent abolitionist, Buxton was no Africanist: as he admitted to MacQueen, 'I am very much out of my element in Geography of any kind, *a fortiori*, in African geography.' Buxton's plans for the Niger's commercial development thus drew not just upon those who had been there – MacGregor Laird, William Allen, and others – but upon the man who had solved the Niger problem without ever visiting the countries he was now advising others about and in whose 1840 *A Geographical Survey of Africa* he had laid out a vision compatible with Buxton's.[17]

Buxton's sentiments cannot be faulted. They nevertheless reflect a particularly patrician attitude of mind towards Africa and Africans, a mix of well-intentioned condescension and moralising disdain. Britain's imperialising negotiations with indigenous peoples in the nineteenth century – whether in India, Afghanistan, British North America, Australia, the Far East, or in Africa – were founded on two presumptions: that the civilising cure for 'native backwardness' lay in blending capitalism with Christianity, and that written agreements which enshrined notions of private property and public ownership had precedence over customary rights and tradition within oral cultures, furthermore that they should replace non-written practices and be understood to do so by all who signed them. The 1841 Niger expedition exemplified this ideology. What Buxton in August 1840 called 'this great national experiment for awakening the people of Africa to a proper sense of their own degradation and misery' would he argued 'otherwise be defeated' without the annexation of African land as British possessions overseas.[18]

For William Allen, Buxton's plans presented an opportunity to visualise the Niger and so add to public interest in the river. Allen had sketched the Niger as part of his survey work during the MacGregor Laird expedition. Engraved, his sketches were published by John Murray in 1840 as *Picturesque Views on the River Niger*. Allen apologised for them 'on the plea of their having been undertaken when he was suffering severely from the effect of the climate', but he took confidence 'from the goodness of the cause', namely, abolitionism. His stated motive was in keeping with the current mood: 'This mission of Peace and Charity – which will redound so much to the true glory of this country – is on the eve of departure; and the deep interest on behalf of Africa – which has never been distinguished in humane minds – will thus derive a fresh stimulus and a more general participation.'

Clapperton's sketch of the rapids at Bussa mark the first visual depiction of the Niger. Allen's picturesque views are a more developed form of topographic representation and in style and content present a peculiarly European and imperial aesthetic. Like many foreign topographical landmarks depicted by British observers in this period, in what have been described as 'strategies of imperial visibility', several of the features on the Niger were named after figures in the British Admiralty and other men of science (Figure 11.3). Maps fixed the Niger in the geographical imagination as their makers charted the course of the river more and more accurately. Allen's sketches helped the British public picture the Niger: constructing images of it as a still unknown river running through 'pestiferous swamps', a tropical world with a pestilential climate and savage inhabitants whose economy harboured a pernicious moral evil (Figure 11.4).[19]

Allen's renewed interest in the Niger was also personal. Like Park, Clapperton, and Richard Lander, he was about to return to west Africa – as commander of the *Wilberforce*, the embodiment in name of Buxton's anti-slavery agenda. The expedition sailed in May 1841 on three steam ships designed by John Laird, MacGregor's brother: the *Albert*, under expedition leader Captain Henry Dundas Trotter, the smaller *Soudan* under Commander Bird Allen, and the *Wilberforce* (Plate 11). A fourth ship, the *Amelia*, acted as tender and was crewed by Africans and West Indians. Each steam ship had a ventilator system designed to circulate fresh air and catch miasmatic air: the scheme was that of David Reid, who in 1840 had designed a similar system for ventilating the new Houses of Parliament. Each captain was appointed a Queen's Commissioner, meaning they had official authority to draw up and sign official treaties: Trotter and Bird Allen on the Niger, William Allen and William Cook on the Chadda/Benue. There were five surgeon-physicians and three chaplains – one, Samuel Crowther, a former Yoruba slave, had been educated in Sierra Leone and ordained in England. A black West Indian, Mr Carr, was responsible for the model farm, and the ships' holds were laden with gifts.

It is untrue to state that there had been nothing like it: Tuckey's expedition had scientists aboard to document and record Africa's natural diversity. Earlier Niger explorers carried gifts, made use of scientific instruments, and commented on the economic botany as they sought news of Park or determined the course of the river. But in its technology, its size, and its aspirations – and at a cost of £80,000, approximately £5 million in today's terms – the 1841 Niger expedition was on a scale quite unseen before.

The ships reached the Niger on 13 August 1841. As the expedition stopped at the principal towns and villages on its way upriver, negotiations were held with local chiefs: end the trade in slaves in the lands under your authority and you will receive the benefits of British trade. Not to do so would be injurious to your interests. Understand the tenets of the Christian faith; dispense with the custom of human sacrifice where still practised and worship the one God.

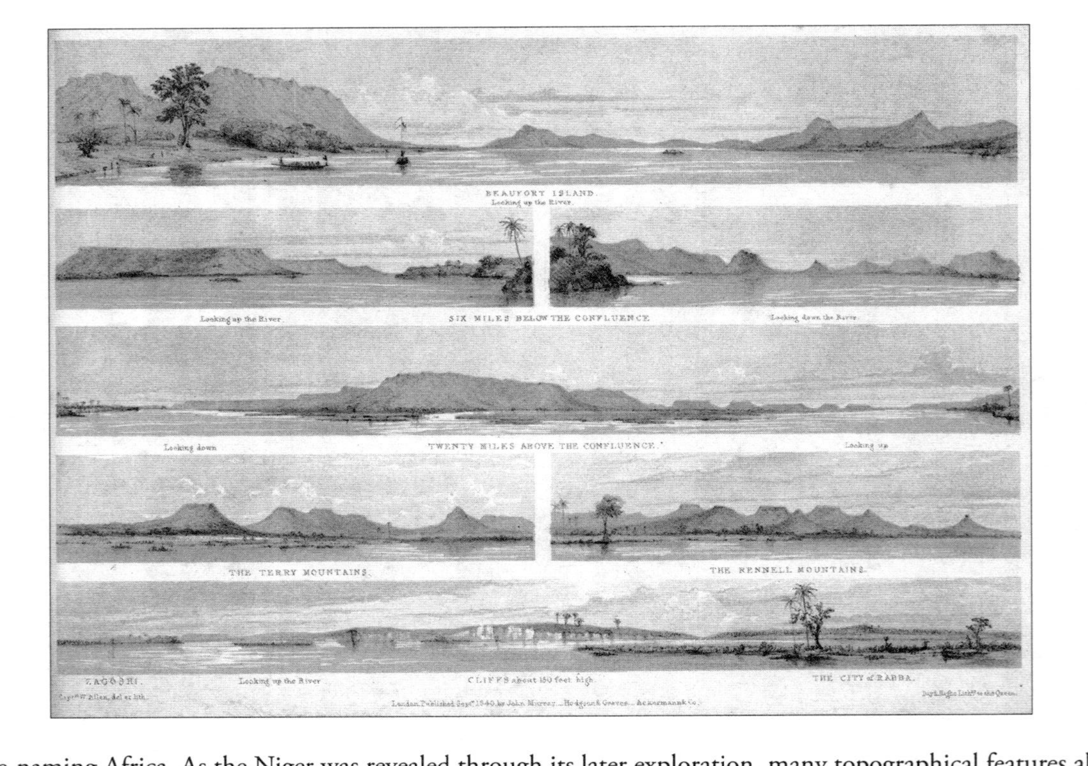

Figure 11.3 Re-naming Africa. As the Niger was revealed through its later exploration, many topographical features along the river's course were named after British scientists and geographers: Beaufort Island after Francis Beaufort, the Admiralty hydrographer; the Rennell Mountains after James Rennell. Allen's mapping and drawing work was at one point interrupted by a large black snake which went to sleep under the legs of his theodolite before it glided away. Source: W. Allen, *Picturesque Views on the River Niger: sketched during Lander's last visit in 1832–33*. Reproduced by kind permission of the National Library of Scotland.

Figure 11.4 The river Nun branch of the Niger. The Nun was the branch of the Niger's delta which the Landers navigated to reach the sea. William Allen here captures the dense and often towering vegetation of the riverbanks in the Niger's lower reaches and the human life there. Source: W. Allen, *Picturesque Views on the River Niger: sketched during Lander's last visit in 1832–33*. Reproduced by kind permission of the National Library of Scotland.

Sign here, or leave your mark, on agreements for the sale of land, for the model farm or to facilitate related interests, although the land in question was communally owned. The land identified for the model farm was a sixteen-mile-long and five-mile-wide stretch along the Niger's east bank at the confluence with the Chadda/Benue. The deal struck for it with the Attah in the town of Idah was concluded for about 700,000 cowries: at the usual rate of exchange of one shilling for 1,000 cowries, this was approximately £35. It was 'an historic occasion, the purchase of the first English [sic] settlement in the interior of Africa, and the first English [sic] foothold in Nigeria'.[20]

For the British and for the Africans with whom they were in contact, the 1841 Niger expedition marks a shift from colonising mission to visions of imperial sovereignty. It may be seen as the moment at which commercial and abolitionist concerns outweighed those of geography – the apotheosis of views about trade and slavery uttered by Banks and others in the 1780s and voiced with increasing emphasis by others in later decades. It might even be considered the moment at which the Niger problem was recast: from a 2,000-year-old mystery regarding the river's direction of flow and its termination as the problem was understood before 1830, to the locus of British colonialism in west Africa.

New beginning as it was, the 1841 Niger expedition shared one distinguishing feature with those explorers and expeditions which preceded it: it was overcome by the geographies it hoped to reveal.

Deaths on board increased as the ships headed upriver. One month after their arrival, sixty-nine men were ill from fever, all bar six white Europeans and most of them British sailors. Seven had died, all British sailors. In total, only fifteen whites out of a total of 146 escaped the effects of fever. Trade, treaties, and travel were hindered by the need to care for the sick and dying.

The three surgeon-physicians differed in their attitudes to fever, in how to treat it, and its root cause. On the *Albert*, James McWilliam believed in 'miasmatous poison' which he considered especially prevalent on the Niger: death rates could be lowered, and symptoms eased, if the ill were removed to the coast or the able bodied were subject to 'intense mental occupation'. One problem was the speed with which the crews were affected, healthy one moment and days later suffering from the fever's rapid impact: 'Fever of a most malignant character broke out in the *Albert*, and almost simultaneously in the other vessels, and abated not until the whole expedition was completely paralysed.'

Proportionately, the death rate was not as high as on the MacGregor Laird expedition. The problem lay in the degree of severity, its intermittent nature, and in the prolonged debilitating effects. Nor was it easily explained. On the *Wilberforce*, Morris Pritchett had no time for miasmatic theory: for him, fever was caused by exposure to the sun and could be treated with castor oil and, as Park and Clapperton had tried, with calomel. On the *Soudan*, Thomas Thomson took a different view: fever was caused by a 'certain peculiarity of atmosphere – call it miasma, call it malaria'. He dosed himself with quinine and so escaped from the fever but did not think to extend the same treatment to the ship's crew.[21]

These differences in medical opinion were paralleled by differences among the captains over what to do as the deaths increased. William Allen was for turning back. He knew from bitter experience the debilitating effects of illness on the Niger. Trotter and Bird Allen were for pressing on. Like the MacGregor Laird expedition and the Bornu Mission, the expedition split in two: Bird Allen and Trotter continued upriver on the *Albert*, and on 21 September 1841, the *Wilberforce* and the *Soudan* headed south. Both were now hospital ships. William Allen was so ill he could hardly stand; Bird Allen died on 25 October 1841. Dispatches between the ships, and from the ships to British officials on Fernando Po, document the mounting death toll and the numbers laid low by fever, difficulties with the ventilator system, frustration at having to withdraw from treaty making, the failure of their abolitionist agenda, even a sense of impending doom.

The *Albert*, at one point under the command of McWilliam as its only fit officer, was assisted in its passage by Becroft on the *Ethiope*. Barrow tried to enlist Becroft into the navy, regarding him as the most knowledgeable authority on the lower Niger delta and on the river's navigation as far as Rabba, but Jamieson would not sanction his release. The model farm reported only limited success. Most of those left alive were stricken by fever, and only a few acres had been cleared. In March 1842, the *Wilberforce* under the command of Lieutenant Webb was despatched to bring the survivors home, though some of the African labourers chose to stay even though the settlement was in chaos: Mr Carr had disappeared and was presumed dead, the jungle was reclaiming what land had been cleared, the first cotton crop had failed, the projected houses had not been built, and the stores were being sold for profit.

Two years in the planning, the 1841 Niger expedition was one of the shortest in duration: the *Albert* spent sixty-four days on the Niger, the *Wilberforce* forty-five, the *Soudan* only forty: 'The best-intentioned and most carefully prepared of all Niger missions, fuelled by Victorian optimism and armed with Victorian morality, ended as a pathetic failure.'[22]

In keeping with the scale of the 1841 Niger expedition, its failure prompted voluminous and prolonged criticism: questions in Parliament and, two years later, a 172-page parliamentary report. The three surviving commissioners William Cook, William Allen, and Henry Dundas Trotter enumerated several reasons why the expedition failed: agreeing a treaty was one thing, enforcing it another; much slave traffic was overland via coffles, not downriver in canoes; they disagreed over the right time of year to navigate the Niger (April, or no earlier than July); climate, especially in the swampy delta, was a key consideration; and ships needed a shallow draft – one reason why the *Ethiope* under Becroft could guide the *Albert* was its draft of only five feet, yet even it grounded.

As the Niger approached the sea, so it was less and less distinct as a single river. The Landers noted this, several times remarking how separate rivers flowed into and out of the Niger in its delta. MacGregor Laird had observed the same of the 160 miles of delta as the Niger approached the coast, the delta extending from Lagos on the west to Old Calabar on the east, where, thirty years earlier, Henry Nicholls had lived and died without piecing together the several parts of the Niger problem. Laird was more assured about the number of rivers as the Niger flowed into the sea, but not at all confident that the delta was fully known: 'Through this delta its waters are discharged into the ocean by twenty-two mouths, the principal of which are the Benin, Warree, Nun, Bonny, and Old Calabar. The Nun branch [Lander's First Brass River] is the only one yet explored.'[23]

William Cook made exactly this point in confessing himself puzzled by the Niger. He thought it impossible 'that a river which Park describes at Sego to be "as broad as the Thames at Westminster"' and which received numerous tributaries could 'dwindle into such an insignificant stream, not having a depth of five or six feet' in the many channels of its delta. Several years later as William Allen and Thomas Trotter, senior assistant surgeon on the *Wilberforce*, looked back on the expedition, they too admitted that 'their knowledge of the channels of the river is at present too imperfect to enable us to give anything like "sailing directions"'.[24] Navigated and increasingly revealed as it was by the 1840s, the Niger did not easily disclose its changing depths.

No recriminations were levelled at the principal participants in the 1841 Niger expedition. No naval careers were ended, or government positions lost. It was not as if the river's mapping was at fault: Trotter singled Allen's Niger map out for praise and used it as the basis for his own. Trotter also counselled over the conduct of the Niger's further exploration – should there be any. His twenty-nine recommendations regarding future travel included regular morning coffee, not hanging up hammocks too early, ensuring clean drinking water and dry clothing, and better medical precautions: 'quinine may be given to the men occasionally in lieu of wine and bark, and its issue may be extended to the whole crew when thought desirable by the surgeon.'[25] This last was prescient advice.

The Niger safely encountered, 1849–1854

One feature of the Niger's exploration is that its in-the-field investigation occurred in episodic bursts, on an almost decadal cycle. Mungo Park followed up his 1795–7 travels in 1805 and 1806. Tuckey departed in 1816, as did the Peddie–Campbell–Gray–Dochard expedition. Heightened interest in the Niger problem, in west African commerce, and in ending slavery prompted a quickening of the expeditionary pulse in the 1820s: Lyon and Ritchie, the Bornu Mission of 1822–4, then Laing, Clapperton, and Lander before, in November 1830, the Landers determined the Niger's termination at first hand. MacGregor Laird's 1832–4 expedition was not repeated until the Niger expedition of 1841.

The failure of the 1841 expedition had several consequences. The momentum of abolitionism was disrupted, and doubts continued over Britain's strategy of naval blockade. Attentive though it was to the evils of slavery, public opinion in Britain shifted from schemes aimed at converting free Africans into land-holding property owners toward more direct forms of colonial improvement – 'factories'. In the west African context this meant small-scale rural workshops under British management with an African labour force rather than the industrial machinofacture then transforming Victorian Britain.

Alarmed at the expedition's expense, the lives lost, and the reputational damage, no one in the Admiralty, the Foreign Office, or Parliament was disposed towards large-scale sea-borne expeditions. But when in 1849 the possibility of a relatively inexpensive Niger expedition was mooted, one that would take the trans-Saharan route to end African slavery, minds changed – and the near-decadal cycle of the Niger's exploration began again.

The initiative came from an evangelical preacher, James Richardson. A disciple of Buxton and a member of the British and Foreign Anti-Slavery Society, Richardson had travelled in the northern Sahara in 1845 and 1846. He published his *Travels in the Great Desert of Sahara* in 1848. The principal object of his travels had been 'to ascertain how and to what extent the Saharan Slave-Trade was carried on'. Showing it to be extensive, organised, and with no signs of cessation – to Richardson 'the human traffic in flesh and blood was the most gigantic system of wickedness the world ever saw' – he sent a copy of his book to Lord Palmerston, Britain's foreign minister and a committed abolitionist.[26] Palmerston responded by suggesting a trans-Saharan Niger expedition on the understanding that Richardson be accompanied by an experienced geographer – or, at least, by someone who could take astronomical observations.

Richardson may have contacted the Royal Geographical Society, and perhaps the Admiralty – surviving records are silent on the matter – but it is difficult to know to whom he might have turned. Becroft may have been approached but his expertise was the Niger from the sea, not across the sand, and between 1849 and 1854 he was British consul for the Bights of Benin and Biafra. William Allen shared Richardson's anti-slavery concerns as is evident in his 1849 *Plan for the Immediate Extinction of the Slave Trade, for the Relief of the West India Colonies*, but would never return to Africa. James MacQueen was an armchair geographer par excellence. Barrow had long since lost interest in the Niger. If articles in the Society's *Journal* in the 1840s are a measure of interest in the Niger, the fact that there were only four in the ten years from 1840 (two by Becroft and two by James MacQueen) suggests that more formal geographical enquiry echoed public disinterest following the 1841 debacle.[27]

Richardson's solution was to turn to Germany – as had Banks to Blumenbach, appeals which produced Friedrich Hornemann and Jean-Louis Burckhardt. Through Augustus Petermann, a German map-maker recently arrived in London following some time spent in Edinburgh with the cartographer Alexander Keith Johnson, and Baron Christian von Bunsen, the German consul in London, Richardson contacted the leading German geographer, Carl Ritter. As had Blumenbach to Banks, Ritter suggested one of his brightest students.

Johann Heinrich Barth was in several ways like Hornemann and Burckhardt. Of humble birth, he strengthened a weak constitution with a regimen of cold baths, exercise, and a strict diet. He learned Arabic (and several west African languages), had interests in comparative philology, history, and what we would now call social anthropology. He took notes on everything. Unlike Hornemann and Burckhardt, he survived his African travels – which lasted five years and produced much new information on north Africa's history and geography. Yet, unlike Park, Clapperton, or the Landers, his published work, in five volumes and over 3,500 disorganised pages, would attract relatively little public acclaim.

Accompanied by Adolph Overweg, a German geologist, Barth joined Richardson in Tripoli in March 1850. They were there met by Frederick Warrington, the diplomat son of Hanmer Warrington. By July 1850, the party reached Murzuq. They resided there, in Ghat, and in oases as they crossed the Sahara. In January 1851, the three separated, planning to meet in Kukawa, the Bornou capital on Lake Chad, in April. Richardson, whose agenda was abolitionism, headed south. Overweg, whose passion was geology, explored the Hausa kingdom of Gobir. Barth, whose concern was exploration, departed for Kano. Richardson however never arrived in Kukawa – he died on 4 March 1851, six days' travel from the rendezvous. Richardson's notes, once recovered by Barth with the remainder of his property, were edited and published in 1853 as *Narrative of a Mission to Central Africa*.

Neither Barth nor Overweg was insensitive to the question of slavery and the slave trade – Barth was appalled at observing the murder and mutilation of 170 captured slaves and the violent separation of children from their mothers by slave traders. But with Richardson's death, what was known as the Central African Expedition lost its humanitarian focus. Overweg and Barth spent the remainder of 1851 and much of 1852 in exploratory excursions from Lake Chad. On one, Barth confirmed that the Chadda/Benue did not have its source in the lake, thus refuting William Allen's belief that the lake and the Niger were connected through this eastern tributary. In August 1852 Overweg died, from a fever caught after duck hunting and not taking off his wet clothes.

Heinrich Barth was now alone in west Africa, the 'sole undaunted European survivor' as he was later termed, an explorer – and an exploring German – in a British-funded expedition whose aim was principally humanitarian. News of Barth's progress did reach London, but only slowly: letters overland took over eighteen months to arrive via Tripoli. In February 1853, Barth asked for assistance (and more thermometers and barometers, having broken most of his stock). Arrangements were made to send out a London-based astronomer as his assistant, Eduard Vogel – another German.

For many in the Royal Geographical Society, the sense that British exploration of the Niger was being hijacked by others was compounded by the fact that news of Barth's progress went first to Petermann and to von Bunsen, and sometimes to Ritter in the Berlin Geographical Society, and only last to British authorities if at all. Petermann published numerous reports on Barth's work under his own name in *The Times* and *The Athenæum*, principally to bolster his emerging reputation as a map-maker. The British geographical community may not have been able to advance anyone to accept Palmerston's invitation to Richardson in 1848, but between 1851 and 1854, it was much exercised by the fact that British money was being used in west Africa by two Germans and, in London, by two other Germans, all to advance their own interests.

Things came to a head in the Royal Geographical Society in London on the evening of 9 January 1854. The evening was devoted to four papers on Africa's exploration. Petermann's 'Latest accounts of the Mission to Central Africa' was the first read. Petermann's talk prompted an 'animated discussion' involving, amongst others, the leading British map-makers John Arrowsmith and Trelawney Saunders, the explorer-artist Thomas Baines, shortly to join the Society-sponsored expedition to northern Australia and later to illustrate David Livingstone's Zambesi expedition, and James MacQueen, a regular attendee at the Society's African nights.

Saunders and the others were irked that good money and expensive instruments had been passed to Vogel, but that little appeared to have come of it. Their principal concern was that Petermann had begun a map showing the progress of the Central African Expedition as early as December 1851, incorporating on various drafts information received from Barth and his companions, but without involving British geographers or map-makers. In a rejoinder to the Society's secretary, Norton Shaw, who had quizzed him on the matter, Petermann said that he had done this to keep costs down. Moreover, continued Petermann, 'I have supplied the information (which no one in this Country, not even Lord Palmerston and the Chev.[alier] Bunsen possesses) – of the General route, and which is of very great interest, – with a readiness as not every one of my colleagues would have done.'[28] To add insult to injury, Petermann incorporated Dixon Denham's route of travel on the Bornu Mission to his map in progress in March 1852 – and billed the Royal Geographical Society for the work involved.

In February 1854, maps based on Barth's correspondence and worked up by Petermann appeared in *An Account of the Progress of the Expedition to Central Africa*. Petermann's name appeared on the title page, despite his having played no part in the expedition or ever visited Africa. Barth, whose travels were shown, knew nothing of the work. The *Account* portrays a particular vision of Africa, and of the explorers of the Central African Expedition and their travels (Plate 12).

One month later, on 18 March, Petermann published a pamphlet entitled *African Discovery: A Letter Addressed to the President and Council of the Royal Geographical Society of London* in which he defended his fellow Germans and his own publication practices and offered further rejoinders over his spoken paper in January. The Royal Geographical Society was slow to publish, protested Petermann, its *Journal* then appearing only once a year. And the Society's officers often sat for a long time on the papers read at its evening meetings – as he felt they had with his – before their publication. In contrast, he pointed out (his emphasis a deliberate rebuke), 'information contained in the *Athenæum* and other *regular* journals, is at once available to every educated person of the civilised world'.[29]

The detail, timing, and location of these exchanges serve to remind us of the wider context to the history and geography of the Niger's exploration: the personal relationships behind explorers' achievements, the hidden histories of map-making and book publishing, the uneven emergence of British and European understanding of the Niger's geography, and the fact that the river continued as a focus of geographical enquiry, in London and in Africa, for decades after the solution to the Niger problem.

Lost as they were, to British officialdom anyway, Barth and Vogel continued in their exploration, oblivious to disputes in London over who had what right to map and to publish news about the Niger. As it became known that

Barth had charted the upper reaches of the Chadda/Benue and was planning to spend time in Timbuktu and the upper Niger, plans were made to send a party to rendezvous with Barth on the river. The Admiralty turned to recognised Niger authorities for the mission. MacGregor Laird oversaw the construction of an iron-built screw steamer, the *Pleiad*, for the mission, and John Becroft was to be its captain. Laird had spoken in January 1854 to the same evening session as Petermann, outlining the qualities of the *Pleiad* and his plans for what, in effect, was a rescue mission.[30] But as the *Pleiad* reached Fernando Po, it was met with bad news: Becroft, long-time Niger navigator, and as captain of the *Ethiope* a key figure in the survival of the 1841 expedition, had died on 10 June 1854.

Several of the principals in the Niger's exploration – Mungo Park most evidently, but also Hugh Clapperton and Walter Oudney – were Edinburgh-trained physicians. So was William Balfour Baikie, the medical officer who took command of the *Pleiad* following Becroft's death. Baikie did not set out to be a Niger explorer; he became one following Becroft's death. His purpose was to rescue Barth and, where he could, to establish trade with west African peoples. In his instructions, MacGregor Laird made it clear that all trading operations were 'subsidiary and auxiliary to the main design of the voyage', which was to ascend the Chadda/Benue 'as high as possible during the rise of the river'. That Baikie later called the *Pleiad*'s voyage an 'exploring expedition' is true, but the term encompassed much: 'Our voyage had points of interest for the utilitarian, the commercial man, the man of science, and the philanthropist.'[31]

With Baikie, the exploration of the Niger is distinguished yet further – but not by his rescue of Barth. In that respect the *Pleiad*'s mission failed. Barth had turned north in May 1855, unaware that he required rescuing, and arrived in Tripoli in September that year. Barth's work was scientifically important, and he was awarded the Royal Geographical Society's Patron's Medal. But he refused their invitation to dine, did not show up as promised to a meeting of the British Association for the Advancement of Science, and was rebuked by the Society for the Extinction of the Slave Trade for returning to Germany with two freed slaves, Dorogu and Abbega, as his servants (humiliated by the matter, Barth later sent them home). The explorer who spent five years in Africa and survived was overlooked because his work was inaccessible and not widely read, and his reputation suffered further because he was antisocial and a German.

Baikie, by contrast, is distinguished for leading the first Niger mission to spend several months on the river and to return without a single fatality. Over half a century after the medically trained Mungo Park had set out, a Scottish physician who did not have exploration as his aim led what is considered the most successful Niger mission by the regular administration of a drug, quinine, known for centuries to guard against fever in the form of chinchona or Jesuit's Bark and known and used in purified form from 1820. And the *Pleiad*'s crew was mostly Africans: twelve Europeans and fifty-three Africans (some of whom had sailed with Becroft): selecting the crew in this way reduced the chances of the mortality experienced by the expeditions of 1832–4 and 1841.

Baikie's experience on the Niger was nevertheless that of his predecessors in several ways. The *Pleiad* regularly grounded during its four months on the river, and the Europeans fell ill with fever. Baikie's experience on a second steam vessel on the Niger was uncomfortably like that of Laird in the early 1830s and that of the 1841 Niger expedition. On 7 October 1857, the *Dayspring* steamer struck a rock at Jebba, about 300 miles north of the confluence of the Niger and the Chadda/Benue, where the model farm had failed and where Samuel Crowther hoped to establish his mission station. The ship stuck fast for almost a year, but with quinine, the crew and passengers recovered, trade was achievable, residence in factories and managed farms was possible, and political ambitions on the abolition of slavery could be realised. During this voyage and its enforced stasis, Lieutenant Glover recovered Park's book of logarithms which, thirty years earlier, Clapperton had declined to part from its African owner. Unlike other steam-driven Niger navigations, immobility was not fatal – Baikie reported that he was 'far from idle', and busy learning two languages and revising geographical works.[32]

In his 1856 published account, Baikie produced a list of the different names by which the Niger was known, distinguishing between the names used for the river above its confluence with the Chadda/Benue, and those below. Eight different indigenous names were listed for the lower Niger, in addition to the 'Niger', 'Kworra', and 'Quorra' names by which the river was known by British and European commentators. Above the confluence, ten different indigenous names were in use, one Arab term '*Bahr Sudán*', and twelve different ones had been identified by European authorities from Ptolemy to Park to Dupuis ('Quorra', 'Niger', 'Kowara', and 'Joliba' being the most used). Baikie does not mention him, but for all the oddity of his proposed sub-Saharan Niger route, Rufane Shaw Donkin may have had a point: looking for *the* Niger hindered the exploration of what, as a single river, was multiply named by African peoples.

William Balfour Baikie's 1854 *Pleiad* mission and the survival of the passengers and crew on the *Dayspring* mark a further shift in the Niger's exploration. The river continued to be of interest to geographers, Baikie speaking several times to the Royal Geographical Society between voyages, and others including Norton Shaw, Francis Beaufort, and Francis Galton discussed and published on the river.[33] Baikie's *Narrative* was accompanied by two other accounts of the *Pleiad*'s voyage: Samuel Crowther's, whose focus was African 'superstition' and missionary work, and by Thomas Hutchinson, the engineer, whose focus was trade and the suppression of disease – through proper clothing, good diet, quinine, and 'appropriate moral influences'.[34]

From the mid-1850s, the river was seen less as a geographical problem and more as a route to spread trade and Christianity, and to the longer-term British settlement of west Africa. Baikie was the first British colonial official to choose to reside on the Niger. He took a deep interest in indigenous languages and comparative philology, abhorred the slave trade, and wrote warmly of the innate capacities of the black African: 'his intellect, when duly cultivated, will rank with that of the white man'. He has been termed, and rightly, 'a pivotal figure of West African exploration'.[35]

Yet he is not remembered as such: one modern scholar has termed his geographical discoveries modest.[36] It is a feature of the Niger's exploration that to be remembered as a Niger explorer, to become famous for being one (especially after one's death), involved more than writing a book about it, drawing a map of the river's course, or failing and dying in the attempt. The status of the explorer as a particular sort of man, and of exploration as a particular set of processes and emotions – physical fatigue and mental anguish, privation, despair, elation, trust in one's fellows, faith in one's instruments – may also be judged by how their work is remembered, how it was understood by their peers and their public, and – if it was – by how and why their life and work was commemorated. Of no one is this truer than Mungo Park.

12

'This celebrated African traveller'

Park commemorated

As he put it to his brother, Mungo Park set off on his first African trip with 'hopes of Distinguishing myself . . . if I succeed I shall acquire a greater name than any ever did'.[1] If not exactly an expression of over-vaulting ambition, it was an honest admission of sorts – that he hoped not only to achieve his goals, and those of Banks and the African Association, but that he might also be recognised for them.

Recognition by one's peers as distinguished, whether as an explorer, an author, as an adventuresome or heroic young man who overcame difficulties to accomplish a given end, is not the same as remembrance. Recognition requires that one's exploratory achievements be noted and understood at the time. Remembrance is both the act of a moment and a longer process, enduring yet uneven. The memory of exploration and of the explorer can be made visible in different forms as it may also be triggered by them: in a statue dedicated to the person's accomplishment and personal fortitude, or by an explorer-author's published narrative and in reviews of it. Memory in these terms matters because of 'the guise of its representation', those practices of commemoration and memorialisation which mark how and why memory is given material form in certain places and particular times, and how it may be contested.

In life, Mungo Park was recognised for his achievements. The dead Park was never out of the public eye: his worldly afterlife distinguished by a strong and varied 'commemorative trajectory'.[2]

Park's words

Park travelled in printed form in different editions and imprints throughout the nineteenth century. Even as his book was published in 1799 by William Bulmer, an unofficial and abridged version appeared that same year.[3] In the 120 years or so between the authorised and unofficial publication of his *Travels in the Interior Districts of Africa* and a 1920 book to that title, over forty different editions of Park's *Travels* were published; and it is still in print. Some retained the original title or shortened it to *Travels in the Interior*, abridging the content. A few combined the greater part of Park's 1799 work with elements of his 1815 *Journal of a Mission*. Others incorporated Park into compendia histories of travel, of Africa in particular, as further exploration brought that continent to the attention of European audiences.[4]

One such was Hugh Murray's *Historical Account of Discoveries and Travels in Africa* (1817). This was an enlarged version of John Leyden's 1799 *A Historical and Philosophical Sketch of the Discoveries & Settlements of Europeans in Northern & Western Africa*. Leyden makes much of Park's work: he confessed to a friend that 'the discoveries of Mungo Park haunted his very slumbers'.[5] Leyden and Park may have known one another. Leyden was from the Scottish Borders, went to the University of Edinburgh, and, like Park, was acquainted with Walter Scott and shared his interests in Border tales and balladry. Leyden later became professor of Hindustani in Calcutta. He may have compiled his 1799 book in the hope of benefitting from Park's success: it is in essence a history of the Niger's exploration up to and including Park's first journey.

Hugh Murray certainly had an eye to those who followed Park in updating Leyden's account: as we have seen, he contributed to and reviewed debates about the Niger's exploration in the periodical press and, in 1830, co-authored a *Narrative of Discovery and Adventure* with the distinguished natural historian Robert Jameson which included Park and other Niger men.[6]

Park's words appeared in cheap chapbook form, in two different editions, in 1816 and in 1818. In 1814, the Quaker philanthropist and educational writer Priscilla Wakefield recast Park's *Travels* in novel form in her *The Traveller in Africa*, adding elements from James Bruce's Abyssinian travels amongst others. Park was a central character in the Rev. Isaac Taylor's *African Scenes for Tarry-at-Home Travellers* (1821). In 1868, Samuel Mossman made Park one of his *Heroes of Discovery* (alongside Livingstone, Franklin, Cook, and Magellan) and, two years later, extended his treatment of him in his *Gems of Womanhood* where, in a section entitled 'Succourers in Days of Distress', the focus is on the African women who helped Park.[7]

As Park's works travelled in these various forms and to different purposes, Park's circumstances were used to comment upon contemporary issues. There is a marked poetic response to Park in the decade between his death and the renewal of Niger expeditions, one stimulated by his 1799 *Travels*, his plight as a lone traveller, and by his romantic sensibilities and sympathetic depiction of Africans, as well as by uncertainty over his death.

Georgiana Cavendish's rewritten version of the African women's words, for example, those sung to Park as he lay weary and dejected on his first trip, became a text for campaigners against the slave trade. The composer and soprano Harriet Abrams took words from Park's *Travels* for her ballad 'The White Man' in 1799, its opening line, 'Let us pity the poor white man' emphasising his plight. In his abolitionist poem 'The West Indies' (1809), the Scottish radical and humanitarian James Montgomery depicted Park as a 'Pilgrim', lost in Africa, but one whose moral stance toward Africans contrasted sharply with that of slavers. In 1811, Mary Russell Mitford, later a successful dramatist, similarly cast Park as a pilgrim in a strange land, his search for the Niger seen as a distinctly Christian quest. Her 'Lines, Suggested by the Uncertain Fate of Mungo Park, The Celebrated African Traveller' was clearly motivated by the half-truths then circulating about Park's death. A ballad entitled 'Lament of the Free Africans for Mungo Park' was current by 1820 if not earlier, its title and lyrics a reversal of the trope of the white European looking disdainfully upon the black African.[8]

Park, it has been suggested, appealed to women poets and authors because his determination and courteous empathy, notably to those African women who offered him food and shelter, helped establish those qualities as distinctively feminine traits in early nineteenth-century culture. Male poets of the standing of William Wordsworth, John Keats, and Robert Southey also acknowledged Park's words and influence. For them, African travel narratives became almost foundational texts for the depiction of the suffering traveller and for the reputational making of Romantic poets. Africa was variously portrayed in their work: an imagined savage space, its northern realms an antique land, as the locus of peoples and places still not understood by Europeans despite the attempts of men like Mungo Park.[9]

One moment in Park's *Travels* received particular attention. Park, we should recall, was at perhaps his lowest ebb during his first African travels in late August 1796 after being robbed, stripped naked, and left for dead: as he later put it, 'I considered my fate as certain, and that I had no alternative, but to lie down and perish.' That he did not was because a small flowering moss caught Park's eye, and realising that he, like the moss, was God's creation he resolved, like the moss, to survive.

The Quaker botanical author Mary Roberts depicted Park's 'moss moment' as the frontispiece to her 1822 *The Wonders of the Vegetable Kingdom Displayed* (Figure 12.1). The scene is Roberts' imagined reconstruction. In contrast to his description of himself, Park is clothed, hatted, armed. The point of the illustration is not a formal depiction of Park's circumstances but is, rather, didactic. In a book whose argument was that God's work was present in all nature – that the proper study and utilisation of plants depended upon

Figure 12.1 Mungo Park's 'moss moment'. At possibly his lowest ebb on his first African travels, Park had his spirits raised after seeing a moss in flower, and he resolved to carry on. The artist here places Park amidst luxuriant and dense vegetation: Park's *Travels* tell a rather different story. Source: M. Roberts, *The Wonders of the Vegetable Kingdom Displayed* (London, 1822). Reproduced with permission from the Library Collection at the Royal Botanic Garden Edinburgh.

'knowledge of their great Artificer' as Roberts put it – Park's moss moment was morally and spiritually instructive.

Similar sentiments motivated the evangelical minister Robert Murray McCheyne in Dundee. In 1836, in which year he began his duties in the city – with news of the failed MacGregor Laird expedition then circulating – McCheyne wrote the poem 'On Mungo Park's Finding a Tuft of Green Moss in the African Desert'. Influenced by Park's tracing 'the mighty Niger's course', the lines 'And ah! Shall we less daring show / Who nobler ends and motives know / Than ever heroes dream / – Who seek to lead the savage mind /

Figure 12.2 Memento Mori illustration of Mungo Park by Giovanni Antonio Sasso (*fl.* 1801–16). Park is shown holding a partly unfurled map of Africa, perhaps to suggest the further Niger work to be done following his death, while his left hand gestures symbolically to the African interior. Image in the Public Domain.

Figure 12.3 Mungo Park's moss (*Fissedens bryoides*) – a small plant with significant implications for Park on his first African journey. Reproduced with permission from the Library Collection at the Royal Botanic Garden Edinburgh.

The precious fountain-head to find / Whence flows salvation's stream?' speak equally to evangelical endeavour in Africa and to the citizens of Dundee, its working-class poor especially. Park's death was marked too by a memento mori illustration, though it is hard to know how widely disseminated this depiction of him was (Figure 12.2).

Park's encounter with the moss had scientific as well as spiritual significance. In November 1887, in thanking the Royal Society for the award of its Copley Medal, botanist Sir Joseph Hooker recounted two circumstances which had helped shape his career. The first, while still a child and after botanising in the Highlands with his father (the botanist William Hooker, a friend to Banks), involved constructing a model of one of the mountains they had ascended: '[I] stuck upon it specimens of the mosses I had collected on it, at heights relative to those which I had gathered them. This was the dawn of my love for Geographical Botany.'

The second came from Park's *Travels*. 'You may remember,' continued Hooker, 'a passage in Mungo Park's travels in search of the source of the Niger, when he describes himself as so faint with hunger and fatigue, that he laid himself down to die,' only to be revived 'by the brilliant green of a little moss on the bank hard by'. Park's achievements and words were sufficiently well known that Hooker could presume upon his audience even though Hooker himself got Park's purpose wrong. 'A scrap of that moss', remarked Hooker, 'was given to my father by Mungo Park, or a friend of his, and was shown to me. It excited in me a desire to read African Travels, and I indulged in the childish dream of entering Africa by Morocco, crossing the greater Atlas (that had never been ascended) and so penetrating to Timbuctoo.'[10] Hooker never did get to Timbuktu, but he did climb the Atlas in the spring of 1871.

Park's moss survives (Figure 12.3). Whether it was actually handed by Park to William Hooker, or to someone else, is uncertain.[11] Park's words affirm its importance to him, as inspiration to continue. Roberts, McCheyne, and Joseph Hooker confirm its wider significance long after Park's death and in contexts entirely different.

Park performed

Park inspired playwrights as well as poets and botanists. In the theatre, however, Park's expeditionary intent and his role in solving one part of the Niger problem was unimportant. What mattered was not geography but his escape from danger, Africa's scenery, the sympathy afforded him by African women, and the attitudes of British audiences toward Africans and slavery.

In *The Africans, Or, The Desolate Island* – oddly described at its opening in 1800 as 'a new Asiatic melo drama' and which featured the character of Omai, the Ra'iatean Pacific islander who had come to London in 1774 and was there much feted by Banks – the scenery was based upon Park's *Travels*. To unknowing metropolitan audiences, one exotic location was much the same as another. In other early nineteenth-century plays, several distinctly anti-slavery in sympathy, Park's book was similarly a key source for depictions of Africa. George Colman's *The Africans; or, War, Love, Duty*, which ran in London in 1808 before moving to Philadelphia, is set in a Foulah village. Colman drew elements of the dialogue from Park's words, and the play takes aim at the horrors of slavery – the involvement of indigenous leaders in dehumanising their fellow Africans and the role of British and European slave traders. The cast of European villains includes the slave traders Marrowbone, Grim, Flayall, and Fetterwell.[12]

A ballet, *Mungo Park; or, The Treacherous Guide*, was performed in May 1819. Its timing and title suggest that it spoke to the rumours then circulating about Park's death and the degree to which trust could be placed in the actions and words of Amadi Fatouma. The reportedly comic spectacle *Mungo Park; or, The Source of the Nile* which ran in 1824 was certainly mistitled – and probably struggled to find much humour in Park's African journeys.[13]

An altogether different Park appeared in northern England in the early 1840s. The three-act play *Mungo Park; or, The Arab of the Niger* by William Bayle Bernard, London-based but American-born, opened in Liverpool's Theatre Royal on 18 February 1841 before moving to Leeds and to Manchester (it later revived in the Royal Surrey Theatre in London in 1862). Advertisements for the play make no mention of the Niger expedition then underway or of the standing of MacGregor Laird in Liverpool and Birkenhead, but the association must have been clear to northern audiences and may account for the play opening in Liverpool.

Where, earlier, Park's words and scenes were used, Park was now the principal character. Laidley was also present as the governor of the Gambia, and six other English characters appear, including John Martyn. Among the five Moors are Karfa Toura, the slave trader who led Park to safety on the return leg of his first journey, Ali who had earlier imprisoned Park, and Ali's daughter, Immalee. Comic relief is provided by the part of Tobey Gander, a naturalist who, he tells us, holds a responsible position in 'the newly-built Museum in my native town of Mud-cum-slushy, Monmouthshire' although all he has collected to date 'has been a pair of Cockchafers that I caught one morning in my coffee, and a curious specimen of a slug that was boring a hole in my ear'. Gander's role as natural historian may be a gentle nod in Banks's direction perhaps or at others, merging in one the botanist, gardener, marine zoologist, and comparative anatomist who accompanied Tuckey on his disastrous venture.

In the play, Park has returned for a second expedition: 'to reach the Niger by the shortest rout [sic], and trace it to its termination' and in the hope that by his actions and the trade that will follow them, the African 'will rise into the dignity of a free enlightened man'. Gander offers his services to Park: 'someone who will explore the Animal Kingdom whilst he's observing human ones'. The party is guided by Karfa Taura.

Act 2 opens on the banks of the Niger. Ali is sitting, watching his daughter who is asleep. The watching Moors report an approaching canoe with five white men aboard. Park appears on the riverbank, bemoaning the loss of his expedition: '31 graves have these hands dug already, and but 7 weeks since their tenants were alive, were marching by my side, all strong with life.' Park is shot by Ali, who has pledged death to the 'Nazarenes', but the bullet hits the handle of Park's knife leaving him wounded but alive, and he is tended by Karfa Taura. Ali captures Park, Karfa Taura, and Gander, and intends that they be torn apart by wild beasts, but Immalee pleads to her father for mercy. Ali relents and the play closes with, centre stage, Park and Karfa Taura, the man who saved him in life and in the theatre.[14]

By the time the play opened in Manchester in July 1843 – a 'highly successful New Melo-Dramatic Spectacle . . . founded upon real events in the

Life of the Great African Traveller' – Park is part of an African animal show. (Figure 12.4). The playbill shows Karfa Taura, a muscular caricature in extraordinary attire, fending off a diminutive lion by sticking his thumb up its nose and his fingers in its eyes while preparing to punch a tiger (not a species native to Africa). In the play, Karfa Taura indeed grapples with big cats, first with a leopard, then a tiger. Park several times encountered or heard lions – but he was never close enough to poke them. The playwright, cast, and audiences may have been aware of the 1841 disaster, and perhaps of the 1843 parliamentary report into it. To them, however, accuracy and real life may not have mattered. This almost pantomime show was the real Park, performing in a work staged to accord with contemporary views about wild Africa and its savage beasts and peoples, and the manly heroism of its explorers.

Park was on stage again in Edinburgh in 2016 and in Denmark before that. The play, *Mungo Park: Travels in the Interior of Africa* was written by two Danes, Martin Lyngbo (who also directed it) and Thor Bjorn Krebs of the Mungo Park Teater Company, a Danish repertory theatre company founded in 1992. Two actors played several parts – including for one Allison Anderson, Joseph Banks, and Lieutenant John Martyn, and by the second James Rennell, Alexander Anderson, and Isaaco amongst others. A third actor played Park. The narrative, a simplification of Park's *Travels* and with elements of his fateful 1815 *Journal*, was enriched by being also based on interviews with modern scholars in Timbuktu and a descendant of King Mansong in Segou. While the play deliberately 'stretched' Park's words, even adding elements of Hollywood-style cinema, Park's story was presented both 'as a cautionary tale as well as one of great bravery and idealism': cautionary in terms of the European colonial project in Africa; brave and idealistic in Park's actions and motivations.

Park was made to confront his own travels (Figure 12.5). The rigours of his epic journey were effectively conveyed by clever use of a wheeled platform on which he trudged. 'Mungo Park', whose life was saved by a slave trader, was used to throw critical light upon slavery, an issue with which Scottish audiences were – and still are – uncomfortable. That the performance openly played with truth and the art of storytelling is not the issue. After all, Park did much the same with his 1799 book as he admitted to Adam Ferguson, Dugald Stewart, and Walter Scott.[15]

Park has performed in different media. Park's *Travels* was used as part of a twenty-minute radio broadcast, number 494 in the Scottish Home Service for Schools, which went out on 6 March 1951.[16] While Africa was his proving ground, Park was archetypically cast as a Scotsman who made good despite his lowly circumstances as the son of a Borders farmer. The programme was listed as one of several 'Stories from Scottish History' rather than as one of African geography.

Figure 12.4 The Advertisement for *Mungo Park, or The Arab of the Niger* for 28 July 1843 as the play opened in Leeds. The play's three acts are described with Act 3 promising to be the most dramatic of all. Image in the Public Domain.

Figure 12.5 Mungo Park in 2016, played by the actor Matthew Zajac, gazes upon the map of his own travels from his *Travels in the Interior Districts of Africa* (London 1799). Reproduced with the permission of Dogstar Theatre and the Mungo Park Teater, Denmark.

In the Footsteps of Mungo Park, a short film made in 1963 and narrated by Ian Cuthbertson, again emphasises Park's background: the film begins and ends with his memorial statue in Selkirk, with scenes from locations on his routes in Africa in between. Park, we are told, overcame the difficulties he had in Africa because he had overcome similar ones in leaving Scotland. This view, that Park's empathy toward Africans and their way of life was born of an innate understanding of the rigours of rural life, is echoed by several of Park's biographers.

Park three times features in novel form, the first in Priscilla Wakefield's *The Traveller in Africa* (1814). There, Park's travels are seen as educationally and morally instructive for the young man in question, Arthur Middleton, whose life was spent in one tour after another. In *Park: A Fantastic Journey*, published in 1932, we are presented with an altogether rather odd Park. The author, Canon John Gray, came late to Roman Catholicism following a career as a poet, translator, and civil servant in the Foreign Office. He was for several years a close friend of Oscar Wilde and may have been the original for Wilde's Dorian Gray, a position he denied. *Park: A Fantastic Story* is his only novel.

In the book, Mungo Park is a Roman Catholic priest and professor of moral theology on a walking tour in the Cotswolds. In a dream – the experience takes place while Park is walking from Burford towards Oxford – Park wakes to find himself in a post-industrial Britain in which the racial stereotypes and social order he encountered in Africa have been inverted. Britain is run by the black Wapami, a Catholic theocracy, whose language is Bapama. They are humane, intellectual, and artistic: qualities that Park saw in the social mores of black Africans during his first travels. Beneath them socially, and literally – for they live underground in networks of tunnels and caves – are the whites, their social systems and morals much degenerated. Motor cars do not exist, trains do. Park is well received by the Wapami chiefs but forbidden to practise his Catholicism. Parts of the dialogue with the chiefs, by name Svillig, Koti, and Dlar, read as if Park was taking notes on the productivity of the country and the lives of their inhabitants, much as the real Park had been instructed to do by Banks and Beaufoy.

It is unclear if this is a statement about Park's actual experiences and his failure to determine the termination of the Niger, or to demonstrate the virtues of Christianity while in Africa. Unsettled by his experiences, Gray's Park wakes from his 'short, deep sleep . . . and a long dream' to find himself back in his England. Much in the novel is unclear. It ends with the words 'Somewhat more elaborate than is usual.' It is hard to know how to describe Gray's *Park*: historical fantasy, grotesquerie, rural reverie, dystopian racial commentary – it is all of these.[17]

In *Water Music* (1980) by T. Coraghessan Boyle, Park is back in his own time, although he might not recognise himself or his peers. Preceding what is a picaresque romp of a novel, often funny and inventive, Boyle offers his readers an 'Apologia'. He makes clear that his motivation was 'principally aesthetic rather than scholarly', that he has been 'deliberately anachronistic' and that he 'strayed from and expanded upon my original sources'. Yet the book is closely attentive to context – to the key role of the African Association and its interest in the Niger, to Park's predecessors Ledyard, Lucas, and Houghton, to Park's captivity by Ali, and much else. It ends poignantly: Park's 'lovely Allie' in 1827, now in her early fifties, holding an unopened letter from her son, Thomas, who had that year gone to Africa to look for his father: 'It was simple, he had it all figured out. He would travel alone, as his father had done on the first expedition, living like the natives, making his way northeast through Ashanti-land and Ibo, striking the Niger at Boussa.'[18] Allison Park had little reason to open the letter: she knew what it said.

Allison Park was used to such news. Her husband's death took years to confirm, and even before it was, her brother John died, in 1809. Her son Alexander died in 1814. Her brother-in-law, Archibald Park, the senior Mungo's brother, died on Mull in May 1820. She was no doubt cheered by the visits north of her son, Adam Park, a military surgeon based in Gravesend. After a visit in the autumn of 1821, he confessed himself melancholy on leaving to return south, 'the country about Selkirk was so intimately associated with the earliest and happiest of my recollections'.[19] Within eighteen months, they would mourn the death in India of the young Mungo Park, the circumstances made worse by the news of the considerable debts he had left behind in London.

Like his father's, Thomas Park's death was a source of some uncertainty. Charles Napier wrote to Thomas Anderson on 31 March 1828: 'you will have seen the melancholy account in the newspapers in reference to our late friend young Park'. Napier had that day been to the Colonial Office to see what further news there might be, and offered Anderson his sincere condolences. By May, he had more news. He had a note 'from old Barrow' who by then had received Thomas Park's effects from west Africa. Barrow confirmed that Thomas's death was not the result of poisoning as had been reported but an accident. Thomas Park 'had sat up in the tree for several hours during a very hot sun, overlooking the natives at their sports, – had drank [sic] several drafts of palm wine & on coming down – <u>a branch gave way</u> which precipitated him from a considerable height and falling on the side of his head – he was observed to put his hands up to his temples at the immediate shock – & the stun bringing on a brain-fever he died in four days – being about 150 miles from the coast'. Barrow was told this by Pascoe, Clapperton's servant and companion, who had just arrived in London from west Africa with Clapperton's papers and news of his death.[20]

Mungo Park's literary afterlife outlasted the lives of several of his children. It drew chiefly upon his *Travels*, hardly at all upon his *Journal*. Park's words were used to give him an agency beyond his 1799 book's original intentions. In the different genres in which Park is made to perform, his travels and his sympathetic depiction of African life are central. His geographical accomplishments are not. In print, Park is made imaginatively different through others' interpretations of what and how he wrote, not for what he did or set out to do regarding the Niger. On stage, he has been differently played – a caricature in the early 1840s, in 2016 a commentator on racial inequality and slavery, and their continued presence in the world.

Memorialising Park

Park lives on in stone as well as on paper. Park's public memorialisation and commemoration began at a meeting in Selkirk town hall on Monday 20 December 1841. The impetus for Park's commemoration came from the townsfolk of Selkirk in a signed petition forwarded to the town's chief magistrate on 1 December 1841 (Figure 12.6). At the meeting, 'various resolutions were adopted with the view of erecting a Monument, or other permanent mark of respect, for the memory of this celebrated traveller'.

Despite the timing, there is no evidence to suggest that these resolutions and the sentiments motivating them reflected concern over news of the 1841 Niger expedition, or even the Niger's exploration at all. At issue was national honour and local pride: 'Too little has been done to connect the name of Mungo Park with the national honour of the Scottish people, and more particularly with the individual sympathies of those inhabiting the district in which he was born.' One newspaper account took the point further:

> As a traveller, as a writer, and as a man, he had the highest claims on his countrymen. From the days of Herodotus down to those of Park, Europe had learned almost nothing of the interior of the great African Continent; and all that has since been done has served merely to stimulate, not to satisfy curiosity . . . Mungo Park was the pioneer who carried the torch so far, and the results of his enterprise demand for him the honours of a philanthropist as well as those of a traveller. Above all, the uncertainty which overhangs his fate invests his memory with the deepest interest.[21]

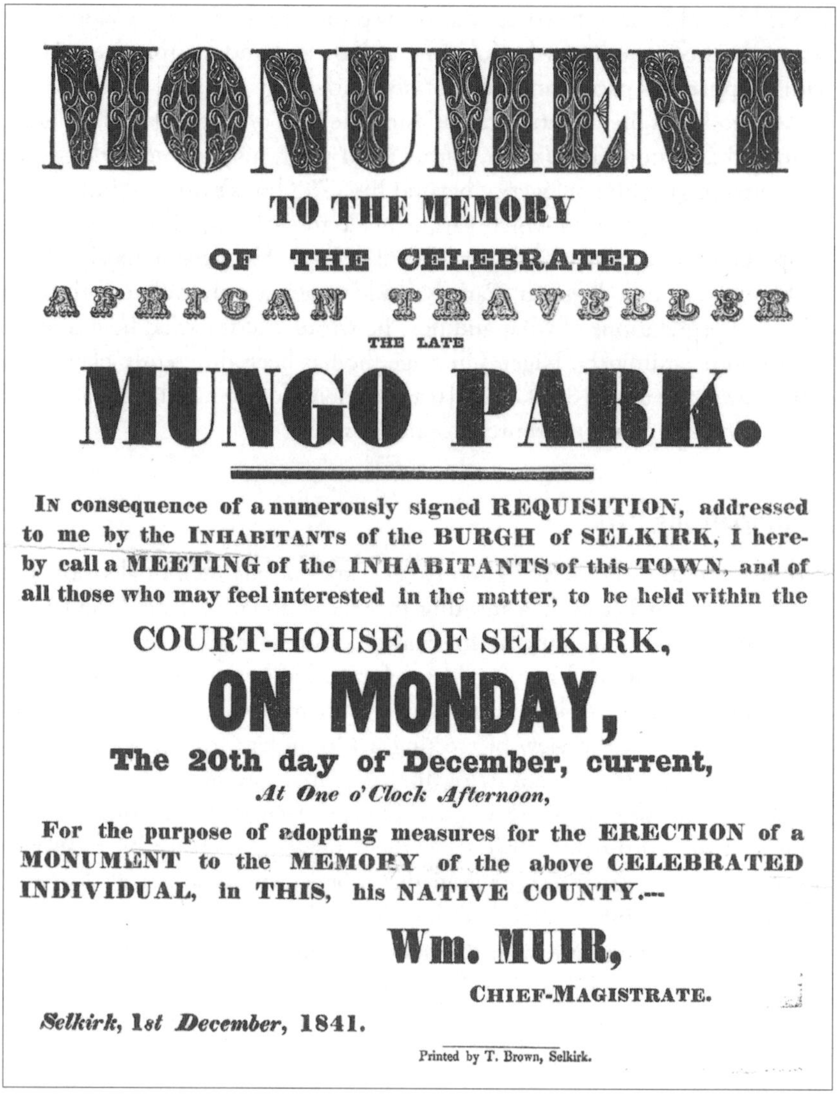

MONUMENT

TO THE MEMORY

OF THE CELEBRATED

AFRICAN TRAVELLER

THE LATE

MUNGO PARK.

IN consequence of a numerously signed REQUISITION, addressed to me by the INHABITANTS of the BURGH of SELKIRK, I hereby call a MEETING of the INHABITANTS of this TOWN, and of all those who may feel interested in the matter, to be held within the

COURT-HOUSE OF SELKIRK,

ON MONDAY,

The 20th day of December, current,

At One o'Clock Afternoon,

For the purpose of adopting measures for the ERECTION of a MONUMENT to the MEMORY of the above CELEBRATED INDIVIDUAL, in THIS, his NATIVE COUNTY.---

Wm. MUIR,

CHIEF-MAGISTRATE.

Selkirk, 1*st December,* 1841.

Printed by T. Brown, Selkirk.

Figure 12.6 A poster of 1841 announcing proposals for a meeting over the erection of a monument to 'the celebrated African Traveller, the late Mungo Park' [MS SC/S/16/12]. With permission from the Collections of Scottish Borders Council administered by Live Borders.

Reports of the 1841 meeting show that Park was considered worthy of commemoration not alone because he was a 'celebrated African Traveller', but because he was, variously, a Scot, a Selkirk man, a Christian, and a tragic figure. In this testimony of remembrance, Park's geographical undertakings – what he did and why – had no apparent place. Park is seen as a local worthy, even an early influence upon Africans' spiritual civilisation, not as someone who helped transform Europeans' geographical ignorance.

Park is not alone in this respect. The citizens of Truro in Cornwall made plans for a memorial statue to Richard Lander within months of his death in 1834. Reporting in May 1836 upon the statue's delayed progress – an earlier mishap had left the master mason bruised and the architect swinging from a poorly fixed ladder before he was rescued – one local paper noted that Lander was being commemorated as the man 'who discovered the source of the Niger'.[22] The historical inaccuracy is beside the point, as it was in Hooker's speech on Park and the moss. Mungo Park was commemorated in Selkirk for much the same reason that Richard Lander was memorialised in Truro – as a local man who died in Africa and whose efforts brought distinction to his townsfolk. The exact circumstances of his death on the Niger, and his purpose in being there, were of secondary importance.

Between December 1841 and June 1858 when a confirmed commission was announced for a statue to Park, to be undertaken by sculptor Andrew Currie in an 'Artistic and Tradesman-like manner', public talks were given about Park in Selkirk and other Borders towns. A memorial fund was established, news about it was carried in the local press, the work of the Park Memorial Committee was regularly reported upon, and local retailers helped collect funds: 'For the benefit of those who may not have had an opportunity of subscribing to the fund for the erection of the proposed structure in honour of the African traveller, we may mention that, in addition to the banks, subscription lists lie in the shops of Mr Dunn, druggist; Mr Turnbull, saddler; Mr Dalgleish, baker; Mr Millar, watchmaker; and Mr Lewis, bookseller.'[23]

Over 120 persons contributed: monies came in – a shilling here, a sixpence there – from seventy-four millworkers from the Ettrick mill in Selkirk, from a Member of Parliament, from Lord Napier who had helped organise the petition to Bathurst over Archibald Park, from local manufacturers, doctors, surgeons, and from several 'Gentlemen' who chose to remain anonymous. The largest single contribution, £50, was from the Duke of Buccleuch. Most donations were from Borders folk, notably from Selkirk and its surrounding area, but they came in too from individuals in Edinburgh, Glasgow, Haddington, Leith, Paisley, and Perth and, furth of Scotland, from Carlisle, Manchester, Newcastle, and Nottingham. The sum of £50 was received from Australia, the bulk from members of the Park family who had emigrated in the early 1850s. Public lectures in Selkirk, Hawick, Peebles, and Melrose brought in over £15.[24]

As the monies came in, plans were made for the completion and inauguration of the monument by March 1859. For one local, who wrote to the editor of the *Southern Reporter* simply as 'A Citizen', it was important that the inauguration go well, unlike the rather pale affair that marked the laying of the statue's base: 'Do not let us go about this matter as if one were ashamed of the great traveller – the monument to his memory – the artist, or ourselves. Let there be no more simple gatherings of "weans" and a few feeble cheers, as at the laying of the foundation stone, but let the welkin ring with the hearty plaudits of Foresters and Souters.' Similar concerns were raised a fortnight later as the inauguration drew near: 'In the event to be celebrated here on Wednesday next the inhabitants of Selkirk and the sons of "The Forest" have a peculiar interest. The distinguished traveller whose memory it is designed to perpetuate may fairly be claimed as a townsman, and one of themselves, and we trust that they will show their appreciation of the honour by making the demonstration as enthusiastic and effective as the occasion demands and deserves.'[25]

Neither correspondent need have worried. The statue to Park was inaugurated on 2 March 1859 'amid demonstrations of the utmost enthusiasm'. As one report of the event noted, 'on every countenance could be traced the desire, overflowing and earnest, to pay a tribute of well-merited honour and gratitude to the memory of the intrepid explorer of Africa'. In unveiling the statue to Park in Selkirk, Mr Johnstone of Alva, a major contributor to the Monument Committee, emphasised Park's meaning for locals: 'We are assembled this day for the purpose of inaugurating this statue to the memory of one of the most illustrious of the children of the Forest – (cheers) – who has cast lustre and honour upon our county, and one, therefore, to whom it is most fitting that we should repay a tribute of memory and of honour (Renewed cheering).' Park's commemoration in this way was fitting, Johnstone continued, because Park had been 'judged by posterity and found worthy . . . the world acknowledges Mungo Park not only as the first who trod that thorny path, but as the hardiest adventurer, and the loftiest spirit of them all'.

To those who doubted the purpose of such a memorial, Johnstone had two responses: 'Some may say or think . . . what sense is there in raising a monument to a man 50 years after he has perished? What good can this do to Mungo Park? . . . There are two objects in raising such records of public service, or great eminence – one, the tribute to the departed great man; the other, the encouragement for great deeds to living lofty souls and rising living talent.'

Impressed as most locals no doubt were with Park's monument and, perhaps, with Johnstone's rhetoric, not everyone felt likewise. One local woman noted, 'I was at the 'nagaration, and sic a crood o' folk I never saw in Selkreik afore. I was 'maiss crushed to death. But what was a' the wark aboot? When they lifted up the claith, the fient a thing could I see to raise sic a steer aboot. Naething but a *stane man*.'[26]

The 'stane man' erected in March 1859 was not what is there today. Two bas relief panels depicting African scenes had been intended in the original design, but the funds were insufficient. The panels were added to the statue years later, in a public ceremony on 10 December 1906. One depicts Park on his ailing horse being assisted in his travels, the other has a prone and bedraggled Park being lent succour by African women (Figure 12.7). However else they may be thought of, this is not the iconography of exploration as triumphalism.

On the evening of the ceremony in 1906, the Africanist and colonial administrator Sir Harry H. Johnston – who would later devote three chapters to Park in his 1912 *Pioneers in West Africa* – gave an address to mark the panels' unveiling. Park was 'the greatest hero of Scottish exploration in the eighteenth century' Johnston noted, a view that would have gone down well with his local listeners. Johnston otherwise taxed his audience's patience and good manners: unnecessarily long, his address became a potted history of the Niger's exploration, a summary of Park's life, and a resumé of Africa's exploration by Scots – and by himself (Johnston was not known for his modesty). As he told the Royal Scottish Geographical Society to whom he lectured that winter on west Africa's exploration, 'I only wished at the time to interest the people of Selkirk in the memory of their great townsman.'[27] They and others had been interested since at least 1841.

In 1911, moves were made to commemorate Mungo Park and Richard Lander jointly in a single monument. Harry Johnston was again involved, the proposal coming from men with whom he was closely acquainted: Lord Curzon, president of the Royal Geographical Society, Sir George Goldie of the African Society, and Lord Scarborough, chairman of the Niger Company. The initial intention was to commemorate the two with a lighthouse sited at the entrance to the Forcados river, a tributary of the Niger, but this plan was abandoned following guidance from Sir Hugh Clifford, governor of Nigeria. Clifford convinced Curzon and the others that the memorial should be built at Jebba, an island in the Niger about eighty miles south of Bussa: it was 'more appropriate to place the memorial near the spot where Mungo Park's work ended and Richard Lander's began'.[28]

Funds slowly began to accumulate, but not everyone was keen. In Scotland, campaigning by the Royal Scottish Geographical Society met

Figure 12.7 The bronze panels on Mungo Park's statue in Selkirk. To the top, Park and his ailing horse are shown being guided by African men. To the foot, on the opposing panel, Park receives succour from African women, as other women and children look on. Photo courtesy of A. Withers.

with a mixed response: Ayr's town council 'did not think it expedient that it should subscribe'; Glasgow's officials were sympathetic but non-committal; Edinburgh's Lord Provost apologised, explaining that the council was taken up with fund raising for the city's memorial to King Edward VII. Funding dried up following the outbreak of the First World War in August 1914 but was renewed following an appeal in 1925 made on behalf of Baroness Conrad-Philips, Mungo Park's great-granddaughter. The monument was completed in September 1929 (although funds were still being sought towards its cost by 1933).[29]

The monument to Park and Lander is impressive, if rather plain and lighthouse-like (Figure 12.8). A commemorative plaque on the obelisk reads 'To Mungo Park 1795 and Richard Lander 1830 who traced the course of the Niger from near its source to the sea. Both died in Africa for Africa.' Other proposals, never realised, intended something different (Plate 14).

Figure 12.8 The Park–Lander memorial on Jebba Island on the Niger. Neither the photographer nor the date of the photograph is known [rgs026141]. © Royal Geographical Society (with IBG)

Figure 12.9 The stone memorial plaque to Mungo Park on the remains of the Park family cottage at Foulshiels. The plaque was placed on the cottage by the Selkirk Antiquarian Society in 1886. Photo courtesy of A. Withers.

If not themselves directly political objects, monuments and statues are easily politicised. Park's statue in Selkirk, like Lander's in Truro, speaks principally to local pride in a fellow townsman. In its origin and situation, the statue to them in Jebba reflects British imperial pride in pioneering geographical achievements – the alpha and omega of a 2,000-year-old geographical problem. It is today promoted as a tourist site, informing locals and visitors alike about the man who solved the first Niger problem and the man who, with his brother, confirmed the second through in-the-field encounter. *In* Africa certainly, less *for* Africa perhaps.

Park's memorialisation was also domestic and destructive, in intention at least. In February 1970, plans were made to commemorate the approaching bicentenary of Park's birth. A Mungo Park Memorial Committee was established to coordinate activities. Its chair was Professor Harold Shepperson, Professor of Commonwealth and American History at the University of Edinburgh. The Park family cottage at Foulshiels, Park's birthplace, had long been maintained by the Selkirk Antiquarian Society, but thoughts now

Plate 8. John Barrow (1764–1848). As Second Secretary to the British Admiralty from 1804 until 1845, Barrow was hugely influential in directing British overseas exploration, not just to the Niger, as he did between 1815 and 1830. Reproduced by courtesy of the National Portrait Gallery, London.

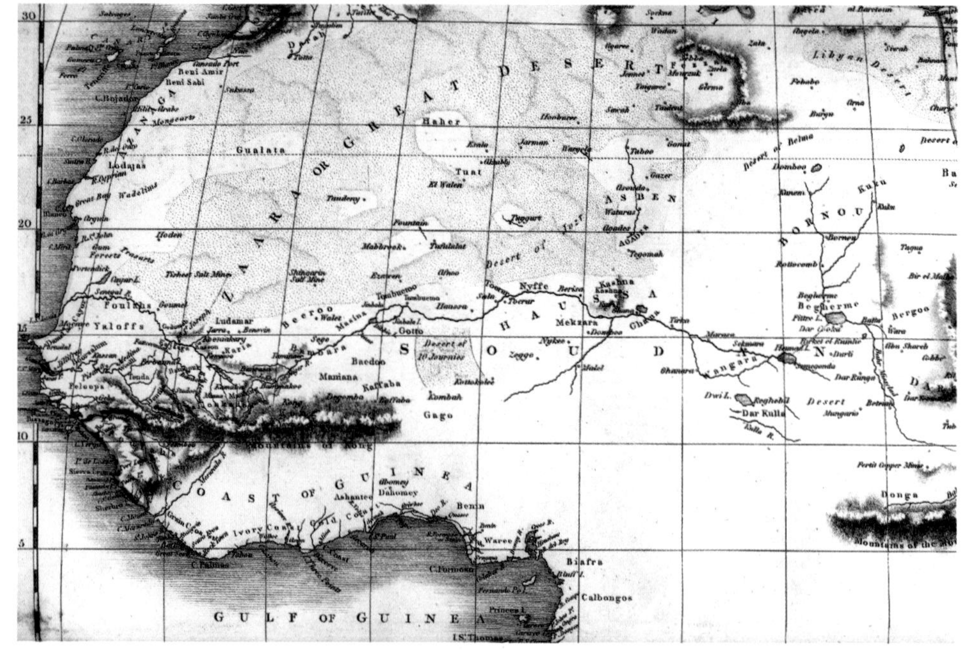

Plate 9. James Playfair would appear to have based his map of north and west Africa on that of James Rennell (1798) or, perhaps, Aaron Arrowsmith (1802) (*cf.* Figures 2.4 and 1.3). Source: J. Playfair, *A New General Atlas: Ancient and Modern* (Edinburgh, 1822). Author's copy.

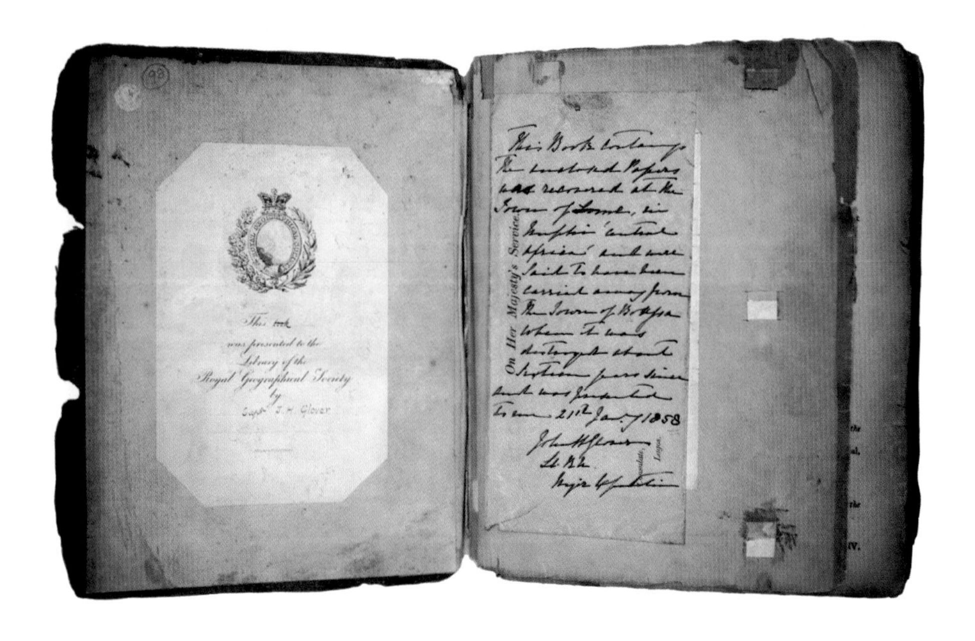

Plate 10. Mungo Park's book of logarithms was saved from the wreckage of 'HMS Joliba' when Park drowned and was brought back to Britain in 1857 by Lt. J. Glover, then of the *Dayspring* expedition. Park used the tables of figures to calculate his position from his longitudinal readings and by using trigonometry (*cf.* Figure 4.4). [rgs800409]

Plate 11. The ships of the 1841–2 Niger Expedition as they assembled off Holyhead prior to departure.
© National Maritime Museum, Greenwich, London.

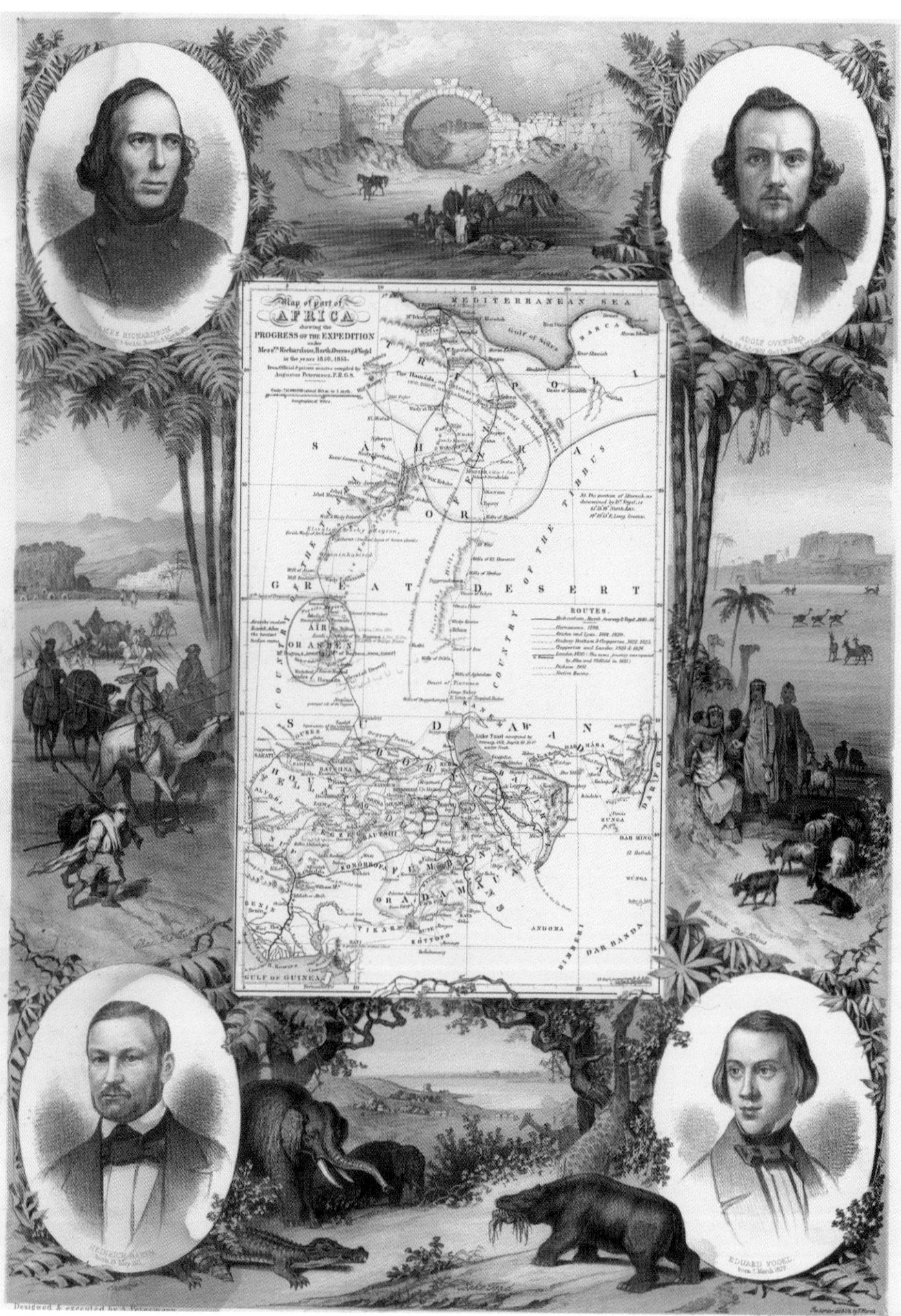

Plate 12. The principals of the Central African Expedition: James Richardson (top left), Heinrich Barth (bottom left), Adolph Overweg (top right) and Eduard Vogel (bottom right). The map, compiled over several years by Petermann, depicts the routes of several Niger expeditions and travellers, from Hornemann to Clapperton and Lander. Source: A. Petermann, *An Account of the Progress of the Expedition to Central Africa* (London, 1854). Reproduced by kind permission of the National Library of Scotland.

Plate 13. The statue to Mungo Park in Selkirk. In his right hand, Park holds a sextant. In his left, he is holding a scroll of paper marked with the words – in the end, prophetic – of his last letter to Lord Camden: 'if I could not succeed in the object of my journey, I would at last die on the Niger.' Photo courtesy of A. Withers.

Plate 14. One of several designs proposed for the Park–Lander memorial statue. The proposal, forwarded in 1928, was by the architect H. Guy Holt. [rgs024638] © Royal Geographical Society (with IBG).

Plate 15. The Park family cottage at Foulshiels today. Photo courtesy of A. Withers.

Plate 16. In these Gambian stamps, issued in 1971 to mark the bicentenary of Mungo Park's birth, Park is shown variously: in his native Scotland; travelling by dugout canoe; and jumping to his death from 'HMS Joliba' as his second expedition comes to an end at 'Busa Rapids'. Author's collection.

Plate 17. Mungo Park, in a portrait by Scottish artist Atkinson Horsburgh (*fl.* 1883–95). With permission from the Collections of Scottish Borders Council administered by Live Borders.

Figure 12.10 Mungo Park commemorated. To the top, Park is shown on board the twin-hulled canoe 'HMS Joliba' on the 20 dalasis coin from the Gambia. To the foot, the Mungo Park Medal. The medal is awarded by the Royal Scottish Geographical Society. The Gambian coin has the correct year for the date of Park's death: the Society's medal does not. Photo courtesy of A. Withers.

turned to making it a Park memorial. However, moves in this direction were hindered by the landowner, the Duke of Buccleuch. As Shepperson informed his committee, 'it looks as though the Estates and Lord Dalkeith are agreed that the best idea is to get rid of the cottage and to "mark the spot"'. Other views were entertained: 'that the form of the cottage should be retained . . . even if it was impossible to get a full reconstruction'. Public pressure proved persuasive. In September 1970, the cottage walls were restored, an earlier memorial plaque repositioned in the cottage's outer wall (Figure 12.9 and Plate 15). Public pressure prevailed but only partly because this was a monument of sorts to Park. Plans to retain the cottage benefitted from an emerging public awareness of Scotland's built heritage and what would be lost by its thoughtless destruction.[30]

Commemorative events in 1971 included public lectures, and an exhibition. Plans for a commemorative postage stamp never came to fruition although Gambia issued Park stamps in 1971 (Plate 16) and has Park on one denomination in its coinage (Figure 12.10). There was even the possibility of a 'Park Pageant' on the Yarrow Water, the river that runs through Selkirk, but this proved too much for the Memorial Committee: 'It costs a lot to produce, has to be done really well and is dependent on weather . . . Mary, Queen of Scots, escaping from Loch Leven Castle is one thing but trying to convert the Yarrow into the Niger is another.' In place of the pageant, Shepperson gave some lectures, schools had special teaching materials prepared on Park and on African exploration, and showings were held of the film *In the Steps of Mungo Park*. Park's memorialisation shows no signs of abating: in 2001, the part of Mungo Park was played by a costumed local – others dressed as Sir Walter Scott and other local worthies – as part of street pageants in Selkirk designed to boost Borders tourism.

Park the explorer

Park achieved distinction through his 1799 book and the events he reported upon there, but he became an explorer fully only in his afterlife. Park's inclusion in Samuel Mossman's 1868 *Heroes of Discovery* is illustrative. Like many such accounts, it offers a particular interpretation of the explorer as a Victorian archetype: someone who overcomes adversity in encountering the unknown, who measures it and renders it knowable in map and other printed form. As a plant hunter in Australia, New Zealand, and the Far East, Mossman was one such himself.

So too was Joseph Thomson. Joseph Thomson had achieved fame in 1879, aged just twenty-one, when he took over the Royal Geographical Society's East African expedition under Keith Johnston (son of the Edinburgh-based map-maker who had employed Petermann) following Johnston's death in Africa. Thomson led a further East African expedition in 1883 and explored the lower Niger in 1885 – but never completed his book on it because he dropped all his expedition notes in the river. He did, however, turn to biographical work upon Park and the Niger.

Thomson's *Mungo Park and the Niger* was published in 1890 in a series entitled 'The World's Great Explorers and Explorations'. The editors were John Scott Keltie, librarian at the Royal Geographical Society, Halford J. Mackinder, Reader in Geography at Oxford, and Ernst Georg Ravenstein, a London-based German geographer, demographer, and cartographer with interests in Africa. When complete – volumes were planned or in press on Marco Polo, Magellan, Barents, Cook, Franklin, Livingstone, and Humboldt amongst others – the series was intended to form a 'Biographical History of Geographical Discovery', and to emphasise 'the life and work of those heroic adventurers through whose exertions the face of the earth has been made known to humanity'.[31] Thomson's account of Park is thoughtful, faithful, scholarly. Park is not depicted as an agent of imperial and racist superiority but as a somewhat reluctant hero: the more heroic, of course, for not having returned from his second journey.

If any one African explorer enjoyed a reputation as a figure of imperial 'derring-do' and for his style of exploration (rather brutal, in truth), it was Henry Morton Stanley, 'discoverer' of Livingstone.[32] Asked to give the inaugural address upon the establishment of the Scottish Geographical Society in 1884 (it was Royal from 1887), Stanley used the occasion to praise Park – and to record the deadly nature of the Niger's exploration:

> It is nearly ninety years ago since Mungo Park left the English factory of Pisania on the Gambia for the Niger. Ten years later he was drowned, while descending the Niger, in the rapids of Bussa, 500 miles from its mouth. Decade after decade witnessed the victims which the river demanded as the price of the knowledge of its course, until Richard Lander, with his brother John, floated from Bussa, down the Niger, to the sea. The geographical knowledge thus dearly bought led the way to the commercial enterprise of Macgregor Laird, in 1832. Scheme after scheme followed, ending in disaster, owing to the general ignorance of how to live under those new climatic conditions, until now the National African Company has established itself on a solid basis, and the enterprise is rewarded by a remunerative profit.[33]

Other geographers thought the same. A founder member of the Scottish Geographical Society and its then editor and secretary, Arthur Silva White used his report to the International Geographical Congress in Paris in 1889 to promote the infant society and laud the achievements of Scotsmen as explorer-geographers. Park to the fore, the Niger's exploration was cast as a distinctly Scottish affair: Park, Clapperton, Oudney, MacGregor Laird, MacQueen, Baikie, Thomson. When the English politician and eugenicist Leonard Darwin, son of Charles Darwin, addressed the Society in 1911, he reported upon the Park–Lander monument proposed for the mouth of the Niger (Figure 12.8) and considered Park 'the most prominent figure amongst the eighteenth-century explorers'.[34]

Others, not geographers, similarly placed Park in the pantheon of great explorers. W. H. Hewitt's *Mungo Park* (1923) is a volume in the 'Pioneers of Progress-Empire Builders' series issued by Sheldon Press. Geographical exploration is taken to be synonymous with Britain's imperial expansion – Park the equivalent of Sir Francis Drake and James Cook. Sir George Scott's book of the same title appeared in a series entitled 'Peeps at Great Explorers' where Park, again in the company of Cook, sits alongside Alexander the Great, Marco Polo, Columbus, Vasco da Gama, and David Livingstone.[35]

Park's death has commonly been taken as 1805 despite the evidence of Amadi Fatouma which points to it being in March or April 1806: 1805 is the date on his statue, and on the Royal Scottish Geographical Society's Mungo Park Medal, first awarded in 1930, for an 'outstanding contribution to geographical knowledge through exploration or adventure in potentially hazardous physical or social environments'. The later date appears on the plaque to Park that marks his house in Peebles (Figure 12.11).

Taking 1805 as the year of Park's death, Selkirk's civic authorities marked Park's 150th anniversary in November 1955 in several ways. Park's statue was floodlit for a week. A united church service was held at the Lawson Memorial Church with an address by the Rev. J. M. Hannah of St John's Episcopal Church. Afterwards, a wreath-laying ceremony was held at the monument. On the evening of 10 November, a talk on Park was given to the Selkirk Antiquarian Society by Ronald Miller, Professor of Geography at Glasgow University. Miller summarised Park's life and, rehearsing the facts of Park's death, concluded that Park's name and fame were inspirational, 'a testimony to the indomitable spirit of man for time to come'.

Sixteen years later, Miller seemed less sure about the need for remembrance. To Miller in 1971, perhaps the most remarkable element of Mungo Park's commemoration and memorialisation was that it happened at all: 'It is astonishing that Park's reputation should have survived the debacle of 1805, for in territorial terms his discoveries of 1795/6 were slender indeed . . . And yet his "Travels" and his likeness are still with us.' For Miller, Park's memorialisation was rooted in an interest in heroic failure: 'Undoubtedly, the world

Figure 12.11 The memorial plaque to Mungo Park on the house in which he lived while a doctor in Peebles. Photo courtesy of A. Withers

loves a loser: "Antarctic" recalls the loser, Scott . . . Would Livingstone be so well known but for the place and manner of his death?'

Shepperson took much the same view: 'However one interprets Park's second expedition to West Africa, his reputation seems to rest on his first journey. It is all very similar to David Livingstone half a century later. What comes later is part of the so-called "tragedy" of both men . . . the dramatic death in Africa which, for Park, as for Livingstone, enhanced his reputation, by creating something of the legend of martyrdom.'

This view has been articulated more recently by those who see heroic failure in exploration and its memorialisation as peculiarly British: 'Mungo Park served as a prototype for British heroic failure in Africa: the lone explorer bravely daring the unmapped territory of the interior and ultimately losing his life in the attempt.' Clapperton's death was 'Mungo Park all over again . . . The exploration of central Africa did not really matter: failure was far more compelling than success.'[36]

Park's commemorative trajectory is diverse and enduring: a Selkirk man, a distinguished Scot, a Christian and an anti-slaver (perhaps more so in death than in life), a heroic explorer, a pioneering African traveller. Park was all these things. He was also medically trained, a competent artist, a loving if somewhat absent husband, a dutiful son, an anxious and ambitious traveller, and a successful author. It is inappropriate to see him as any one thing. It is equally inappropriate to interpret his success on his first journey or his failure on his second in territorial terms: that was not his aim. What principally motivated him and those who sent him was European ignorance – about Africa, the Niger, its direction of flow, its route, and its end point. Questions about African society and culture and the possibility of trade were of secondary importance. They became more important after Park's death, at much the same time that Park's afterlife began.

Park's afterlife and its different expression – in print, on stage, in statuary, on stamps, and on coins – does not mean we lose sight of the real Park. Simply, different contemporaries saw him differently. Exploration, explorer, and author are not fixed labels.

13

'Cool, impassioned, cowardly, courageous'

Afterlife – Park's biographers

What we know of Mungo Park, of his achievements and their significance, comes from his letters, books, and biographers. There is no autobiography – *I, Mungo Park* is an adaptation for younger readers of Park's 1799 *Travels*, the book that stands as Park's own words on his African journey influenced by Bryan Edwards and James Rennell though we know it to be. Park's 1815 *Journal of a Mission* is the posthumous record of his second expedition and Park's first biographical memoir, containing as it does John Whishaw's 'Account of the Life of Mungo Park' with its several appendices. Other biographies – in 1835, 1898, with two in 1934 preceding Lupton's 1979 account and Duffill's in 1999 – confirm Park as an explorer of importance.[1]

Biographical writing is a changing art. Nineteenth-century biography is imbued with a kind of moral passion, its fundamental motive being to describe an admirable role model – admirable because crowned with achievement (usually) and with noble failure if not. More recent practice in telling lives is disciplined yet subjective, balancing scholarly attention to source materials and so to authoritativeness with imaginative creativity. Biography as itself the subject of study – 'metabiography' – is thus instructive in revealing how a subject's life has been written about and in highlighting the purpose of different biographers.[2]

Understanding what Park did, with what result, even knowing what sort of man Park was will always reflect different purposes and the demands of given audiences. Nigerian authors for example see Park differently. In Nwelme and Okeh's *Saving Mungo Park*, Park is not the Niger's discoverer, but a strange white man rescued from the river by locals. Written for African children, the book plays with Park's 1799 narrative to raise questions over Park's purpose and about exploration as the first step to colonialism. In *Formations*, a popular history of Nigeria, Park is first among equals of those 'mad men and missionaries' who helped explore the Niger and shape that nation.[3]

Recognising that Park's life, his geographical achievements, and his uncertain death have been interpreted differently is to understand that geographical exploration – the events and personnel themselves – and exploration history – what we say about those events and personnel – have significance beyond the facts themselves. Mungo Park merits the biographical scrutiny afforded him for what he did and for his continuing afterlife. The halting solution to the 2,000-year-old Niger problem – rooted as it was in an ignorance which in Britain and Europe occupied politicians, traders, humanitarians, and the geographically minded for over half a century – began with a man many considered fitted and fated for the task.

Explaining Park the man

John Ledyard was, Henry Beaufoy being impressed by his manliness and 'the inquietude of his eye'. Daniel Houghton was, Beaufoy noting a 'natural intrepidity of character and constitutional good humour'. Friedrich Hornemann was, being reportedly a 'robust man of an athletic bodily constitution'. So too was Mungo Park – physically constituted, that is, for the rigours of exploration.

Park's early biographers certainly saw Park thus. For Whishaw, Park was 'perfectly well proportioned', his 'frame active and robust, fitted for great exertions and the endurance of great hardships'. The anonymous 'H.B.' in 1835 took issue with Whishaw's account of Park, seeing it 'in many respects deficient' because it was a memoir and not a fuller 'Life of Park' informed by later circumstances regarding the Niger's exploration and additional personal information – like Whishaw, 'H.B.' was able to incorporate information gleaned from Park's relatives in his biography. Yet he too thought that Park was similarly fitted for exploration: 'Mungo Park was admirably qualified, by the constitution both of his mind and of his body, for that life of enterprise to which, fortunately for his own fame and for the cause of science, he early attached himself. His person was tall and muscular, and united great bodily strength with an extraordinary capacity for enduring fatigue.' These physical capacities were matched by his mental and moral qualities: 'With regard to the nobler qualifications of a traveller, intelligence, mental energy, accurate observation, fidelity, sagacity, and prudence, no one has ever surpassed, and few have equalled Mungo Park.' Maclachlan in 1898 was of the same view, seeing Park as 'tall and broad-chested, muscular, firmly knit, and of excessively hardy constitution'.[4]

Both Park biographies published in 1934 interpret his capacity for exploration as an almost racial matter – not in terms of a presumed and innate white European superiority but from Park being Borders 'stock', shaped by his physical environment and by his forebears. Recognising, as do I, that 'our concern must be with what the man did rather than with what he was', Stephen Gwynn nevertheless considered Park an example of the Borders type: 'His fitness for the task lay really in the fact that he came from one of the hardiest stocks in the world.'

To Lewis Grassic Gibbon – who had little time for Maclachlan's 1898 biography but thought better of 'H.B.'s' 1835 account, 'sincere and serene in the conventions of camouflage peculiar to its age' as it was – Park, like his father, 'came from the soil, up out of the darkness with that horde of men of like kind, the dour even-tempered peasants of Scotland'. Mungo Park the Niger's explorer was wrought from a hard life in a rural Scotland long since vanished:

> For the matrix that produced Mungo Park, the now ruined whinstone house, mantled with ivy and set with a neat little tablet [see Figure 12.9], is unconvincing enough unless one remembers the quality of his father and mother, and the life they led. To cram a family of thirteen, not to mention an altogether improbable maidservant, inside the compass of those narrow walls, and imagine it a winter night with the elder Mungo coming laired from the fields, the cruse-lamp with a murky splutter on the earthen floor and the plain walls, the scrubbed table and the shining dresser, is to reach a type of life now remote from anything in Scotland or outside the huts of the Esquimaux.[5]

These views of Park, like all views of Park, are products of their time. Gwynn's 'Borders type' reflects ideas, still current in the 1930s, about 'inferiorist' racial and ethnic types, a form of physical anthropology that drew connections between skull shape, skin colour, and intellectual capacity. In Britain, they were directed at the rural and Celtic populations especially. Gibbons' is of a piece with his novels about rural Scotland, with the steadfastness and iron-tender morality of its farming folk that he depicted in his *Sunset Song*, the first volume of his *Scots Quair* trilogy written at the same time as his Park biography.[6]

Such pseudo-racial and environmentally determinist ideas, never universally held even when prevalent, are now discredited. Yet Sanche de Gramont subscribed to a variant in 1975 when explaining not just Park but also the place in the Niger's exploration of Hugh Clapperton, Walter Oudney, Alexander Gordon Laing, and William Balfour Baikie. Late eighteenth-century Scotland, he explains, was 'a land of proverbial poverty, where the peasants were as thin-ribbed as their cattle and life as pinched and grim as the weather . . . The explorers' requirements, physical endurance and spiritual forbearance, were the Scotsman's only birthright . . . Thus, a land of proverbial poverty and periodic calamity . . . provided more famous African explorers than England itself, precisely because its barren soil and migratory habits were ideal for shaping adventurous and determined men.'[7]

Park may have been muscular, strong, and capable of enduring fatigue. He is hardly likely to have been deemed suitable by Banks had he not been the 'right stuff'. In explaining him as an explorer, Park's nineteenth-century biographers emphasise his physical qualities as a particular sort of man – constitutionally suited, of the Borders, of farming folk – long after the events themselves. He is an explorer post hoc. That is the image captured in Horsburgh's portrait of him painted sometime in the early 1880s (Plate 17). With the subject not available to sit for him, Horsburgh probably used Edridge's frontispiece portrait as his model (*cf.* Plate 7): even allowing for the greatcoat, the impression is of a powerful physical presence.

Park is here explained and portrayed as an explorer – was almost predestined to be an explorer – because of his own bodily strength and from the poverty and nature of his home environment. Both allowed him to overcome his experiences in Africa on his first journey. Yet his background was hardly one of poverty. It is a biographical view which, if widely held in the nineteenth century, contrasts sharply with Park's place as a student in Enlightenment Edinburgh, with his description of himself in Africa, and with later accusations about his conduct toward Africans on his second expedition.

Robert Chambers included Park in his 1835 *Biographical Dictionary of Eminent Scotsmen* for his achievements, not his attributes. Art critic and social commentator John Ruskin paid Park's physicality no mind but took issue with his moral judgement. Park's declaration that 'he would rather brave Africa and all its horrors rather than wear out his life in long and toilsome rides over the hills of Scotland' as a doctor was seen by Ruskin as an absence of the instinct of personal duty, disregard for his family, and a want of interest in his profession. For Ruskin, Park would have done more good as a doctor 'among Fair Scotch laddies in a day . . . than among black Hamites in a lifetime'.[8]

In truth, Park was moulded much more by his social world than by any physical capacities. Park's modern biographers emphasise the importance of his schooling and education. Whishaw is the only one of his nineteenth-century biographers to do so, praising the virtues of Scotland's parochial schools in one appendix to his 1815 'Account' to support his view. Mungo was one of three sons supported in their education – a fact several biographers explain as the father's far-sightedness and commitment to learning. That may be so. It might also be a practical consequence of inheritance: as the seventh child and third son Mungo Park would not have inherited the farm. Even if he had wanted to farm – and neither farming nor the ministry appealed – Foulshiels was leased from the Duke of Buccleuch who would have had the final say over its tenancy.

Park's education, formal and informal, did not fit him for African exploration any more than did his religious beliefs or his moral values, but these were the bases to that world of intellectual enquiry and social patronage which did. Selkirk was more important for his apprenticeship under Thomas Anderson, his friendship with Alexander Anderson, and in meeting his future wife than it was for his schooling. Late Enlightenment Edinburgh introduced Park to the power of the mind, to critical thinking, to formal instruction in a range of subjects, and to altogether broader intellectual and social horizons.

Where early biographers depict Park as reserved, a diligent if somewhat introvert student, Duffill's student Park is altogether livelier: 'a young man with an active social life, who mixed with the aspiring young *literati* of the city, participated in debates at The Pantheon, a famous Edinburgh debating society, attended the theatre, made fishing expeditions with friends and wrote poetry'.[9] Botanical field trips in the Pentlands with John Walker, and in the Highlands with James Dickson, notebook and sketch pad to hand, doubtless demonstrated the connections between observation and accuracy in the field, and between on-the-spot writing and its later refinement. But Park became an African traveller through opportunism and others' patronage, and did so in London, not from what he was taught from pulpits and classrooms in Selkirk or learned in lecture theatres and medical societies in Edinburgh.

Mungo Park was first put on the road to Africa, in May 1794, by Joseph Banks and Henry Beaufoy and then because others had failed in the task he undertook to accept. Ledyard and Houghton died in Africa. Lucas got no further than Tripoli. Willis spent the money but never took ship. Park's Sumatran voyage may have whetted his appetite for travel, and it helped convince Banks that he had found the African Association's next suitable Niger man.

Park returned to Africa in 1805 because the opportunity again presented itself and competing Niger theories needed to be tested, because he presented Banks and Lord Camden with a credible scheme for the Niger's exploration, because he was fretful and bored as a doctor in Borders Scotland, and, not least, because he had failed to complete what had been asked of him a decade earlier.

Understanding Park's achievements

Mungo Park did not discover the Niger. Neither did Daniel Houghton who saw it flowing eastwards years before Park did. The Niger had already been discovered.

The 2,000-year-old Niger problem and its attempted resolution by Park and by those who preceded and followed him was a problem rooted in European ignorance. Those Africans who lived on or near the Niger's banks knew the river as a source of livelihood and a route to longer-distance trade. For different parts of its 2,600-mile course, the river was known by these peoples by different names. True, no one, amongst those who worked on the river and managed their lives according to its seasonal rhythms had a comprehensive understanding of the Niger's geography from origin to termination. They had no need. 'Discovery' is appropriate only if we wish to privilege European perspectives and to ignore indigenous knowledge, and to dismiss the agency of locals in shaping what explorers did.

Mungo Park did not discover the Niger but he certainly explored the Niger. He was the first European to do so and live. His significance, then and now, rests in his providing the first detailed account of west Africa and its peoples, their customs and moral geography, and something vital about the Niger's course. He established by direct observation the eastwards direction of flow of the Niger. Park is important for being both explorer and author – a lone white traveller, a first-hand credible witness, a sympathetic writer, the man who settled the first part of the Niger problem and who survived to tell others.

Park achieved fame for his success in 1796 and for returning to Africa in 1805 to make good his failure, from which second trip he did not return. A distinction should be made then between the Park who explored west Africa between 1795 and 1797 accompanied by only a few others and who returned to produce *Travels in the Interior Districts of Africa*, and the Park who explored west Africa in 1805 and early 1806 as the head of a large expeditionary party, whose *Journal of a Mission* is a partial account, and who died there in mysterious circumstances.

Park's achievements on his first journey were significant despite the circumstances of their undertaking. He was often ill. His instruments did not always work properly. Park provided descriptions of African society and of many places and their location despite his bodily frailty and despite his compass being shattered and his sextant being faulty and then stolen – one reason why James Rennell felt it necessary to adjust the positions of some places as he prepared Park's maps and his own *Geographical Illustrations*. We should recall too as Rennell put it, that 'a great part of Mr. Park's geographical memorandums are totally lost'. What came back was nevertheless sufficient for Rennell to produce a map of Park's travels. This map revised Rennell's work with Beaufoy, replaced the earlier work of d'Anville (overly reliant upon the ancients as Rennell held it to be) – and, in Rennell's insistence upon the Mountains of Kong, misled future Niger explorers for decades. These mountains remained an errant feature on western maps of Africa until the early twentieth century – a continued reproach regarding European uncertainty about the continent.

Park at first listened to and learnt from locals. It is unfair, perhaps, to say that he benefitted from being ill in Pisania at the start of his first journey, but he certainly used the months of his illness wisely: learning Mandingo, sketching, making notes on economic botany, listening to the conversations of *slatees*. Most of his biographers note that Park was in some measure prepared beforehand for the fact of the Niger flowing eastward because of what he had garnered on the subject after he left the coast. Yet none comment on Park's confession that he was confused on the matter from the very start of his travels despite the guidance of Rennell and of Beaufoy, and that he even believed that its course was the other way.

Park's achievements include admitting to his own uncertainty. They also include a willingness to accept the halting and interrupted nature of his journeying and to acknowledge that he owed his survival to the assistance afforded him by others – local women on more than one occasion, and to slave trader Karfa Taura for some weeks on his return journey.

Mungo Park achieved status as a pioneering African explorer – possibly the greatest of them all to judge from nineteenth-century commentators and biographers – because his findings were of value to the African Association and because of that body's view of its mission. Becoming an explorer depended upon his becoming an author. Becoming an author required that Park accept the assistance of others: the patronage of Banks and the close attentions of Bryan Edwards over his words and of James Rennell on his maps. That Park's book was not wholly his own does not diminish the substance of its content – although we must recognise, as several of his biographers do not, that Park confessed to leaving things out so as not to test the credulity of his readers. Park was not the first to have his words worked with by others and in the process find himself a feted author. In the context of the Niger's exploration, he was certainly not the last.

Park's second expedition sought to address the shortcomings of his first: not visiting Timbuktu as instructed or identifying the possibilities for furthering trade with the west African kingdoms, and not solving the termination of the Niger. As a contribution to a fuller geographical understanding of the region and its peoples, the expedition was unproductive and quite unlike his first journey: its limited geographical value lay, as Whishaw notes, in correcting Park's earlier astronomical observations and so moving the line of his former route further north.

We know about Park's second expedition because his notes from it survived, thanks to Isaaco and to Park's foresight, to become his posthumous

Journal of a Mission. As Whishaw also notes, allowance must be made for the fact that Park's 1815 *Journal* is incomplete and is, largely, in note form: 'A work, thus imperfect, and which the unfortunate fate of its author has prevented from being brought to a completion, is entitled to peculiar indulgence.' To Whishaw, the book's very incompleteness and its terse prose could bring the reader closer to knowing what exploration was really like: 'the very nakedness and simplicity of its descriptions and its minute details of petty circumstances, may be thought by some readers to convey a more accurate and distinct conception of the process of an African journey, and of the difficulties with which such expeditions are attended, than a more elaborate and polished narrative'.[10]

Whishaw's remarks serve to emphasise the importance of book history to exploration history – of knowing about the events and the persons behind the sometimes elaborate and often polished narratives of exploration. Park's achievements on his second expedition did not include adding significantly to the geography of the Niger. Following conversations in Scotland with George Maxwell – and no doubt more disagreeably with Rennell in Brighton and with Banks in London both of whom thought otherwise – Park set out again for Africa in the belief that the Niger and the Congo were one. He set off down the Niger in September 1805 unsure, but, as he put it in his last letter to Camden, 'I am more and more inclined to think that it can end nowhere but in the sea.'

The Park that emerges from his *Journal* is a rather uncertain, even contradictory, figure. Lewis Grassic Gibbon certainly thought so in a further work of 1934: 'Cool, impassioned, cowardly, courageous, imperturbable, Mungo Park's character in analysis after a hundred and fifty years disintegrates into fragments seemingly irreconcilable enough.'[11] Park's later achievements, hardly at all geographical, were his determination and his commitment to his companions: taking notes and longitudinal measurements as his expedition fell apart, offering succour to dying men, laying bare and for the second time his anguish and loneliness in Africa upon the death of Alexander Anderson. Park's *Journal* portrays the difficulties of exploration – several of them self-inflicted. We are shown a Park who, it seems, was prepared to travel down the Niger without stopping to pay the necessary tribute to local chiefs and who, with John Martyn, opened fire upon local people as they tracked 'HMS Joliba' downstream. Park's determination to succeed where he had earlier failed is admirable, but it would be the death of him: 'but though all the Europeans who are with me should die, and though I were myself half dead, I would still persevere; and if I could not succeed in the object of my journey, I would at last die on the Niger'.[12]

Park's death: questions of credibility

Mungo Park's death was not known for certain until, on 1 September 1811, Isaaco passed his report to Colonel Maxwell. Five years had passed since the events reported upon had occurred. The facts were then not more widely known until the November 1814 issue of Thomas Thomson's *Annals of Philosophy* looked to steal John Murray's thunder, but it was chiefly through Whishaw's 1815 'Account' that the facts became public.

Park's death was not a simple matter of fact. Lawyers working to settle his affairs encountered difficulties since, while Park's death was accepted, no one knew exactly when it had occurred. This made awkward the timing, and thus the amount, of the annuities due to his widow. Writing on 7 March 1825 to Andrew Lang, a solicitor in Selkirk and one-time Sheriff's clerk to Walter Scott, Thomas Anderson noted what he suggested should be Park's legal date of death: 'The time of his Death is uncertain and it may be better to consider it as having taken place about Whitesunday [sic] or Martinmas 1814'. Anderson reported too that Mr Buchanan – the family's solicitor Alexander Buchanan and Allison Park's brother-in-law – had dated his inventory of Park's estate the first of November 1814. The monies in Park's navy stocks were transferred to his widow in August 1815. Buchanan and Allison Park were exchanging letters to much the same end at the same time. Allison Park was gathering notes on the childrens' expenses to send to Buchanan, including the cost of their clothes – and confessed herself worried about the strapping young Mungo: 'for he has got so much of the bean about him that he will hardly wear amended thing [sic] when he is at Foulshiels'. In a further letter to Lang on 6 April 1825, Anderson indicated that 'old Dickson' – his mentor James Dickson who had involved himself in Park's financial plans in the event of his death in Africa – had thought Park 'alive long after 1806' but, by 1813, had begun to act to support Allison Park on the assumption that Mungo Park was now dead.[13]

For those concerned not with the timing of his legal demise but with the facts of Park's death on his second expedition, doubts were expressed over the cause of Park's reported drowning and that of his remaining companions, about the integrity of Amadi Fatouma, Isaaco's informant and Park's former guide, and about the testimony of the sole survivor of the expedition, the slave from whom Amadi Fatouma was given the information he recounted to Isaaco. Park's death was a matter of credibility in several senses: over its timing and cause, regarding those who reported it, and in terms of how it was reported, by word of mouth or in writing.

Park's biographers discuss these several things variously. Whishaw and 'H.B.' accept Isaaco's report of Amadi Fatouma's words, the latter in tones typical of his day and directed at the niggling remarks of the anonymous reviewer in the *Edinburgh Review* in February 1815: 'The journal of a semi-barbarous African trader, who reports the oral communications of men wholly illiterate, concerning events which had occurred five years before, cannot be expected to be without those little blemishes, which assume such magnitude in the microscopic eye of hypercritical reviewers.' Maclachlan was likewise quite prepared to accept that 'for the circumstances of Park's death we depend upon the word of an unknown slave'. Neither Gwynn nor Gibbon give the issue much attention. Gwynn is virtually silent on the matter. Gibbon doubts only Amadi Fatouma's description of the rocks and the narrowness of the rapids at Bussa given that many accounts emphasise the Niger's breadth. Only Lupton deals with the tales of Park's death as they were recounted to the explorers who came after him.[14]

Doubts over the credibility of Amadi Fatouma's account diminished as later explorers turned to Park's death. In 1826, Hugh Clapperton's visit to Bussa allowed him to see for himself where Park and Martyn had perished. In 1830, the Lander brothers visited the same spot. On both occasions, indigenous testimony corroborated what Isaaco had reported in 1811 although discrepancies remain over whether Park was being warned by locals to avoid the rapids – the explanation given to Clapperton – or, as was reported to the Landers and is commonly believed to be what happened and why, that he and his party were attacked because they had failed to pay tribute to local chiefs and had killed several men who sought to follow their canoe.

Park was relatively silent in his *Travels* on slavery as a moral and political question. Yet slavery was a central element of his exploration and of his afterlife. On the outward leg of his first trip, he was accompanied by two slaves, Johnson and Demba. On the return leg, Park survived thanks to a slave trader, heading for home in 1797 on an American slave ship. Accounts of his death depend upon the testimony of a slave, and his posthumous *Journal* was published by an institution concerned to end the slave trade. Abolishing the slave trade was part of the wider objectives of many of the men who followed Park, as they sought to determine the facts of his death and to explore the Niger to its end.

Exploring the Niger

Exploration gives continents their dimensions, seas their depth, mountains their range and height, and rivers their course. Explorers encounter, measure, describe, and depict the world, whether alone and on foot or accompanied by others, working by ship or on land, listening to others' words or reading their instruments and seeing for themselves. The world has the geography it does because exploration at one time or another and in one way or another has revealed it to be so – and continues, of course, via remote sensing and satellite imagery: 'ocular demonstration' at one remove.

Interpreted thus, the exploration of the river Niger – what between about 1788 and 1830 and to Europeans was the world's greatest geographical problem – is one revelatory chapter in the world's geographical unveiling, and a revelation uneven in its timing and in its consequences. The exploration of the Niger was a distinct and complex process directed at knowing where the river ran, its direction of flow, its course, its end, and even its variable width and depth. The motives behind the Niger's exploration went beyond knowing the river's physical geography, mystifying though that geography was: did the river connect with the Nile? Did it run into the sands of Africa? Where did it end? Solve the Niger problem and west Africa's interior might open up – to merchants, abolitionists, and evangelists as well as to map-makers and settlers.

In confirming that the river ran eastwards, into the African interior, Mungo Park's encounter with the Niger in 1796 was a moment of exploration by first-hand observation. A triumph for personal perseverance, it solved by ocular demonstration the first part of a 2,000-year-old problem. His death a decade later confirmed the dangers intrinsic in such exploration: natural hazards, disease, the effects of climate, and the uncertainties which could follow when meeting cultures other than one's own and ignoring the courtesies demanded of travel in lands strange to the traveller. Because he failed in solving the second part of the Niger problem, Park's death set in train further exploration.

Between August 1815, when Goulburn wrote to Barrow to sanction its renewed exploration, to November 1830 when the Landers traced the river to the sea, the Niger was subject to an expeditionary impulse without parallel in the world's exploration. The expeditions of Peddie–Campbell–Gray–Dochard, Tuckey, Lyon and Ritchie, Clapperton, Oudney, and Denham, Laing, and Clapperton with Richard Lander, read as a record of institutional cupidity, personal stupidity and considerable intrepidity, significant but uneven achievement, and death – all without solving the problem left unfinished by Park. In this fifteen-year period of exploratory activity, the emphases behind the Niger's exploration shifted. Geographical questions remained important, precisely because Park had left the river's later course and termination hanging. Questions of commerce and trade increased in their significance, as means to extend Britain's political and colonial reach, realise its humanitarian intentions, and guard against French interests.

Between 1794 and 1805, Park, Hornemann, Nicholls, and Burckhardt were all sent into the field by the African Association, but as individuals in search of geographical fame, with an interest in the world and, perhaps, to make a name for themselves. The expeditions between 1815 and 1830 were each shaped and planned as collective endeavours, in London, by the Colonial Office, the Foreign Office and the Admiralty, and by Lord Camden, Lord Bathurst, and John Barrow, and in Tripoli by Hanmer Warrington and the Bashaw or his lieutenants. Expedition leaders were instructed before departure. Most had expensive compasses, sextants, and other instruments to plot the location of places and fix their own position. Several spoke Arabic or could get by in one or more west African dialect or language. They sought out and valued indigenous knowledge, seeing it as key to information about Park and to the Niger's several names if not always to the river's true course.

Maps commonly followed their endeavours. The maps in the narratives published in the wake of these expeditions were based upon instrumental readings taken in the field – either to corroborate or to correct those of earlier explorers – but they were compiled by others, later and elsewhere. Only Hugh Clapperton among the Niger's explorers, and then on his second expedition, entered Africa with a map in hand to guide him. In the map evidence produced by Clapperton in particular (*cf.* Figures 8.6 and 10.3), and by the Landers in their 1832 account (Figure 10.5), the Niger 'emerged' on paper as the Niger problem was solved by direct observation in the field.

Given the involvement of these individuals and expeditions and biographers' later attention to the Niger's exploration as in-the-field enquiries initiated by Park, it is ironic that the Niger problem was solved decades before the Landers by persons who never visited Africa, who relied not upon scientific instruments and a robust constitution, but upon indigenous information, comparative analysis, textual exegesis, and critical reason.

Park's biographers emphasise his capacities on his second expedition as an explorer, observing that Park set off in 1805 to test the views of George Maxwell, a Scottish African trader, over the Niger–Congo idea, not to test the hypothesis of Christian Reichard, a German town clerk. Several ignore Reichard altogether. The later Niger explorer and Park biographer, Joseph Thomson, did not. He gives credit where it was due: 'Never was a better instance of a mental discovery of a geographical fact. Reichard's hypothesis is a graphic description of the middle and lower Niger.'[15] None of Park's biographers mention William Hutton who, like Reichard, successfully answered the problem, nor do they note the text-based and wrong-headed enquiries in the 1820s of John Dudley, Joseph Dupuis, and Rufane Shaw Donkin. One figure, however, does feature.

To Maclachlan, James MacQueen's 1821 *Geographical and Commercial View of Northern and Central Africa* was 'truly marvellous': 'Not only was the author right in his contention that only in the Bight of Benin should they look for the termination of the Niger; he was, in general, correct in laying down the details of the course of the river, and in his description of the salient features of the countries through which it passed. He was a man of extraordinary insight.' Thomson is of like mind. MacQueen had 'extraordinary genius and industry, and admirable clear-sightedness and judgement'. In solving the Niger problem without ever seeing that river and doing so over a decade before the Landers, MacQueen was the archetype of a critical or armchair geographer – a 'stay-at-home geographer' as Thomson termed him: 'Never was a piece of arm-chair geography worked out more admirably. In its broad outlines it was perfectly correct. To M'Queen [sic] it was as much a certainty as if he had actually explored and mapped it on the spot.'[16]

Before 1830, there were, then, two strands to the Niger's exploration. Both had especially clear expression and exponents in the 1820s. For John Barrow of the Admiralty, the architect of British expeditionary enquiry, and for the colonialist and 'stay-at-home geographer' James MacQueen, the exploration of the Niger was a personal conflict of interests over the appropriate methods. As an establishment figure, Barrow could afford to dismiss the text-based philological enquiries of Dudley and of Donkin. But when, in November 1830, the Landers confirmed in the field what MacQueen and Hutton had argued in 1821, and Reichard in 1803, Barrow's views on the Niger were refuted through the very methods he had long advocated. His position and reputation as a geographical authority was undermined at just the moment that he and others established the Royal Geographical Society and in doing so brought the African Association to an end.

After 1830, the Niger problem was re-cast. Questions about where the river ended were replaced by questions about the river's navigability and as a route for trade. What for 2,000 years had been a geographical puzzle over the river's direction and termination became a matter of politics, commerce, and humanitarianism. The Niger became the focus of technological and political investment – steamship expeditions geared toward the eradication of slavery and the development of trade agreements with west Africa's kingdoms. This investment is embodied in twin moments of exploratory disaster and British imperial hubris, the 1832–4 MacGregor Laird expedition and the 1841 Niger expedition.

As British sailors and expeditionary leaders lay dying in Africa, their ships aground on the Niger's sandbanks, their uncertainty over disease theory reflected in rising death tolls, Mungo Park was being given a new afterlife. Until now, Park's biographers have paid this no heed. Subscriptions toward his memorial statue in Selkirk began in December 1841: the *Soudan*, the *Albert*, and the *Wilberforce* had departed for the Niger in May that year. By the time Parliament had debated the 1841 disaster and produced its 1843 report upon it, Park was on stage in Liverpool and in Manchester. A few years later, Park's logarithm book was recovered from west Africa, a fitting symbol of the locational data required to fix places geographically and of the rigours of exploration.

In London in March 1859, William Balfour Baikie, Scottish physician-turned-African-explorer, was preparing his notes on his successful Niger expedition for an address to the Royal Geographical Society in which he would urge the river's continued exploration. Within weeks, the Jamaican-born teacher and printer Robert Campbell would set off in charge of the grandly named Niger Exploring Party, with the aim of colonising parts of west Africa with African American immigrants. As he did so, back in London, James MacQueen turned his expertise in critical geography to bear upon the claims of the explorer John Hanning Speke as British geographers, now broadly satisfied with their understanding of the Niger (notwithstanding Baikie's exhortations), turned to their attention to the Nile.[17]

In Selkirk, in March 1859, Scottish physician-turned-African-explorer Mungo Park – a 'geographical missionary' as Joseph Banks had described him over sixty years earlier – was being immortalised in stone: a local man and explorer-author whose exploratory life and mysterious death is synonymous with the majestic Niger.

Abbreviations

AA	Abbotsford Archives
BCA	Borders Council Archives
BL	British Library
BMNH, DTC	British Museum of Natural History, Dawson Turner Collection
BPP	British Parliamentary Papers
CA	Chatsworth House Archives
ECA	Edinburgh City Archives
NLS	National Library of Scotland
RGS–IBG	Royal Geographical Society (with the Institute of British Geographers)
RSGS	Royal Scottish Geographical Society
TNA	The National Archives
UoE, CRC	University of Edinburgh Library, Centre for Research Collections

Notes

Introduction

1 The 'Plan of the Association' was written by the secretary to the Association, Henry Beaufoy, in 1788. I have drawn from its later reprinting in *Proceedings of the Association for Promoting the Discovery of the Interior Parts of Africa*, two vols (London: W. Bulmer and Co., 1810), vol. I. The quotes are from pp. 3, 8 and 9. The founding members of the African Association were: Sir Joseph Banks, naturalist and patron of the natural sciences, who sailed with Cook on the *Endeavour* (1768–71) and who was president of the Royal Society between 1778 and 1820; Andrew Stuart, a lawyer who, with Adam Ferguson, served on the Board of Trade; Lord Rawdon (Francis Edward Rawdon-Hastings), politician and military commander; General Conway, the military commander Henry Seymour Conway; Sir Adam Ferguson, Scottish advocate and politician; Sir William Fordyce, a surgeon-physician; Mr Pulteney, the Scottish advocate, landowner, and politician, born William Johnstone, who changed his name to Pulteney following his marriage in 1767 and who was knighted in 1794; Mr Henry Beaufoy, a Quaker and, in 1788, MP for Great Yarmouth; the Bishop of Llandaff, Richard Watson, who in addition to being Archdeacon at Ely Cathedral, was at one time Regius Professor of Divinity and of Chemistry at the University of Cambridge; Lord Carysfort, the parliamentarian John Proby, 1st Earl of Carysfort; and Sir John Sinclair, politician, agricultural improver and pioneering social statistician.

2 Anders Sparrman, a disciple of the Swedish botanist Carl Linnaeus, published his work on southern Africa as *A Voyage to the Cape of Good Hope, towards the Antarctic Polar Circle and round the World: But chiefly in the Country of the Hottentots and Caffres, from the year 1772 to 1776*. The Scottish freemason and African traveller James Bruce 'discovered' the origins and course of the Blue Nile during his lengthy travels in North Africa and in what is now Ethiopia: in effect, he confirmed the work of early seventeenth-century Spanish and Portuguese Jesuit missionaries, the first Europeans to encounter the Blue Nile. Bruce's book, *Travels to Discover the Source of the Nile; in the years 1768, 1769, 1770, 1771, 1772, and 1773* was well received despite lingering doubts as to its credibility (Bruce's claims were later established to be true).

3 Rennell's remarks are from his discussion on the 'Construction of the Map of Africa' in the *Proceedings of the Association for Promoting the Discovery of the Interior Parts of Africa* (London: W. Bulmer and Son, 1798), pp. 211–12.

4 Browne, W. G., *Travels in Africa, Egypt and Syria from the Year 1792 to 1798*, xxiv. See also Hallett, R., *The Penetration of Africa: European Enterprise and Exploration, Principally in Northern and Western Africa up to 1830. Volume 1 to 1815*, pp. 125–33; Eze, E. C. (ed.), *Race and the Enlightenment: A Reader*; Oguejiofor, J. O., 'The Enlightenment gaze: Africans in the minds of western philosophy', *Philosophia Africana* 10 (2007), pp. 31–6. On Africa's nature being seen as 'Other', particularly the continent's tropical regions, see Arnold, D., *The Problem of Nature: Environment, Culture and European Expansion*, pp. 141–3.

5 De Almeida Gebara, A. L., 'The British search for the course of the Niger River: from a geographical problem to potential possession', *História* (São Paulo) 31 (2012), pp. 146–70.

6 Lupton, K., *Mungo Park African Traveler*; Duffill, M., *Mungo Park West African Explorer*; Duffill, M., 'Notes on a collection of letters written by Mungo Park between 1790 and 1794, with some remarks concerning his later published observations on the African slave trade', *Hawick Archaeological Society Transactions* [no vol.] (2001), pp. 35–55; Williamson, K., and Duffill, M., 'Mungo Park, man of letters', *Scottish Literary Review* 3 (2011), pp. 55–79; Hallett, R., *Records of the African Association 1788–1831*.

7 Bovill, E. W., *The Niger Explored*; Bovill, E. W. (ed.), *Missions to the Niger*. Edward Bovill died in 1966, his 1968 work being completed from surviving notes by his widow: it is perhaps for this reason that this work is not as thorough as his earlier studies and, in places, lacks clear attribution to the relevant sources in contrast to the detail characteristic of his edited *Missions to the Niger*. On Bovill's contribution to the history of African exploration, see Laithwaite, G., 'Obituary: E. W. Bovill', *Geographical Journal* 133 (1967), pp. 280–1. Bovill's knowledge stemmed from having served in north and west Africa during the First World War. The quote on Park is from *The Niger Explored*, p. 1.

8 Sattin, A., *The Gates of Africa: Death, Discovery and the Search for Timbuktu*; De Gramont, S., *The Strong Brown God: The Story of the Niger River*. At about the time as he wrote *The Strong Brown God*, Sanche de Gramont rejected his French aristocratic background and changed his name to Ted Morgan (an anagram of 'De Gramont'), under which name, and as an American citizen, he wrote biographies of Franklin D. Roosevelt and Winston Churchill. The quote is from *The Strong Brown God*, p. 13. The many other works on the Niger are considered throughout what follows. Among several commentators who, like de Gramont, record the Niger's exploration from Park onwards together with some attention to the Niger, see, in the later nineteenth century, Richardson, R., *Story of the Niger: A Record of Travel and Adventure from the Dogs of Mungo Park to the Present Time*, and, among modern authors, Brent, P., *Black Nile: Mungo Park and the Search for the Niger*, and Hudson, P., *Two Rivers: Travels in West Africa on the Trail of Mungo Park*.

9 On John Wesley Powell and the Colorado river, see Worster, D., *A River Running West: The Life of John Wesley Powell*, and Dolnick, E., *The Great*

Unknown: John Wesley Powell's 1869 Journey of Discovery and Tragedy through the Grand Canyon. On Schomburgk in Guyana, Britain's sole South American colony, see Burnett, D. G., *Masters of All They Surveyed: Exploration, Geography, and a British El Dorado.*

Chapter 1

1. [John Cory], *A Geographical Historie of Africa, written in Arabicke and Italian, by John Leo a More, borne in Granada and brought up in Berberie*, p. 3. On early Islamic map-makers' image of the African interior, see Shea, R. W., and Bell, D., 'Charting the unknown: Islamic cartography and visions of Africa in the "Abbasid era"', *History in Africa* 46 (2019), pp. 37–56.
2 On the European mapping of Africa, see Relano, F., *The Shaping of Africa: Cosmographic Discourse and Cartographic Science in Late Medieval and Early Modern Europe*; Tooley, R. V., *Collectors' Guide to Maps of the African Continent and Southern Africa*; Stone, J. C., *A Short History of the Cartography of Africa.* The reference to Labat is Labat, J.-B., *Nouvelle relation de l'Afrique Occidentale*, and from Moore is Moore, F., *Travels into the Inland Parts of Africa*, p. xi.
3 D'Anville, J. B. D., 'Mémoire concernant les rivières de l'interieur de l'Afrique, sur les notions tirées des Anciens & des Modernes', *Mémoires de Litterature de l'Académie Royale des Inscriptions et Belles-Lettres* 26 (1759), pp. 64–81; Stone, *Short History of the Cartography of Africa*, pp. 23–46; Kennedy, D., *The Last Blank Spaces: Exploring Africa and Australia*, pp. 12–16. Rennell's remarks upon d'Anville's 1749 map appear in chapter XII, the 'Construction of the Map of Africa', in *Proceedings of Association for Promoting the Discovery of the Interior Parts of Africa* (London: C. MacRae, 1790), vol. I, p. 214.
4 Kennedy, *The Last Blank Spaces*, p. 7. On this point more generally, see Shapin, S., *A Social History of Truth: Civility and Science in Seventeenth-Century England*, especially chapters 3 and 5; and Outram, D., 'On being Perseus: new knowledge, dislocation, and Enlightenment exploration', in Livingstone, D. N., and Withers, C. W. J. (eds), *Geography and Enlightenment*, pp. 281–94.
5 Hallett, R., *Records of the African Association 1788–1831*, pp. 52, 55. Ledyard's account of his voyage with Cook was published as *A Journal of Captain Cook's Last Voyage to the Pacific Ocean*. On Ledyard, see Gray, E., *The Making of John Ledyard: Empire and Ambition in the Life of an Early American Traveler.*
6 Hallett, *Records of the African Association*, p. 61.
7 Hallett, *Records of the African Association*, pp. 123–6. For Beaufoy's description of Houghton, see *Proceedings of the African Association* (1792) vol. I, p. 15. On pre-circulated queries and the solicitation of knowledge in the eighteenth century, see Sloan, P., 'The gaze of natural history', in Fox, C., Porter, R., and Wokler, R. (eds), *Inventing Human Science: Eighteenth-Century Domains*, pp. 112–51; Fox, A., 'Printed questionnaires, research networks and the discovery of the British Isles, 1650–1800', *Historical Journal* 53 (2010), pp. 593–621.

8 For what is known of Houghton, see Hallett, *Records of the African Association*, pp. 120–40 from which this summary is derived, and Sattin, A., *The Gates of Africa: Death, Discovery and the Search for Timbuktu*, pp. 91–109, 113–16, 120–4 (who takes much of his account of Houghton from Hallett).
9 *Proceedings of the African Association* (1792), vol. II, p. 17.
10 Hallett, *Records of the African Association*, pp. 141–2.
11 On the use of the term 'explorer', see Bourguet, M.-N., 'The explorer', in Vovelle, M. (ed.), *Enlightenment Portraits*, pp. 257–315 and, on the idea of the term as a 'back-formation', Craciun, A., 'What is an explorer?', *Eighteenth-Century Studies* 41 (2011), pp. 29–51. For recent summaries of the reinterpretation of exploration, see Kennedy, D., 'Introduction: reinterpreting exploration', in Kennedy, D. (ed.), *Reinterpreting Exploration: The West in the World*, pp. 1–18; Driver, F., *Geography Militant: Cultures of Exploration and Empire*; and Martin, P. R., and Armston-Sheret, F., 'Off the beaten track?: critical approaches to exploration studies', *Geography Compass* 14 (2020), pp. 1–14.
12 Lockhart, J. B., 'In the raw: some reflections on transcribing and editing Lieutenant Hugh Clapperton's writings on the Borno Mission of 1822–25', *History in Africa* 26 (1999), pp. 157–95. On writing in the field/on the move, see Keighren, I., Withers, C. W. J., and Bell, B., *Travels into Print: Exploration, Writing, and Publishing with John Murray 1773–1859*, especially pp. 34–67, and, on this point for several explorers in later nineteenth-century Africa, Kennedy, *The Last Blank Spaces*, pp. 214–15. On field notebooks as a portable archive, see Bourguet, M.-N., 'A portable world: the notebooks of European travellers (eighteenth to nineteenth centuries)', *Intellectual History Review* 20 (2010), pp. 377–400.
13 For study of the relationships between exploration, authorship, and exploration narratives, see Keighren, Withers and Bell, *Travels into Print*, passim. On these questions for the Arctic, see Cavell, J., *Tracing the Connected Narrative: Arctic Exploration in British Print Culture, 1818–1860* and Craciun, A., *Writing Arctic Disaster: Authorship and Exploration*. For later African exploration and the work of David Livingstone for example, see Driver, F., '*Missionary Travels*: Livingstone, Africa and the book', *Scottish Geographical Journal* 129 (2013), pp. 164–78; and, on the African explorer John Hanning Speke, Finkelstein, D., *The House of Blackwood: Author-Publisher Relations in the Victorian Era* and Finkelstein, D., 'Unraveling Speke: the unknown revision of an African exploration classic', *History in Africa* 30 (2003), pp. 117–32.
14 These issues are the subject of Barnett, C., 'Impure and worldly geography: the Africanist discourse of the Royal Geographical Society, 1831–1873', *Transactions of the Institute of British Geographers* 23 (1998), pp. 239–51; Driver, F., and Jones, L., *Hidden Histories of Exploration: Researching the RGS-IBG Collections*; Konishi, S., Nugent, M., and Shellam, T. (eds), *Indigenous Intermediaries: New Perspectives on Exploration Archives*; Schaffer, S., Roberts, L., Raj, K., and Delbourgo, J. (eds), *The Brokered World: Go-Betweens and Global Intelligence, 1770–1820*; Kennedy, *The Last Blank Spaces*, pp. 159–94.

On exploration and empire in relation to travel writing in the nineteenth century, see Thompson, C., 'Nineteenth-century travel writing', in Das, N., and Youngs, T. (eds), *The Cambridge History of Travel Writing*, pp. 108–24.

15 On this point, see Edney, M. H., *Mapping an Empire: The Geographical Construction of British India, 1765–1843*.

16 On these questions, see Schaffer, S., 'Easily cracked: scientific instruments in states of disrepair', *Isis* 102 (2011), pp. 706–17; Baird, D., *Thing Knowledge: A Philosophy of Scientific Instruments*; Baker, A., '"Precision", "Perfection" and the reality of British scientific instruments on the move during the 18th century', *Material Culture Review* 74–75 (2012), pp. 14–29; Bourguet, M.-N., Licoppe, C., and Sibum, H. O. (eds), *Instruments, Travel and Science: Itineraries of Precision from the Seventeenth to the Twentieth Century*; Fleetwood, L., 'No former travellers having attained such a height on the Earth's surface: instruments, inscriptions in the Himalaya, 1800–1830', *History of Science* 56 (2018), pp. 3–34; Goodman, M., 'Proving instruments credible in the early nineteenth century: the British Magnetic Survey and site-specific experimentation', *Notes and Records of the Royal Society of London* 70 (2016), pp. 1–18; Rae, E., Souch, C., and Withers, C. W. J., '"Instruments in the Hands of Others": the life and liveliness of instruments of British geographical exploration, *c.*1860–*c.*1930', in MacDonald, F., and Withers, C. W. J. (eds), *Geography, Technology and Instruments of Exploration*, pp. 139–61; Raj, K., 'When human travellers became instruments: the Indo-British exploration of Asia in the nineteenth century', in Bourguet, M.-N., Licoppe, C., and Sibum, H. O. (eds), *Instruments, Travel and Science: Itineraries of Precision from the Seventeenth to the Twentieth Century*, pp. 156–88; Wess, J., 'The role of instruments in exploration: A study of the Royal Geographical Society, 1830–1930', (Unpublished PhD, University of Edinburgh, 2017); Wess, J., and Withers, C. W. J., 'Instrument provision and geographical science: the work of the Royal Geographical Society, 1830–*ca*1930', *Notes and Records of the Royal Society of London* 73 (2019), pp. 223–42; Withers, C. W. J., 'Science, scientific instruments and questions of method in nineteenth-century British geography', *Transactions of the Institute of British Geographers* 38 (2013), pp. 167–79; Withers, C. W. J., 'Geography and "thing knowledge": instrument epistemology, failure and narratives of 19th-century exploration', *Transactions of the Institute of British Geographers* 44 (2019), pp. 676–91.

17 Bovill, E. W., *Missions to the Niger, II: The Bornu Mission, 1822–25 Part I*, p. 75.

18 Dritsas, L., 'Expeditionary science: conflicts of method in mid-nineteenth-century geographical discovery', in Livingstone, D. N., and Withers, C. W. J. (eds), *Geographies of Nineteenth-Century Science*, pp. 255–77. On the use of fieldwork and different methods in geographical enquiry in historical context, see the essays in Nielsen, K. H., Harbsmeier, M., and Ries, C. J. (eds), *Scientists and Scholars in the Field: Studies in the History of Fieldwork and Expeditions*. On different methods and the mapping of east central Africa in the mid-nineteenth century, see also Withers, C. W. J., 'Map making, defamation, and credibility:

the case of the *Athenaeum*, Charles Tilstone Beke, and W. & A. K. Johnston's *Edinburgh Educational Atlas* (1874)', *Imago Mundi* 73 (2021), pp. 46–63.

19 Kennedy, *The Last Blank Spaces*, pp. 5–6.

20 Fleming, F., *Barrow's Boys: A Stirring Story of Daring, Fortitude and Outright Lunacy*. Fleming briefly treats of Barrow's Niger men in chapter 16, 'The Riddle of the Niger', pp. 253–73.

21 On metabiographical studies, see Russell, P., *Prince Henry 'the Navigator': A Life*; Rupke, N. A., *Alexander von Humboldt: A Metabiography*; Higgitt, R., *Recreating Newton: Newtonian Biography and the Making of Nineteenth-Century History of Science*; Jeal, T., *Stanley: The Impossible Life of Africa's Greatest Explorer*; and Livingstone, J., *Livingstone's 'Lives': A Metabiography of a Victorian Icon*. On the biography of scientific instruments, see Daston, L. (ed.), *Biographies of Scientific Objects*. The idea of the 'life' and 'liveliness' of scientific instruments in geographical exploration is explored by Rae, Souch and Withers, '"Instruments in the Hand of Others"', passim.

22 On Mungo Park presented to children as a famous Scottish Africanist, see Ganeri, A., *Stunning Scotland*, p. 63. On Mungo Park as a subject in Nigerian dance and rap music, see Korede Bello, https://www.youtube.com/watch?v=il7vIMecTaI&index=7&list=PLPoWxokyjhFSLJaBkZn4dBF7b265ZCsxD&t=0s&app=desktop where, as a schoolmaster asks his class 'Who discovered the Niger?' a girl answers, 'Mungo Park'. Another pupil [Bello] asks 'But who discovered Mungo Park?' This prompts mayhem and dancing. Rap musician Burna Boy calls Mungo Park a 'fool' in lyrics which also state 'fuck Mungo Park': Burna Boy, https://falseto.com/2020/08/17/burna-boy-i-want-to-inspire-a-revolution/ [Last accessed October 20, 2020].

Chapter 2

1 Some biographies and biographical entries (including the first edition of the *Oxford Dictionary of National Biography*) give his birthdate as 10 September, perhaps from misinformation given by the Park family: the date of 11 September 1771 is given in the parish registers: see Lupton, K., *Mungo Park: The African Traveler*, p. 5 (and in the second, 2004, edition of the *Oxford Dictionary of National Biography*).

2 Robertson, T., 'Parish of Selkirk', in Sinclair, J. (ed.), *The Statistical Account of Scotland: Drawn up by the Communications of the Ministers of the Different Parishes*, vol. II (1792), pp. 434–48. Robertson's account was published in 1792; I have assumed its authorship to be the previous year. There are no firm figures for farm income or rental in Park's day, but Lupton notes that Foulshiels generated an annual rental of £74 in 1766 (that sum payable to the Duke of Buccleuch) and had 720 sheep and 10 cattle. The Scotland- and Europe-wide harvest failure of 1782–3, which was exacerbated by the effects of earlier 'short' years, is discussed in Flinn, M. (ed.), *Scottish Population History from the 17th century to the 1930s*, pp. 233–7.

3 Robertson, 'Parish of Selkirk', p. 443; Lupton, ibid., p. 6.
4 The description of Newark Castle is from Robertson, 'Parish of Selkirk', p. 437. On geographical education in eighteenth-century Scotland, see Withers, C. W. J., *Geography, Science and National Identity: Scotland since 1520*, pp. 112–57, especially pp. 128–39. For Guthrie's comment on the Niger, see Guthrie, W., *A New Geographical, Historical, and Commercial Grammar*, p. 516. For Salmon's, see Salmon, T., *A New Geographical and Historical Grammar*, twelfth edition, p. 540. Robert Burns' admission in 1787 over the influence of Salmon's and Guthrie's geographical grammars – and a letter of 1791 in which he refers to 'the latest edition of Guthrie's Geographical grammar' as among 'the valuable books' he was soon to be sent – is cited in Roy, G. R. (ed.), *The Letters of Robert Burns*, vol. I, pp. 138, 140; vol. II, p. 66.
5 Park, M., *The Journal of a Mission to the Interior of Africa, in the Year 1805*, p. 163.
6 The literature on the Enlightenment in Scotland and the Scottish Enlightenment is extensive. For summaries, see the essays in Broadie, A. (ed.), *The Cambridge Companion to the Scottish Enlightenment*; Withers, C. W. J., and Wood, P. (eds), *Science and Medicine in the Scottish Enlightenment*; and Wood, P. (ed.), *The Scottish Enlightenment: Essays in Reinterpretation.*
7 These general remarks about the medical degree at the University of Edinburgh are taken from Duffill, M., *Mungo Park West African Explorer*, pp. 14–15. Unaccountably, given the depth of his knowledge on Park, Duffill neglects to mention that Park also took John Walker's natural history class. On Park in Walker's class list of students in April 1791, see UoE, CRC, MS. Dc.1.18/9, f. 73, 18 April 1791.
8 Rutherford's botanical teaching does not survive in detail. These remarks are based on UoE, CRC, Coll-1929, Gen.786, 'Notes taken at Dr. Rutherford's Botanical Lectures' [May 1790, by a person unknown]. The notes reflect a rather dry and technical listing of the structure of plants, botanical terms, and in Lecture 30 (f. 50), engagement with Linnaeus's binomial nomenclature and the ordering of plants.
9 For details of Walker's teaching, Park's classmates, and the connections between subjects such as chemistry and mineralogy as taught in Edinburgh's medical school, see Eddy, M. D., *The Language of Mineralogy: John Walker, Chemistry and the Edinburgh Medical School, 1750–1800.* Walker's geological lectures (themselves wide-ranging) are the subject of Scott, H. W. (ed.), *Lectures on Geology by John Walker.*
10 Duffill, *Mungo Park West African Explorer*, p. 16. Chirurgo-Physical Society, *Laws and Regulations of the Chirurgo-Physical Society*, p. 15 (the regulation), p. 25 (Park's membership). The difference between an extraordinary member (Mungo Park) and an ordinary member (Alexander Anderson) seems to have been whether one intended to graduate as a physician or surgeon (ordinary member) as opposed to simply taking one or more of the required diet of courses for the medical degree.

11 The Park–Laidlaw correspondence is the subject of Duffill, M., 'Notes on a collection of letters written by Mungo Park between 1790 and 1794, with some remarks concerning his later published observations on the African slave trade', *Hawick Archaeological Society Transactions* (2001), pp. 35–55, especially pp. 35–9.
12 The evidence for Park's probable involvement in the Pantheon debate of 14 April 1791 on Ramsay and Fergusson is discussed in Duffill, ibid., p. 38.
13 On James Anderson and *The Bee*, see Mullett, C. F., 'A village Aristotle and the harmony of interests: James Anderson (1739–1808) of Monks Hill', *Journal of British Studies* 8 (1968), pp. 94–118; Anderson's 'Hints respecting the study of Geography', appears in *The Bee* 6 (December 1791), pp. 208–14, and is continued, under the title 'Philosophical Geography', on pp. 335–41. On Anderson, *The Bee*, and Park's poetry, see Williamson, K., and Duffill, M., 'Mungo Park, man of letters', *Scottish Literary Review* 3 (2011), pp. 55–79: the claim that Park, in his poem respecting Cullen, was seeking the approbation of his peers is made on p. 66.
14 UoE, CRC, Dc. 2. 29, f. 38.
15 The quotes are from Dickson, J., 'An account of some plants newly discovered in Scotland', *Transactions of the Linnean Society* II (1794), pp. 286–91. Dickson recounts further Scottish plant finds in the society's *Transactions* in vol. I, pp. 181–2 and in vol. II, pp. 343–5. Dickson's principal work was his *Fasciculus Plantarum Cryptogamicarum Britanniae.*
16 Duffill, 'Notes on a collection of letters', p. 43.
17 Duffill, *Mungo Park West African Explorer*, pp. 22–6; Duffill, 'Notes on a collection of letters', pp. 43–4; Marsden, W., *The History of Sumatra, containing an account of the government, laws, customs, and manners of the native inhabitants, with a description of the natural productions, and a relation of the ancient political state of that island*, pp. iii, viii.
18 Park, M., 'Descriptions of eight new fishes from Sumatra', *Transactions of the Linnean Society* 3 (1797), pp. 33–8.
19 Park also sketched other, more common, reef fish species which did not appear in his paper to the Linnean Society (probably because they were common). These, together with moss specimens collected from west Africa and other material, were later discovered in papers belonging to Alexander Park, Mungo's brother: Hancock, E. G., and Wheeler, A., 'An addition to the archival record of Mungo Park (1771–1806), African explorer', *Archives of Natural History* 15 (1988), pp. 311–15.
20 NLS, MS 10782, ff. 180–1, Mungo Park to Alexander Anderson, 20 May 1794; Lupton, *Mungo Park: The African Traveler*, p. 36.
21 Hallett, R., *Records of the African Association 1788–1831*, pp. 137–8; Lupton, *Mungo Park: The African Traveler*, p. 37.
22 Hallett, ibid., pp. 141–56. Hallett's description of the Willis affair as a 'fiasco' is from p. 156.
23 NLS, MS Acc. 13149, Nevil Maskelyne to Mungo Park, 4 October 1794.

24 On Rennell's life and work, see Rennell, J., *Memoir of the Map of Hindoostan*; Markham, C. R., *Major James Rennell and the Rise of Modern English Geography*; Baker, J. N. L., 'Major James Rennell, 1742–1830, and his place in the history of geography', in Baker, J. N. L., *The History of Geography*, pp. 130–57; Cook, A. S., 'Major James Rennell and the Bengal Atlas (1780 and 1781)', in *India Office Library and Records Report for the Year 1976*, pp. 5–42; Edney, M. H., *Mapping an Empire: The Geographical Construction of British India, 1765–1843*.
25 Rennell, J., *The Geographical System of Herodotus, Examined, and Explained, by a Comparison with those of other Ancient Authors, and with Modern Geography*, pp. 4–5.
26 The remark about the committee of the African Association receiving 'new and interesting intelligence' is from a page of manuscript, in Rennell's hand, held together with a copy of his 1790 map in the British Library's King's Topographical Collections: BL, Maps K. Top. 117. 24.1.c., f. 1. The letter from Beaufoy to Barnard is undated but which must date from about May 1793 is held in the British Library in its map collections: BL, Maps K. Top. 117.24.1.g.
27 This information is taken from *Elucidations of the African Geography; from the Communications of Major Houghton, and Mr Magra; 1791*, a copy of which is bound, as a separate publication, in a British Library copy of the *Proceedings of the African Association*, with the shelfmark BL, 454, ff. 15 and 16. The quotes here are from pp. 5 and 22, respectively. The reference to 'Mr. Magra' in the title is to Perkins Magra, then British consul in Tunis. Although no author is given, the author is James Rennell. That this copy of the publication also states 'Jos: Banks' suggests that it was Banks's personal copy. On the conjunction of map and memoir in Enlightenment map-making, see Edney, M. H., 'Reconsidering Enlightenment geography and map making: reconnaissance, mapping, archive', in Livingstone, D. N., and Withers, C. W. J. (eds), *Geography and Enlightenment*, pp. 165–98.
28 Rennell, J., 'On the rate of travelling, as performed by camels; and its application, as a scale, to the purpose of geography', *Philosophical Transactions of the Royal Society of London* 81 (1791), pp. 129–45. The quote is from p. 141.
29 The quote by Beaufoy on Ledyard and the map is from *Proceedings of the Association for Promoting the Discovery of the Interior Parts of Africa* (London: Printed by C. MacRae for the Association, 1790), p. 18.

Chapter 3

1 Park, M., *Travels in the Interior Districts of Africa*, p. 7. The quotes that follow from Park are all from his 1799 *Travels*, unless otherwise stated. Where more than one quote is used, the citation identifies the respective quotes by the page or pages in question.
2 The idea (and myth) of the 'chameleon fieldworker', perfectly attuned despite being foreign to the social world he/she is examining, is developed in Geertz, C., *Local Knowledge: Further Essays in Interpretive Anthropology*.

3 Park, *Travels*: 'in my chamber', p. 7; 'delirium', p. 8; 'those who have heard it' and 'fertility and abundance', p. 9.
4 Fabian, J., *Out of Our Mind: Reason and Madness in the Exploration of Central Africa*.
5 Park, ibid.: 'afterwards very rapidly' and 'season for travelling', p. 12; 'unknown to me' and 'approved my determination', p. 13.
6 Park, ibid., p. 33.
7 The work here referenced is Richardson, J., *A dictionary, Persian, Arabic, and English*. Written in part with William Jones, the Orientalist, Richardson's work laid the foundation for later Orientalist studies of Persian grammar.
8 Park, ibid.: 'new to him', p. 54; 'afraid to avow', p. 75; 'interior of Africa', p. 95.
9 Park, ibid., p. 102.
10 Park, ibid.: 'Moorish country', p. 103; 'left to perish', p. 104.
11 The term 'Moor' was used loosely by Park and others to include those of Islamic faith (the principal association in Park's eyes), desert-edge peoples, and, occasionally, Moroccans.
12 Park, ibid.: 'their apprehensions', p. 113; 'a shower of rain', p. 116.
13 Park, ibid.: 'in case of accidents', p. 114; 'cattle, and goats', p. 121; 'human being', p. 122; 'buttoning and unbuttoning', p. 124; 'disagreeable inmate', p. 123.
14 Park, ibid.: 'were artificial', p. 56; 'the matter jocularly', 'satisfy her curiosity', and 'for my supper', p. 132; 'very indecent', p. 133.
15 Park, ibid.: 'insult and irritation', 'plague the Christian', and the extended quote are all from p. 125. The emphasis is in the original.
16 Park, ibid.: 'fresh distresses', p. 138; 'insufferably hot' and 'sensible pain', p. 135.
17 Park, ibid.: 'office of *barber*' and 'commencement of the operation', p. 126; 'fortunate circumstance' and 'recovering my liberty', p. 127; 'painful sensation' and 'received nourishment', p. 142.
18 Park, ibid.: 'Daman Jumma', p. 164; 'strangers to its dictates', p. 163; 'refusing to proceed', 'myself understood', and 'worse than either', p. 168.
19 Park, ibid., pp. 186–87.
20 On this point, see Haddadd, E. A., 'Body and belonging(s): property in the captivity of Mungo Park', in Harper, G. (ed.), *Colonial and Post-Colonial Incarceration*, pp. 124–43.
21 Park is depicted in heroic mould by Leask, N., *Curiosity and the Aesthetics of Travel Writing 1770–1840*, p. 131. On Park as a sentimental 'anti-hero', see Pratt, M. L., *Imperial Eyes: Travel Writing and Transculturation*, pp. 74–85. The description of Park as a 'distinct anomaly' I take from Kate Marsters' 'Introduction' to Park's *Travels in the Interior Districts* (2000), p. 3.
22 Park, *Travels*: 'stroke of thunder' and 'entrusted him', p. 171; 'apparent unconcern', p. 173, 'the Desert looked pleasant', p. 175.
23 Park, ibid.: 'come to an end', p. 179; 'Moorish Sultan', p. 185; 'merriment to the Bambarrans', p. 193; 'price of a button', 'early the next day', and 'shutting my eyes', p. 194.

24 Park, ibid.: 'see the water' and '*to the eastward*', p. 194 (the emphasis is in the original); 'endeavours with success' and 'same manner', p. 195.
25 Park, ibid.: 'weary and dejected', p. 197; 'our tree' and 'I could give her', p. 198; 'large red lion', p. 208; 'fatigue and hunger', p. 209; 'completely obstructed', pp. 209–10; 'perish with me', p. 212; 'the geography of countries', p. 214.
26 Park, ibid.: 'to avoid me', p. 225; 'for all travellers' and 'poor fellows', p. 236; 'serious nature', p. 239; 'distant mountains', p. 240.
27 Park, ibid.: 'kept my memorandum', 'lie down and perish', and 'influence of religion', p. 243; 'caught my eye' and 'surely not!', p. 244. The modern commentator is Leask, *Curiosity and the Aesthetics of Travel Writing 1770–1840*, p. 81, fn. 88.
28 Park, *Travels*: 'broken to pieces' and 'could not repair', p. 249; 'faint and sickly', p. 251; 'yellow from sickness', 'extreme poverty', and 'some Arab in disguise', p. 253.
29 Park, ibid.: 'thought proper' and 'satisfy him', p. 254.
30 Park, ibid.: 'until sunset', p. 330; 'risen from the dead', 'venerable incumbrance', and 'into a boy', p. 358; 'Black men are nothing', p. 359; 'a mind *above his condition*' (emphasis in the original) and 'rudeness to refinement', p. 360. On this point, see Juengel, S., 'Mungo Park's artificial skin; or, the year the white man passed', *The Eighteenth Century* 47 (2006), pp. 19–38.
31 Park, *Travels*: 'remainder of the voyage', p. 361; 'exertion at the pumps', p. 362. Park's letter in which he informs Banks that he, Banks, will be liable for the cost of Park's homeward voyage is in BMNH, DTC, vol. 10, f. 209, Mungo Park to Joseph Banks, 13 November 1797.

Chapter 4

1 On the several stages which exploration narratives go through between writing in the field and publication, see MacLaren, I. S., 'Exploration/travel literature and the evolution of the author', *International Journal of Canadian Studies* 5 (1992), pp. 39–68; MacLaren, I. S., 'In consideration of the evolution of explorers and travellers into authors: a model', *Studies in Travel Writing* 15 (2011), pp. 221–41. The idea of the explorer being produced through authorship is taken from Craciun, A., 'What is an explorer?', *Eighteenth-Century Studies* 41 (2011), pp. 29–51, and Craciun, A., *Writing Arctic Disaster: Authorship and Exploration*. See also Keighren, I., Withers, C. W. J., and Bell, B., *Travels into Print: Exploration, Writing, and Publishing with John Murray 1773–1859*.
2 Park's words are from his *Travels in the Interior Districts of Africa*, p. viii; the Association's words are from *Proceedings of the Association for Promoting the Discovery of the Interior Parts of Africa* (London: Printed for the Association by W. Bulmer and Son, 1798), p. 1.
3 *Proceedings of the Association for Promoting the Discovery of the Interior Parts of Africa* (1798), pp. 20 and 27.
4 The quotations are from Rennell's '*Geographical Illustrations of Mr Park's Journey,*

and of North Africa at Large', in the *Proceedings of the Association* (1798), pp. 51, 52, 63, 64, 79, 86, and 143, respectively. The persistent presence in cartography of the mythical mountains of Kong is explored by Bassett, T. J., and Porter, P. W., '"From the Best Authorities": the Mountains of Kong in the cartography of west Africa', *Journal of African History* 32 (1991), pp. 367–413. On eighteenth-century beliefs in the physical (and climatic) determination of moral qualities, see Withers, C. W. J., *Placing the Enlightenment: Thinking Geographically about the Age of Reason*, pp. 139–63.

5 On these points, see Bravo, M., 'Precision and curiosity in scientific travel: James Rennell and the Orientalist geography of the new imperial age (1760–1830)', in Elsner, J., and Rubieś, J.-P. (eds), *Voyages and Visions: Towards a Cultural History of Travel*, pp. 162–83; Mayhew, R., 'The character of English geography *c.*1660–1800: a textual approach', *Journal of Historical Geography* 24 (1998), pp. 385–412; Mayhew, R., 'Mapping science's imagined community: geography as a Republic of Letters, 1600–1800', *British Journal for the History of Science* 38 (2005), pp. 73–92.

6 This section is shaped by my preliminary discussion of these issues: Withers, C. W. J., 'Geography, enlightenment and the book: authorship and audience in Mungo Park's African texts', in Ogborn, M., and Withers, C. W. J. (eds), *Geographies of the Book*, pp. 191–220.

7 The words to what is titled 'A Negro Song, from Mr. Park's Travels' appears in Mungo Park, *Travels in the Interior Districts of Africa*, following p. xxviii [but without pagination]. The song, and the musical score, is entitled 'Song from Mr. Park's Travels'. While the version in Park's *Travels* clearly states its 'author' as the Duchess of Devonshire, there is no record of this in her contemporary surviving papers. There are, however, a few minor variations between the manuscript version in related papers in the Chatsworth Archives CA, DF 12/1/1/5, which version, interestingly, is not in the Duchess's hand, and that printed in Park's *Travels*, and appears to be later than 1799.

8 Park, *Travels*, pp. 15–18.

9 Blouet, O. M., 'Bryan Edwards, F.R.S., 1743–1800', *Notes and Records of the Royal Society of London* 54 (2000), pp. 215–22.

10 *Proceedings of the African Association for Promoting the Discovery of the Interior Parts of Africa* (London: W. Bulmer, 1810), II, p. 8. On Park to his sister, see NLS, MS 10290, ff. 1–2v, 21 September 1798.

11 *Proceedings of the African Association for Promoting the Discovery of the Interior Parts of Africa* (London: W. Bulmer, 1810), II, p. 2.

12 Shapin, S., *A Social History of Truth: Civility and Science in Seventeenth-Century England*, pp. 193–242.

13 Park, *Travels*, pp. viii–ix.

14 At this time, different prime meridians were used by different nations and different scientific communities for different purposes, a fact which led, as contemporaries recognised, to considerable geographical confusion: on this, see Withers, C. W. J., *Zero Degrees: Geographies of the Prime Meridian*.

15 Park, *Travels*, p. ix (emphasis in the original).

16 The surviving Young–Park exchanges appear as Appendix III in Park, M., *The Journal of a Mission to the Interior of Africa, in the Year 1805*, pp. cix–cxi.

17 Park, *Travels*, p. ix (emphasis in the original). The possibility that Park is quoting Shakespeare's *Othello* is noted by Regard, F., 'Introduction: articulating empire's unstable zones', in Regard, F. (ed.), *British Narratives of Exploration*, p. 7. On the trope of the modest author in exploration narratives, see Keighren, I., Withers, C. W. J., and Bell, B., *Travels into Print: Exploration, Writing, and Publishing with John Murray 1773–1859*, pp. 100–7.

18 The observation by Scott appears in an undated manuscript letter, from Scott via John Murray the publisher to John Whishaw, in or about 1814. The letter is affixed in a copy of Park's *Travels in the Interior Districts* held by Selkirk Library Archives (Borders Council) as SC/5/56 [Letter 10 in a set of interleaved letters]. The quotes are from f. 2 and f. 2v of this letter. This copy of Park's book bears the bookplate of Patrick Miller of Dalswinton; possibly the owner of the estate of that name in the Nith valley in Dumfriesshire, and friend to the poet Robert Burns.

19 Grieve, H. J. C. (ed.), *The Letters of Sir Walter Scott*, 4, p. 52.

20 This claim of Park's *Travels* is made by Pratt, M. L., *Imperial Eyes: Travel Writing and Transculturation*, p. 75.

21 These points are from Sher, R. B., *The Enlightenment & the Book: Scottish Authors & Their Publishers in Eighteenth-Century Britain, Ireland & America*, pp. 185–7, 217. On Nicol's plea to Banks, see Isaac, P., *William Bulmer: The Fine Printer in Context*, p. 95.

22 On the subscribers, see Park, *Travels in the Interior Districts of Africa* [second edition of 1799], pp. xxi–xxviii.

23 These sales figures, for about a decade from May 1799, are from the Bell and Bradfute records, ECA, SL138/7/9 [17 May 1799] to SL138/7/14 [11 October 1808]. Park's purchase (and return) of the Damberger volume is at SL138/7/10, 6 January 1801. Intriguingly, on 29 May 1799, one 'P. Miller of Dalswinton' is listed as a purchaser: it is possible that this is the same Patrick Miller in whose copy of Park's *Travels* is placed Sir Walter Scott's letter to John Whishaw on Park's admission concerning the incomplete nature of his published work: see note 18 above.

24 Park, M., *Travels in the Interior Districts of Africa* (Philadelphia: Humphreys, 1800), pp. ix–xv.

25 On these questions, see Topham, J. M., 'Science, print, and crossing borders: importing French science books into Britain, 1790–1815', in Livingstone, D. N., and Withers, C. W. J. (eds), *Geographies of Nineteenth-Century Science*, pp. 311–44; Martin, A. E., *Nature Translated: Alexander von Humboldt's Works in Nineteenth-Century Britain*. On Monrad, see Monrad, H. C., *Efterretninger om det indre Afrika, uddragne af de af det i London oprettede afrikanske Selskab udgivne Oplysninger om Houghtons og Parks Reiser, som Foreløber for denne sidstes større Reisebeskrivelse* [Reports on inner Africa, abstracted from information published about Houghton's and Park's Travels by the African society

established in London, as a forerunner to the latter's more extensive travel account]. On Danish interest in west Africa, see Hopkins, D., 'Books, geography and Denmark's colonial undertaking in west Africa, 1790–1850', in Ogborn, M., and Withers, C. W. J. (eds), *Geographies of the Book*, pp. 221–46. The anonymous 1812 text based on the works of Park and of John Barrow is *Travels in the Interior of Africa, by Mungo Park. And in Southern Africa, by John Barrow. Interspersed with notes and observations, geographical, commercial, and philosophical. By the author of the* New System of Geography.

26 These claims are based on a book census survey of copies of Park's *Travels* held in UK university and college libraries, and from examining copies elsewhere as I could. The Spitalfields Society copy is held in the Fisher Rare Book Library, University of Toronto. Bryan Edwards' copy is in the University of Birmingham Library, Banks's in the British Library, Dugald Stewart's in the University of Edinburgh Library, Sir Walter Scott's in Abbotsford. Rennell's annotated copy is held by the Royal Geographical Society (with IBG). F. R. Rodd claims that 'Rennell's notes and his emendations on his own map of Park's travels were made from Major Laing's book, "Travels in . . . Western Africa," which was published by John Murray in 1825': Rodd, 'Rennell's comments upon the journeys of Park and Laing to the Niger', *Geographical Journal* 86 (1935), pp. 28–31, quote from p. 29. This cannot be so since Rennell dates his emendations at 1805 and clearly had sight of Park's notes and calculations from his second journey before the publication of Laing's 1825 book.

27 *The Edinburgh Magazine, or Literary Miscellany* (1799) 14, p. 33; *Anti-Jacobin Review and Magazine* (September 1799) 4, p. 14; *New London Review* (July 1799) 2, p. 31.

28 Duncan's lectures on medical jurisprudence emphasised not just the different causes of death – infanticide, drowning, suffocation, and so on – but upon the places in which they occurred and the institutions such as the medical police or other authorities who had oversight of them. See Duncan, A., *Heads of Lectures on Medical Jurisprudence, or the Institutiones Medicinae Legalis*, and UoE, CRC, Coll-1587, 'Notes from Dr Duncan's lectures on medical jurisprudence taken down by Charles Anderson, 1792–93.'

29 Park, *Travels*, pp. 297–8.

30 Duffill, M., 'Notes on a collection of letters written by Mungo Park between 1790 and 1794, with some remarks concerning his later published observations on the African slave trade', *Hawick Archaeological Society Transactions* (2001), pp. 35–55, see especially pp. 45–7.

31 Edwards, B., *History, Civil and Commercial, of the British Colonies in the West Indies*, 2, pp. 34, 72.

Chapter 5

1 BMNH, Banks MSS. A: 3; p. 52, Joseph Banks to Mungo Park, 10 September 1795.

2 The material on Friedrich Hornemann's *Proposal* and the Banks–Blumenbach correspondence is from Bovill, E. W. (ed.), *Missions to the Niger, I: The Journal of Friedrich Hornemann's Travels and The Letters of Alexander Gordon Laing*, pp. 1–39.

3 Godlewska, A. M. C., *The Napoleonic Survey of Egypt: A Masterpiece of Cartographic Compilation and Early Nineteenth-Century Fieldwork*.

4 BMNH, DTC, vol. 11, f. 269, Friedrich Hornemann to Joseph Banks, 19 August 1799.

5 The quote from Hornemann, from a letter to Banks of 29 November 1799, is from Bovill, E. W. (ed.) *Missions to the Niger*, I, p. 29. On disguise and trust, see Withers, C. W. J., 'Disguise – trust and truth in travel writing', *Terrae Incognitae* 52 (2021), pp. 48–64.

6 Hornemann, F., *The Journal of Frederick Hornemann's Travels: From Cairo to Mourzouk, the Capital of the Kingdom of the Fezzan, in Africa, in the Years 1797–8*. Hornemann's work was first published, in German, in 1801. The English edition was published by the African Association the same year as part of its *Proceedings* and includes maps and commentaries by James Rennell.

7 Bovill, *Missions to the Niger*, I, p. 37; Hallett, R. (ed.), *Records of the African Association 1788–1831*, p. 30, and, on Hornemann in more detail, pp. 171–90. The quotations from Hornemann are from pp. 175 and 186. Hornemann is sympathetically treated by Sattin in his *The Gates of Africa*, pp. 188–229, and, on his death, pp. 254–56.

8 BMNH, DTC, vol. 11, ff. 71–2, Mungo Park to Joseph Banks, 14 September 1798; f. 79, Mungo Park to Joseph Banks, 20 September 1798.

9 BMNH, DTC, vol. 11, f. 80, Joseph Banks to Mungo Park, 21 September 1798; f. 82, Banks to Robert Moss, 21 September 1798 on Park as 'this fickle Scotsman'; ff. 86–7, Banks to Park, 25 September 1798; ff. 88–9, Park to Banks, 26 September 1798. The Robert Ross letter of 24 September 1798 is inserted as the first item in the copy of Park's *Travels in the Interior Districts of Africa* owned by Patrick Miller of Dalswinton, now held in Borders Council (Selkirk Library) Archives.

10 Schwartz, J., *Robert Brown and Mungo Park: Travels and Explorations in Natural History for the Royal Society*.

11 On Robert Brown as a student of John Walker, see UoE, CRC, Dc. 1. 18, ff. 74–74v, May 1792. On Banks's and wider Enlightenment interests in Australia, see Gascoigne, J., *The Enlightenment and the Origins of European Australia*. Flinders and the *Investigator* voyage have been the subject of considerable attention: see Morgan, K., *Matthew Flinders, Maritime Explorer of Australia*.

12 The letter from Mungo Park to his wife on 10 January 1800 is NLS, MS 5308, f. 425. On the Macartney Mission, see Bickers, R. A. (ed.), *Ritual and Diplomacy: The Macartney Mission to China, 1792–1794*.

13 NLS, Acc. 12055, item 3, Mungo Park to Allison Park [no date, but probably January 1801]

14 NLS, Acc. 6207, Mungo Park to Joseph Banks, 13 October 1801.

15 On Rennell's 1802 essay, see Hallett, R., *Records of the African Association 1788–1831*, pp. 254–5. BMNH, DTC, vol. 12, ff. 296–303, James Rennell to William G. Browne, 31 December 1801. Brief mention of Watt and Winterbottom and their ventures in 1794 appears in Park, M., *Journal of a Mission to the Interior of Africa*, p. cviii.

16 On the Association's memorandum, see TNA, CO 267/10. On Banks to Lord Liverpool, President of the Board of Trade, see BL, Add. MSS. 38233, f. 94, Banks to Liverpool, 8 June 1799.

17 Lupton, K., *Mungo Park: African Traveler*, pp. 134–5.

18 BMNH, DTC, vol. 14, f. 162, Mungo Park to Joseph Banks, 20 October 1803. The wording 'Exploration pure and simple' is from Lupton, ibid., p. 135.

19 The 'Memoir *delivered by* Mungo Park, *Esq. to* Lord Camden, *on the 4th of* October, 1804', is in BMNH, DTC, vol. 15, ff. 232–41.

20 While it is clear that Park subscribed to the Niger–Congo theory by 1804, the precise substance of his conversations with Maxwell were not widely known until 1811 and were not published until 1820: see 'Account of the Correspondence between Mr Park and Mr Maxwell, respecting the identity of the Congo and the Niger. By the Rev. William Brown, D. D. Minister at Eskdalemuir', *Edinburgh Philosophical Journal* 3 (1820), pp. 102–14 and pp. 205–18. It is clear, however, that Rennell knew of Park's belief in the Niger and the Congo being the same river as he, Rennell, tried to persuade Park otherwise before Park departed for Africa for the second time.

21 This discussion of the contrasting ideas on the Niger's termination is taken from Appendix IV in Park, M., *The Journal of a Mission to the Interior of Africa, in the Year 1805*, pp. cxii–cxxiii. Although Park is the nominal author, the book was compiled from several sources by John Whishaw after Park's death.

22 Reichard, C. G., 'Ueber die vermuthung des Dr. Seetzen zu Jever, dass sich *Niger* in Afrika vielleicht mit dem Zaire vereingen konne', *Zach's Monatliche Correspondenz* (May 1802), pp. 409–15; Reichard, C. G., 'Ueber den angekündigten, nun bald erscheinenden, *Atlas des ganzen Erdkreises*', *Allgemeine Geographische Ephermeriden* 12 (1803), pp. 129–70, especially pp. 155–67.

23 As Hallett shows, Rennell certainly knew of Reichard's theory by 1815. In two letters to Banks that year, he laid out his reasons why Reichard's ideas should be rejected: evaporation rates made his [Rennell's] argument possible, but, principally, because of the Mountains of Kong: 'The improbability of the *Joliba*'s penetrating the ridge of Kong at all,' Hallett, R., *Records of the African Association 1788–1831*, p. 255. It seems likely that Rennell knew of Reichard's ideas before then, but this evidence of 1815 is the earliest in the English language to engage with the German's ideas.

24 'Memoir to Lord Camden', BMNH, DTC, vol. 15, f. 241.

Chapter 6

1 NLS, MS 20311, f. 43, 'A List of Mathematical Instruments and other articles provided by directions from Mr Sullivan, with receipts for the same, 24 January 1804.' On payments to Sidi Ombark Bouby between 1 January and 31 October 1804, see NLS, MS. 10708, f. 143. On Sidi Ombark Bouby and Andrew Anderson's night-time adventures in London, NLS, Acc 9834/2, Andrew Anderson to Thomas Anderson, 8 December 1804.

2 NLS, Acc 12055, items vii–xiii. Letters of Mungo Park to Allison Park, 9 October 1804; 28 October 1804, 14 November 1804, 28 November 1804; 10 December 1804; 26 December 1804; 3 January 1805.

3 NLS, Acc.12055, item xiv, Mungo Park to Allison Park, 28 January 1805. Park's letter of the same date to Thomas Anderson is NLS, MS 5319, f. 220.

4 The mention by Park of Allison's picture 'in fine stile' is from NLS, Acc 12055, item viii, Mungo Park to Allison Park, 28 October 1804. I have been unable to identify who the artist was or when it was done, or whether the original painting survives. It is highly likely that Mungo Park had his portrait painted by the Scottish artist and portraitist Henry Raeburn, probably in 1798 when Park had returned from his first African trip and was in Scotland visiting family and preparing his notes for his *Travels*. It is possible that Allison would have been painted then as well. Evidence for the Raeburn work on Mungo Park comes from an article in *Country Life* in 1968. There, a portrait, by an 'unknown artist of an unknown sitter', is presented for comment by a Miss Margaret Ritchie of St Boswells. The painting had long been in her family's keeping. While the editor confirms that the sitter was Park and suggests a date of 1798, the attribution of the artist as Raeburn was only tentative: *Country Life* 143, Issue 3719 (13 June 1968), pp. 1576–7. Comments from expert art historians and curators confirm the portrait as Park and the artist as Raeburn and suggest that the date of 1798 is very likely. Despite extensive enquiries, neither I nor the many others who have helped look have been able to trace the painting's whereabouts or to know what happened to Miss Ritchie's belongings at her death. No trace has been found of the painting of Allison Park.

5 BMNH, DTC, vol. 15, ff. 242–8. The quotes are from ff. 242, 244, 248, respectively. The mention of 'miasmata' is a reference to contemporaries' belief in miasmas/miasmatic theory, that diseases were caused by an accumulation of 'bad' air. This view, widely held until it was displaced by germ theory, was debated by many of the Niger's explorers in the 1830s and on the 1841 Niger expedition in particular: see chapters 10 and 11.

6 This phrase regarding Nicholls is from de Gramont, S., *The Strong Brown God: The Story of the River Niger*, pp. 102–3. Nicholls is treated briefly but affectionately by Sattin, A., *The Gates of Africa*, pp. 248–51, 253, and more fully by Hallett, R., (ed.), *Records of the African Association 1788–1831*, pp. 191–210, 268–9.

7 As Lupton notes, there is a date error of a day in Park's 1815 printed account. His letter to Banks makes clear that the expedition departed Kayee on 27 April;

his 1815 *Journal* notes 26 April. For consistency when citing from the *Journal*, I have kept to Park's original dating.

8 No unit or scale was given here but these must be Fahrenheit readings (although the Celsius scale was in use following its invention by Anders Celsius in 1742). In °C, the temperatures listed here would be, respectively, 38°C, 50°C, and 25°C.

9 NLS, Acc 12055, f. xvi, Mungo Park to Allison Park, 26 April 1805. The letter to Thomas Anderson is NLS, MS 581, f. 424, Mungo Park to Thomas Anderson, 26 April 1805.

10 These paragraphs are taken from the 'Copy of Mungo Park's Notes for his Journal, preserved by Sir J. Banks in his own handwriting', BMNH, DTC, vol. 15, ff. 357A–63.

11 Park, M., *The Journal of a Mission to the Interior of Africa, in the Year 1805*: 'interior of the country', 'about an hour after', and 'the immersion was correct', are from pp. 7, 21–2, and 24, respectively. The letter to Banks is BMNH, DTC, vol. 15, f. 365.

12 BMNH, DTC, vol. 15, ff. 356–7, Mungo Park to Joseph Banks, 26 April 1805.

13 Park, *Journal of a Mission*: 'at a moment's notice', p. 30; 'puzzled how to act', p. 31; 'recovered our guide', p. 32.

14 NLS, Acc 12055, f. xvii, Mungo Park to Allison Park, 29 May 1805.

15 BMNH, DTC, vol. 16, ff. 39–40, Mungo Park to Joseph Banks, 28 May 1805.

16 Park, *Journal of a Mission*, p. 54 (emphasis in the original). The quotes 'at the same time', 'an end to our journey', and 'to defend ourselves' are from pp. 37 and 47, respectively.

17 Park, ibid.: 'this place', p. 66 (emphasis in the original); 'wealth in their possession', p. 73; 'going into confusion', p. 74; 'to the last', p. 75.

18 Park, ibid.: 'to the ranks', p. 85; 'asses on the road', p. 86; 'very much lacerated', p. 89.

19 Park, ibid., p. 120.

20 Sattin, *Gates of Africa*, p. 239.

21 Park, *Journal of a Mission*, p. 75.

22 Park, ibid.: 'branches of the tree', p. 103; 'arrive at the Niger', p. 117; 'to walk or sit upright', p. 131; 'forwards to meet them', p. 133.

23 Park, ibid.: 'who spoke Mandingo', p. 137; 'stream along the plain', p. 140 (emphasis in the original).

24 Park, ibid., pp. 143, 144, respectively. Lupton gives the names, and the date and cause of death of the expedition's members in his Appendix II: Lupton, K., *Mungo Park African Traveler*, pp. 222–3.

25 Park, *Journal of a Mission*: 'violence all night', p. 143; 'sleep for six days', p. 145. Calomel, mercurous chloride, was widely used at this time to treat a variety of diseases. In large doses – such as that self-administered by Park – it could be dangerous, often worsening the symptoms it aimed to cure. Park came close at this moment to 'doing a Ledyard'.

26 Park, ibid.: 'seen any of them', p. 150; 'came into Bambarra', p. 151; 'of the white people' and 'if Mansong wishes it', p. 152; 'if I reach the salt water', p. 153.
27 Park, ibid.: 'count my cash', p. 159; 'water when loaded', p. 163.
28 Park, ibid.: 'wilds of Africa', p. 163; 'morning, or evening', p. 164.
29 NLS, Acc 12055, f. xix, Mungo Park to Allison Park, 19 November 1805 (emphasis in the original); NLS, MS 3278, ff. 60–61v, Mungo Park to Thomas Dean, 19 November 1805. The statement 'deranged in his mind' about an un-named private soldier is from Park, *Journal of a Mission*: p. lxxx.
30 BMNH, DTC, vol. 16, ff. 159–60, Mungo Park to Joseph Banks, 16 November 1805.
31 Park, *Journal of a Mission*, p. lxxx.
32 Park, ibid.: 'dead by their stings', p. 181; 'they are all dead', p. 207.

Chapter 7

1 Park, *Journal of a Mission*: 'ready for action' and 'great number of men', p. 209; 'near the river side' and 'attempting to escape', p. 214. The estimate of Park's death at four months after leaving Sansanding is given on p. 218.
2 Park, ibid., pp. lxxxv–lxxxvi; 'on his oath', pp. 217–18. On Park's words regarding Isaaco, see NLS, MS 3278, ff. 60–61v, Mungo Park to Thomas Dean, 19 November 1805, quote from f. 60v.
3 [Anon.], 'Review of *The Journal of a Mission to the Interior of Africa in the Year 1805, by Mungo Park: Together with other Documents, Official and Private, relative to the same Expedition: To which is prefixed, an Account of the Life of Mr. Park,*' London, John Murray, 1815, *Edinburgh Review* 24 (1815), pp. 471–90, quote from p. 483.
4 Park, ibid., p. lxxxvi.
5 *The Charleston Courier*, 1 July 1806. On the slave trade in this area at the time of Park's death, see Mouser, B. L., 'The trial of Samuel Samo and the trading syndicates of the Rio Pongo, 1797–1812', *International Journal of African Historical Studies* 46 (2013), pp. 423–41. The Rio Pongo flows into the Atlantic Ocean near Boffa in modern Guinea.
6 *The Times*, 12 January 1808; 19 May, 14 June, and 8 November 1810; 4 October 1811; 4 April 1812; *Edinburgh Evening Courant*, 1 January 1808.
7 NLS, MS 40908, no foliation.
8 RGS–IBG, MS 8/8.37, Letter from H. S. N. Edwardes to the secretary of the Royal Geographical Society, 6 February 1921. I am grateful to Francis Herbert, former Map Librarian at the RGS, for drawing this letter and sketch map to my attention.
9 De Gramont, S., *The Strong Brown God: The Story of the River Niger*, pp. 100–1.
10 Bovill, E. W., *The Niger Explored*, p. 31.
11 Burckhardt, J. L., *Travels in Nubia*, p. v.
12 Burckhardt, ibid., p. xxvi.

13 Burckhardt, ibid., p. lviii.

14 In addition to his *Nubia* volume, see Burckhardt, J. L., *Travels in Syria and the Holy Land, by the late John Lewis Burckhardt published by the Association for Promoting the Discovery of the Interior Parts of Africa*. The anglicisation of Burckhardt's name was a marketing ploy by John Murray.

15 Burckhardt, *Travels in Syria and the Holy Land*, p. ii. The extracts from the instructions to Buckhardt are taken from Hallett, R. (ed.), *Records of the African Association 1788–1831*, pp. 218–23. The quote from Rennell regarding Burckhardt is from BMNH, DTC, vol. 18, p. 288.

16 Hallett, *Records of the African Association 1788–1831*, pp. 226–7.

17 Adams, R., *The Narrative of Robert Adams*, p. xii. The comments on the book being set to the same format as the Park volumes come from the anonymous reviewer in the *Quarterly Review* 14 (January 1816), p. 473.

18 For Jomard's view of Adams, see Heffernan, M., '"A Dream as Frail as Those of Ancient Time": The in-credible geographies of Timbuctoo', *Environment and Planning D: Society and Space* 19 (2001), pp. 203–25, quote from p. 209.

19 'Account of the captivity of Alexander Scott, among the wandering Arabs of the Great African Desert, for a period of nearly six years. Drawn up by T. S. Traill, M. D., F.R.S.E. With geographical observations on his routes, and remarks on the currents of the ocean on the north-western coast of Africa, by Major Rennell, F.R.S.', *Edinburgh Philosophical Journal* 4 (1821), pp. 38–54, 225–34, and 235–41. A further full account of the Scott case, differently titled and with Rennell's name given as 'Russell' but otherwise with the same essential content, appears in the Traill papers held at the National Library of Scotland (NLS, MS 19390).

20 Riley, J., *Loss of the American Brig Commerce*, pp. x, xiv, 605. For a fuller discussion of the Riley volume and its American predecessor, see Keighren, I., Withers, C. W. J., and Bell, B., *Travels into Print: Exploration, Writing, and Publishing with John Murray 1773–1859*, pp. 117–21.

21 This phrase is from the *Rules and Regulations of the African Institution: Formed on the 14th April 1807*, p. 9.

22 These cover a brief history of education in Scotland (Appendix I); extracts from the work of d'Anville, Rennell, and Watt and Winterbotham on west Africa (II); a summary of the exchanges between William Young and Park over the former's imputation regarding Edwards' assistance to Park on his *Travels* (III); the different ideas then current respecting the Niger's termination (IV); a note on the botanical specimens collected by Park and sent to Banks (V); and a note on the increase of trade between Britain and Africa since the abolition of the slave trade in 1807 (VI).

23 BCA, SC/5/56, Item 7, ff. 1–2v, John Whishaw to Lord Camden [undated but *c*. January 1814].

24 'Advertisement', to Park, *Journal of a Mission*, [no pagination but final page of the Advertisement].

25 NLS, MS 40909, John Murray to John Whishaw, 6 February 1814; NLS, MS 41263, Letter 18, John Whishaw to John Murray, 14 February 1814.

26 NLS, MS 786, ff. 11–12v, Walter Scott to John Whishaw, 24 April 1815.
27 [Thomas Thomson], 'Isaaco's Journal of a Voyage after Mr. Mungo Park, to ascertain his Life or Death', *Annals of Philosophy* (November 1814), Article XI, pp. 369–86.
28 NLS, MS 41908, ff. 370–2, John Murray to John Whishaw, 9 November 1814.
29 BCA, SC/5/56, Item 8, John Whishaw to Joseph Banks, 26 November 1814. On the alteration in timing to Park's family, see BCA, SC/5/56, Item 9, John Whishaw to Joseph Banks, 3 December 1814.
30 NLS, MS 41263. On Murray to Bulmer over the paper shortages and the implications for cost, see NLS, MS 40908, f. 3, John Murray to William Bulmer, 7 December 1814.
31 NLS, MS 20311, ff. 3–6, Indenture between John Thornton, treasurer to the African Institution, Adam Park, then a military surgeon in Gravesend, and Archibald Buchanan, a Glasgow accountant and brother-in-law to Allison Park, 7 June 1816. On the costs incurred and profits made in publishing *The Journal of a Mission*, see NLS, MS 42725, ff. 1–2.
32 NLS, MS 41263, [no foliation], Letter 11, John Whishaw to John Murray, 9 December 1814.
33 NLS, MS 41263, [no foliation], Letter 12, John Whishaw to John Murray, 18 November 1815.
34 [Henry Brougham], [Review of *The Journal of a Mission to the Interior of Africa, in the Year 1805, by Mungo Park*], *Edinburgh Review* 24 (1814–15), pp. 471–90, quotes from pp. 490 and 472, respectively.
35 [John Barrow], [Review of *The Journal of a Mission to the Interior of Africa, in the Year 1805, by Mungo Park*], *Quarterly Review* 13 (1815), pp. 120–51.
36 On Adam Park's remarks to Whishaw, see NLS, MS 41263, [no foliation], Letter 14, John Whishaw to John Murray, 18 May 1815. On Whishaw's remarks about controversy potentially assisting sales, see MS 41263, [no foliation], Letter 17, John Whishaw to John Murray, 10 April 1815.

Chapter 8

1 Barrow, J., 'Introduction', in Tuckey, J. H., *Narrative of an Expedition to Explore the River Zaire*, pp. ii–iii. Bovill is insistent regarding Barrow's role on the Niger: 'Nobody contributed more to the solution of the Niger problem than John Barrow but for whom British interest in African exploration might well have withered on the vine in the early years of the century. By his enthusiasm and persistent advocacy of an unpopular cause he prevented the continuance of the work begun by Joseph Banks and the African Association passing into foreign hands.' (Bovill, E. W., *The Niger Explored*, p. 245.) On Barrow's role in orchestrating British exploration more widely in the first half of the nineteenth century, see Fleming, F., *Barrow's Boys*.
2 The full incident, taken from John Ross's unpublished memoir [dated 1848], is

cited in Keighren, I. M., Withers, C. W. J., and Bell, B. (eds), *Travels into Print: Exploration, Writing, and Publishing with John Murray 1773–1859*, pp. 93–4.

3 On Barrow as reviewer in the *Quarterly*, see Cutmore, J. B., 'Sir John Barrow's contributions to the *Quarterly Review* 1809–24', *Notes and Queries* 41 (1996), pp. 326–8, and Cameron, J. M. R., 'Sir John Barrow as a *Quarterly* reviewer, 1809–1843', *Notes and Queries* 43 (1998), pp. 34–7.

4 The first part of this subheading is from Bovill's description of the Peddie–Campbell and Tuckey expeditions: Bovill, *The Niger Explored*, pp. 33–40.

5 TNA, ADM 1/4234, Henry Goulburn to John Barrow, 2 August 1815.

6 On Peddie acknowledging receipt of the Adams volume from Goulburn, see TNA, CO 2/5, f. 42, James Peddie to Henry Goulburn, 3 August 1816. On Cock's offering advice to the Colonial Office, see TNA, CO 267/41 [no folio]. Gray, W., and Dochard, D., *Travels in Western Africa, in the Years 1818, 19, 20, and 21, from the River Gambia, through Woolli, Bondoo, Galam, Kasson, and Foolidoo, to the River Niger*.

7 The description of 'profitless' and a 'tedious volume' is from Bovill, *The Niger Explored*, p. 34; of a 'record . . . and misunderstanding' from Mouser, B. L. (ed.), *The Forgotten Peddie/Campbell Expedition into Fuuta Jaloo, West Africa, 1815–17*; 'lost in place' from Kennedy, D., 'Lost in place: two expeditions gone awry in Africa', in Mayhew, R. J., and Withers, C. W. J. (eds), *Geographies of Knowledge: Science, Scale, and Spatiality in the Nineteenth Century*, pp. 235–54. See also Mouser, B. L., 'Forgotten expedition into Guinea, West Africa, 1815–1817: an editor's comments', *History in Africa* 35 (2008), pp. 481–9.

8 Gray and Dochard, *Travels in Western Africa*, p. 349 (on the Niger), on meeting and journeying with Isaaco, pp. 239–41.

9 The Tuckey papers are principally held as TNA, ADM 1/2617 (Bundle T34): Tuckey's journal, instructions to the expedition members, and other relevant papers.

10 Barrow, 'Introduction', in Tuckey, *Narrative of an Expedition to Explore the River Zaire*, pp. xxxvi, and xli–xlii.

11 On thanking the editor, see the anonymous review in *British Critic* 10 (August 1818), pp. 154–5; on awareness of Barrow's insertions and editorship, see the anonymous review in *Monthly Review* 86 (June 1818), pp. 113–29.

12 Barrow, 'Introduction', in Tuckey, *Narrative of an Expedition to Explore the River Zaire*, p. xx (on Reichard), p. xxii (on Park's death), p. xliii (on Eyre). NLS, MS 40555, f. 70, John Barrow to John Murray, 8 May 1817. On Tuckey rejecting the Niger–Congo hypothesis, see TNA, ADM 1/1267, T34, James Tuckey to John Barrow, 30 August 1815. On Tuckey's papers reaching the Admiralty, see TNA, ADM 1/2617, 6 March 1817.

13 The Ritchie–Lyon 'Mission to Fezzan' is discussed in Bovill, *The Niger Explored*, pp. 55–70.

14 Bowdich, T. E., *Mission from Cape Coast Castle to Ashantee, with a Statistical Account of that Kingdom, and Geographical Notices of other Parts of the Interior of Africa*, 'last places he visited', p.11; 'realising beyond my expectations', p. 86.

15 Bowdich, *Mission from Cape Coast Castle to Ashantee*, pp. 90–1.

16 Salamé, A., *A Narrative of the Expedition to Algiers in the Year 1816, under the Command of the Right Hon. Admiral Lord Viscount Exmouth*; Jackson, J. G., *An Account of the Empire of Marocco, and the District of Suse*; and Jackson, J. G., *An Account of Timbuctoo and Housa, Territories in the Interior of Africa*. The term 'Go-Between' is from Schaffer, S., Roberts, R., Raj, K., and Delbourgo, J. (eds), *The Brokered World: Go-Betweens and Global Intelligence, 1770–1820*.

17 For the two translations, see Bowdich, *Mission to Cape Coast Castle to Ashantee*, pp. 478–81. On their later scrutiny, see [Anon.], 'Remarks on the two accounts of the death of Mungo Park', *Asiatic Journal and Monthly Miscellany* 11 (1821), pp. 453–6.

18 Bowdich, *Mission to Cape Coast Castle to Ashantee*, 'most satisfactory manner', p. 186. This summary is distilled from his 'Geography' chapter, pp. 161–228.

19 [John Barrow], 'Review of 1. *Mission from Cape Coast Castle to Ashantee, with a Statistical Account of that Kingdom, and Geographical Notices of other Parts of the Interior of Africa*. By T. Edward Bowdich, Esq. Conductor. 1819. 2. *The African Committee*. By T. Edward Bowdich, Esq. Conductor of the Mission to Ashantee. 1819', *Quarterly Review* 22 (1820–1), pp. 273–302, quote from p. 292. Hugh Murray, 'Observations on the information collected by the Ashantee Mission, respecting the course of the Niger, and the interior of Africa', *Edinburgh Philosophical Journal* 1 (1819), pp. 163–70, quotes from pp. 164 and 170.

20 TNA, F.O. 8/7, Lord Bathurst to Joseph Ritchie, 1 February 1818. The view of the Fezzan as a forward base for African discovery is articulated in a letter to Goulburn from the Treasury in December 1817: TNA, CO 2/8, 3.

21 TNA, CO 2/8, 11, Hanmer Warrington to Lord Bathurst, 7 March 1818.

22 Lyon, G. F., *A Narrative of Travels in Northern Africa, in the Years, 1818, 19, and 20; accompanied by Geographical Notices of Soudan, and the Course of the Niger. With a chart of the routes, and a variety of coloured plates, illustrative of the costumes of the several natives of Northern Africa*: 'some have supposed', p. 146; 'southward of Dongola', p. 148.

23 [John Barrow], 'Review of *Voyage dans l'Intérieur de l'Afrique aux Sources du Sénégal et de la Gambie, fait en 1818, par ordre du Gouvernement Francais*. Par. G. Mollien. (Paris, 1820)', *Quarterly Review* XXIII (1820), pp. 225–44, map on p. 237.

24 Bovill, E. W. (ed.), *Missions to the Niger, I: The Journal of Friedrich Hornemann's Travels and the Letters of Alexander Gordon Laing*; Bovill, E. W. (ed.), *Missions to the Niger II: The Bornu Mission, 1822–25 Part 1*; *Missions to the Niger III, The Bornu Mission, 1822–25 Part 2*; *Missions to the Niger IV: the Bornu Mission, 1822–25 Part 3*. Bovill summarised the Bornu Mission as chapter V in his *The Niger Explored*, pp. 71–148. The quotes are from this summary work, pp. 145 and 146, respectively. On Denham's notebooks from which Murray omitted sections of the original narratives, see RGS–IBG, DD/14–DD/16, DD/17, and DD/18, 1–4.

25 On the Bornu appointments, see TNA, CO 2/13, especially ff. 1–292; FO 8/8, ff. 1–90.

26 Denham, D., Clapperton, H., and Oudney, W., *Narrative of Travels in Northern and Central Africa, in the years 1822, 1823, and 1824, by Major Denham, F.R.S., Captain Clapperton, and the late Dr Oudney*, 1, pp. 14.

27 These contradictory ideas were discussed by John Barrow as he reviewed Robert Adams' account of his African experiences. [Barrow, J.], 'Review of Adams, R., *Sketches taken during Ten Voyages to Africa*', *Quarterly Review* 29 (1823), pp. 508–24. As Barrow stated there (p. 522), 'where is the celebrated Niger?'

28 Bovill, *Missions to the Niger III*, pp. 382–3.

29 Denham, Clapperton, and Oudney, *Narrative of Travels in Northern and Central Africa,* Appendix No. VIII, pp. 147–8.

30 The classic example is that of Jean-François de Galaup, compte de Lapérouse, investigating Sakhalin in the 1780s. On making landfall on the beach, an indigenous Ainu man drew a map in the sand in attempting to answer the Frenchman's query whether Sakhalin was an island or a peninsula: Bravo, M., 'Ethnographic navigation and the geographical gift', in Livingstone, D. N., and Withers, C. W. J. (eds), *Geography and Enlightenment*, pp. 199–235.

31 [Barrow, J.], 'Prefatory Notice to the Narrative of Captain Clapperton's Journey from Kouka to Sackatoo', in Denham, Clapperton, and Oudney, *Narrative of Travels in Northern and Central Africa*, p. 2: unpaginated, but preceding Clapperton's 'Journal'.

32 [Barrow], 'Prefatory Notice to the Narrative of Captain Clapperton's Journey. On Barrow's review of the Bornu Mission, see [Barrow, J.], 'Recent discoveries in Africa, made in the years 1823 and 1824, by Major Denham, Captain Clapperton, R.N. and the late Dr Oudney, extending across the Great Desert to the tenth degree of northern latitude, and from Kouka in Bornou to Sackatoo, the capital of the Soudan Empire', *Quarterly Review* 33 (1826), pp. 519–49. Given Barrow's admission to Murray that he had 'trimmed' Clapperton's journal, his praise in the review (p. 530) for Clapperton's 'unaffected and manly style' is mischievous.

33 [Barrow], ibid., *Quarterly Review* 33 (1826), p. 549.

34 TNA, CO 2/14, f. 31, John Barrow to Robert Wilmot-Horton, 11 November [no year date given but 1824].

35 Laing, A. G., *Travels in the Timannee, Kooranko, and Soolima countries in western Africa.*

36 Bovill, *Missions to the Niger*, I, p. 294.

37 Bovill, *Niger Explored*, p. 182.

38 Barrow, J., 'Art VI. Journal d'un Voyage à Temboctoo et à Jenné, dans l'Afrique Centrale, &c.', *Quarterly Review* 42 (1830), p. 459.

39 TNA, CO 2/20, Alexander Laing to Hanmer Warrington, 21 September 1826.

40 Laing's argument that the Niger ran into the Bight of Benin, which he principally based upon Clapperton's work and Sultan Bello's oral account, is laid out in a document he produced in Gademis on 28 September 1825: Gordon

Laing, 'Cursory remarks on the course and termination of the Great River Niger': TNA, CO 2/115, ff. 119–30.

41 Heffernan, M., '"A Dream as Frail as those of Ancient Time": the in-credible geographies of Timbuctoo', *Environment and Planning D: Society and Space* 19 (2001), pp. 203–25.

42 Bovill, *Niger Explored*, p. 181. The sense that Laing was attuned to Sahelian customs and life and so able to benefit from his time in Timbuktu is clear in Wise, C., 'Writing Timbuktu: Park's hat, Laing's hand', in Wise, C. (ed.), *The Desert Shore: Literatures of the Sahel*, pp. 175–200.

43 TNA, CO 323/144, John Wilson Croker to Robert Wilson Hay, 13 February 1826.

Chapter 9

1 On 'critical geography' as it concerned the nineteenth-century exploration of east central Africa, see Dritsas, L., 'Expeditionary science: conflicts of method in mid-nineteenth-century geographical discovery', in Livingstone, D. N., and Withers, C. W. J. (eds), *Geographies of Nineteenth-Century Science*, pp. 255–77, and Bridges, R. C., 'W. D. Cooley, the Royal Geographical Society and African geography in the nineteenth century', *Geographical Journal* 142 (1976), pp. 27–47 and pp. 274–86. On Rennell as a Classical critical geographer, see Koelsch, W., *Geography and the Classical World: Unearthing Historical Geography's Hidden Past*, pp. 106–16. On geography's subject content and social status in nineteenth-century Britain, see Withers, C. W. J., *Geography and Science in Britain, 1831–1939: A Study of the British Association for the Advancement of Science*. On professionalisation in modern science in which context a curriculum, disciplinary identity and a salary were part, see Morrell, J. B., 'Professionalisation', in Olby, R. C., Cantor, G. N., Christie, J. R. R., and Hodge, M. J. S. (eds), *Companion to the History of Modern Science*, pp. 980–9.

2 Dudley, J., *A Dissertation showing the Identity of the Rivers Niger and Nile; chiefly from the Authority of the Ancients*, p. 95.

3 Dudley, ibid., pp. 92–3.

4 On Barrow's review of Dudley, which he incorporated in a fuller discussion of Lyon's 1821 *Narrative*, see [Barrow, J.], 'Review of *A Narrative of Travels in Northern Africa, in the Years, 1818, 19, and 20; accompanied by Geographical Notices of Soudan, and the Course of the Niger. With a chart of the routes, and a variety of coloured plates, illustrative of the costumes of the several natives of Northern Africa*. By G. F. Lyon', *Quarterly Review* 25 (1821), pp. 25–50: Dudley is discussed on pp. 46–50. William Roberts' review appears in *The British Review and London Critical Journal* 18 (1821), pp. 198–202.

5 Donkin, R. F., *Dissertation on the Course and Probable Termination of the Niger*, pp. 3–4, 5.

6 For fuller discussion, see Withers, C. W. J., *Zero Degrees: Geographies of the Prime Meridian*.

7 Donkin, *Dissertation on the Course and Probable Termination of the Niger*, p. 61.
8 Donkin, ibid.: '*prima facie* evidence', p. 109; 'to the very last', p. 122; 'mountain base of Central Africa', p. 123.
9 BL, Add MSS 34614, ff. 110–11, John Barrow to Charles Napier, 26 July 1829.
10 [Barrow, J.], 'Review of *Dissertation on the Course and Probable Termination of the Niger*. By Lt.-Gen. Rufane Shaw Donkin, London, 1829', *Quarterly Review* 41 (1829), pp. 226–39. On Barrow's encouragement of Beechey, following the coastal survey work of W. H. Smyth, see TNA, CO 2/12, ff. 107–8, John Barrow to Henry Goulburn, 6 February 1821. The Beechey survey was published as Beechey, F. W., and Beechey, H. W., *Proceedings of the Expedition to Explore the Northern Coast of Africa, from Tripoly Eastward; in MDCCCXXI. and MDCCCXXII. comprehending an account of the Greater Srytis and Cyrenaica; and of the Ancient Cities comprising the Pentapolis.*
11 [Donkin, R. F.], *A Letter to the Publisher of the Quarterly Review*, p. 3.
12 BL, Add MSS 34614, ff. 187–8, John Barrow to Charles Napier, 28 September 1829.
13 Hutton, W., *A Voyage to Africa: including A Narrative of an Embassy to one of the Interior Kingdoms in the Year 1820; with remarks on the course and termination of the Niger, and other principal rivers in that country*, p. 3.
14 Dupuis, J., *Journal of a Residence in Ashantee*, p. xxvi. Dupuis's geographical chapter is chapter VIII of his *Journal* where, without explanation, pagination starts afresh, in lower case Roman numerals, all preceding chapter pagination being in Arabic figures.
15 Dupuis, ibid., p. xxvii.
16 Hutton, *Voyage to Africa*, pp. 416, 390.
17 Hutton, ibid., pp. 390, 391.
18 MacQueen, J., *A Geographical and Commercial View of Northern Central Africa: containing a Particular Account of the Course and Termination of the great River Niger in the Atlantic Ocean*, p. v.
19 Robertson, G. A., *Notes on Africa: Particularly Those Parts which are Situated between Cape Verd and the River Congo*. On one occasion, Robertson describes being told by one Ado Gheese about his encountering four white men, two of whom were very ill, descriptions of one matching Park, but at a time later and in a location well to the south of other reports concerning Park's second expedition and reported death. Robertson noted this but in closing remarked 'After all, this account must be received with great caution, there being nothing in Mr. Park's account to corroborate it; nor in Isaaco's Journal.' Robertson, ibid, p. 187.
20 Lambert, D., *Mastering the Niger: James MacQueen's African Geography & the Struggle over Atlantic Slavery*, especially pp. 31–57, and, on MacQueen's reading aloud from Park's *Travels*, pp. 105–16.
21 In his 1821 book, MacQueen wrote that his ideas concerning the Niger running into the Atlantic Ocean were first articulated in a 'treatise' or 'pamphlet' (he used both terms) which was published in Edinburgh in 1816: 'it is nearly

five years, since, in a small treatise, I pointed out that, in the Bights of Benin and Biafra, the Niger certainly entered the Ocean': MacQueen, *Geographical and Commercial View of Northern Central Africa*, p. vii. This *c.*1816 work has resisted all attempts to find it in major libraries and map catalogues: Lambert makes no mention of it in his helpful listing of 'Publications by or attributed to James MacQueen': Lambert, *Mastering the Niger*, pp. 274–6. The fact that the map by MacQueen is dated 6 June 1820 makes it unlikely that this *c.*1816 work was accompanied by a map, and unlikely too that his arguments changed between then and 1821.

22 NLS, MS 4005, f. 201, James MacQueen to William Blackwood, December 1820. MacQueen's commercial vision for Africa is discussed by Curtin, P. D., *The Image of Africa: British Ideas and Action, 1780–1850.* On this description of MacQueen, see Lambert, *Mastering the Niger*, p. 4.

23 MacQueen, *Geographical and Commercial View of Northern Central Africa*, p. vii.

24 NLS, MS 4011, f. 9, James MacQueen to William Blackwood, 9 April 1823; MacQueen, J., 'M'Queen on the course and termination of the Niger', *Blackwood's Edinburgh Magazine* 13 (1823), pp. 417–32; Barrow, J., 'Review of *A Geographical and Commercial View of Northern Central Africa: containing a particular Account of the Course and Termination of the Great River Niger in the Atlantic Ocean*. By James MacQueen, Edinburgh, 1821', *Quarterly Review* 26 (1822), pp. 51–81.

25 NLS, MS 4017, f. 217, James MacQueen to William Blackwood, 8 May 1826.

26 [MacQueen, J.], 'Geography of Central Africa: Denham and Clapperton's Journal', *Blackwood's Edinburgh Magazine* 19 (1826), pp. 687–709.

27 NLS, MS 4017, f. 225, James MacQueen to William Blackwood, 21 November 1826.

28 NLS, MS 4017, f. 223, James MacQueen to William Blackwood, 12 November 1826. [MacQueen, J.], 'The River Niger – Termination in the Sea', *Blackwood's Edinburgh Magazine* 30 (1831), pp. 130–6.

29 [Barrow, J.], 'Review of *A Geographical and Commercial View of Northern Central Africa*', *Quarterly Review* 26 (1822), p. 55 and p. 56, respectively. MacQueen's view of the close connections between the *Quarterly Review* and the Government is apparent in a note to Blackwood in August 1827: 'The Quarterly Review <u>was</u> the Channel through which the Colonial Office spoke on our Colonial subjects': NLS, MS 4019, f. 309, James MacQueen to William Blackwood, 5 August 1827. The clear implication of the underlining for emphasis was that MacQueen hoped that *Blackwood's Edinburgh Magazine* would replace the *Quarterly* in this respect.

30 [Barrow, J.], 'Recent discoveries in Africa, made in the years 1823 and 1824, by Major Denham, Captain Clapperton, R.N. and the late Dr Oudney, extending across the Great Desert to the tenth degree of northern latitude, and from Kouka in Bornou to Sackatoo, the capital of the Soudan Empire', *Quarterly Review* 33 (1826), pp. 519–49, quotes from pp. 543 and 544.

31 This is particularly so of Barrow and the *Quarterly Review*, a point stressed by Cameron, J. M. R., 'John Barrow, the *Quarterly*'s imperial reviewer', in Cutmore, J. (ed.), *Conservatism and the Quarterly Review*, pp. 133–49.

32 Wilson, S. A., *Real Stories: Taken from the Narratives of Various Travellers*, pp. 62–3, p. 106.

33 The 'Various opinions' quote is from Playfair, J., *A System of Geography, Ancient and Modern*, VI: pp. 226–7. On Playfair as a geographical author and teacher, see Withers, C. W. J., 'James Playfair (1738–1819)', *Geographers Biobibliographical Studies* 24 (2005), pp. 79–85. On ideas around the geography of reading, see Secord, J. A., *Victorian Sensation: The Extraordinary Publication, Reception, and Secret Authorship of* Vestiges of the Natural History of Creation, and James A. Secord, *Visions of Science: Books and Readers at the Dawn of the Victorian Age*.

34 NLS, Acc 9834/6 [no date, but January 1823]. The memorial was first drawn up in November 1822 but revised as various names, including leading figures from Selkirkshire and other counties, were added as memorialists.

35 NLS, Acc 9834/7, Adam Park to Thomas Anderson, 1 December 1824.

36 MacQueen, 'Geography of Central Africa', p. 705.

Chapter 10

1 TNA, CO 2/16, ff. 191–207, Lord Bathurst to Hugh Clapperton, 30 July 1825.

2 Bovill only suggests that Barrow was involved: Bovill, E. W., *The Niger Explored*, p. 192. The evidence that Barrow drafted Clapperton's instructions but that they were issued under Bathurst's name is clear in the correspondence from Barrow to Hay on 25 July 1825: TNA, CO 323/144.

3 The suggestion that Columbus might travel ahead was made by Barrow in a letter to Hay in the Colonial Office before Clapperton's instructions were issued by Bathurst: TNA, CO 2/17, f. 13, John Barrow to Robert Hay, 5 July 1825. This detail is significant on two counts: Barrow was preparing Clapperton's instructions throughout July 1825; Hay was helping supervise African affairs before his formal appointment.

4 TNA, CO 2/17, ff. 29–30, John Barrow to Robert Hay, 24 August 1825.

5 TNA, CO 323/144, John Barrow to Robert Hay, 21 April 1826 and 22 May 1826.

6 [MacQueen, J.], 'Geography of Central Africa: Denham and Clapperton's Journal', *Blackwood's Edinburgh Magazine* 19 (1826), pp. 687–709, quote from p. 698.

7 Clapperton's writing, and his life and travels, is excellently served in Lockhart, J. B., *A Sailor in the Sahara: The Life and Travels in Africa of Hugh Clapperton, Commander R.N.*

8 Clapperton, H., *Journal of a Second Expedition into the Interior of Africa, from the Bight of Benin to Socatoo, to which is added The Journal of Richard Lander*

from Kano to the Coast, partly by a more eastern route, p. 84. On the belief that Europeans bought slaves to eat them, see Bovill, *The Niger Explored*, p. 28.

9 Clapperton, ibid., p. 100.

10 Clapperton, ibid., pp. 134–5.

11 TNA, CO 392/3, 44–6, Robert Hay to Hanmer Warrington, 22 July 1825.

12 Clapperton, *Journal of a Second Expedition*, pp. 316–17.

13 [John Barrow], 'Introduction', in Clapperton, ibid., pp. xviii, xix.

14 Barrow, ibid., p. xx.

15 Barrow, ibid, p. xix.

16 On this claim, see Lockhart, *A Sailor in the Sahara*, p. 264.

17 [Barrow, J.], 'Review of *Journal of a Second Expedition into the Interior of Africa, from the Bight of Benin to Soccatoo*', *Quarterly Review* 39 (1829), pp. 143–83, quotes from p. 177 and p. 179, respectively.

18 TNA, CO 2/17, f. 378, James Rennell to Robert Hay, 26 September 1826. Barnes, apparently, knew of the course of trade upriver from Timbuktu and beyond. There is no evidence to show that Barrow knew of Rennell's disclosure to Hay, but it is probable that he did.

19 TNA, CO 324/78, ff. 80–1, Robert Hay to William Martin Leake, 3 March 1827.

20 Barrow, 'Review of *Journal of a Second Expedition into the Interior of Africa*', *Quarterly Review* 39 (1829), p. 180.

21 Barrow's comments on Lander's 'Wanderings in Africa' appear in NLS, MS 40057, ff. 33–34v [no day/month/year but 1829]. The book was published as Lander, R., *Records of Captain Clapperton's Last Expedition to Africa*, the subtitle describing Lander as Clapperton's 'faithful attendant, and the only surviving member of the expedition'. Reviewers understood that Lander's style in his contribution to Clapperton's 1829 *Journal* was a product of circumstance, not least Lander's own illness, and that Lander was engaged in a second expedition: see, for example, the *Monthly Magazine, or, British Register* 9 (1830), pp. 462–3, which review ends by noting that 'Lander is already on his way back to Africa' (p. 463).

22 Lander, R., and J., *Journal of an Expedition to Explore the Course and Termination of the Niger; with a Narrative of a Voyage down that River to its Termination*, II, quotes from pp. 3, 6, and 12–13, respectively.

23 Lander, R., and J., ibid., II: 'no where in existence', p. 45; 'a noble river', p. 258; 'pride and enthusiasm', p. 279.

24 Lander, R., and J., ibid., II, 'covetous prince', pp. 301–2. John Lander's description of Park's tobe is from his 'Diary kept by John Lander on the Niger', NLS, MS 42326, ff. 370–1.

25 Lander, R., and J., ibid., I, p. 205.

26 Lander, R., and J., ibid., III: 'amazement and terror', p. 33; 'object we had in view', pp. 45–6; 'from the sea coast' and 'anticipations vanished', p. 133; 'astonishing rapidity' and 'saw him no more', p. 134; 'world of ours', p. 233; 'of considerable magnitude', p. 171.

27 Lander, R., and J., ibid., III: 'knowledge of our language', p. 187; 'satisfaction and rapture, p. 270; 'my powers of description', p. 244.
28 Bovill, *The Niger Explored*, p. 244.
29 Hallett, *The Niger Journal of Richard and John Lander*, p. 33.
30 Lander, R., and J., ibid., III, pp. lxiii–lxiv.
31 NLS, MS 40668 [no folio], John Lander to John Murray [no date but 1831].
32 NLS, MS 4030, James MacQueen to William Blackwood, 16 June 1831. The Ross here referred to is probably Sir James Clark Ross, with whom Barrow was on good terms. Barrow and the Arctic explorer John Ross did not get on.
33 [Barrow, J.], 'Review of *The Journal of the Royal Geographical Society of London for the Year 1830–1831*', *Quarterly Review* 46 (1832), pp. 55–80, quotes from pp. 78 and 79.
34 Barrow, ibid., pp. 79 and 80; Bovill, *The Niger Explored*, p. 246. On the publication of the Landers' 1832 book, see Hallett, R., *The Niger Journal of Richard and John Lander*. Barrow's view that the name 'Niger' should not be used to mark the whole course of this single river also features in his introduction to Clapperton's 1829 book: 'The name of Quorra, or Cowarra, by which it is known universally in Soudan, and probably also to the westward of Timbuctoo, ought now, therefore, to be adopted on our charts of Africa.' (Clapperton, *Journal of a Second Expedition*, p. xix.)
35 [MacQueen, J.], 'Geography of Africa – Quarterly Review. Letter from James MacQueen', *Blackwood's Edinburgh Magazine* 31 (1832), pp. 201–16, quote on p. 211.
36 NLS, MS 4030, f. 153v, James MacQueen to William Blackwood, 29 May 1831.

Chapter 11

1 De Gramont, S., *The Strong Brown God: The Story of the Niger River*, p. 191.
2 [MacQueen, J.], 'Geography of Africa – Quarterly Review. Letter from James MacQueen', *Blackwood's Edinburgh Magazine* 31 (1832), p. 216.
3 Leake, W. M., 'Is the Quorra, which has lately been traced to its Discharge into the Sea, the same River as the Nigir of the Ancients?', *Journal of the Royal Geographical Society of London* 2 (1832), pp. 1–28, quotes from pp. 1 and 28, respectively.
4 TNA, CO 324/83, ff. 196–7, Robert Hay to John Barrow, 19 August 1831.
5 Lander, R., and J., *Journal of an Expedition to Explore the Course and Termination of the Niger; with a Narrative of a Voyage down that River to its Termination*, III, p. 314.
6 Laird, M., and Oldfield, R. A. K., *Narrative of an Expedition into the Interior of Africa, by the River Niger, in the steam-vessels Quorra and Alburkah in 1832, 1833 and 1834*, I: 'in every possible way' and 'effort to shake off', p. 122; 'not half a ton of ivory', p. 143.
7 Laird and Oldfield, ibid., p. 173.

8 RGS–IBG, JMS 1/29, MacGregor Laird to Roderick Murchison, 23 April 1852.
9 Laird and Oldfield, *Narrative of an Expedition into the Interior of Africa*, p. v.
10 Nicholls, E., Calabar, D., and Coulthurst, C. H., 'Failure of another expedition to explore the interior of Africa', *Journal of the Royal Geographical Society* 2 (1832), pp. 305–12.
11 Becroft, J., 'On Benin and the upper course of the River Quorra, or Niger', *Journal of the Royal Geographical Society of London* 11 (1841), pp. 184–192.
12 RGS–IBG, JMS 1/3, Edward Nicholls to R. W. Hay, 1 July 1833.
13 Allen, W., 'On a new construction of a map of a portion of West Africa, showing the possibility of the Rivers Yéu and Chadda being the outlet of the Lake Chad', *Journal of the Royal Geographical Society of London* 8 (1838), pp. 289–307, quotes from pp. 291, 292, and 295 respectively.
14 On Allen and Oldfield and their differing opinions over the river systems making up the Niger's delta, see Oldfield, R. A. K., 'A brief account of the ascent of the Old Calabar river in 1836', *Journal of the Royal Geographical Society of London* 7 (1837), pp. 195–8; Allen, W., 'Is the Old Calabar a branch of the River Quorra', *Journal of the Royal Geographical Society of London* 7 (1837), pp. 198–203.
15 Washington, J., 'A brief account of a Mandingo, native of Nyáni-Marú, on the River Gambia, in western Africa', *Report of the Eighth Meeting of the British Association for the Advancement of Science*, p. 97.
16 Buxton, T. F., *The African Slave Trade and its Remedy*, pp. 368 and 429, respectively.
17 This paragraph is derived from Lambert, D., *Mastering the Niger: James MacQueen's African Geography & the Struggle over Atlantic Slavery*, pp. 176–206; MacQueen, J., *A Geographical Survey of Africa: Its Rivers, Mountains, Production, States, Population, &c, with a Map on an Entirely New Construction to Which is Prefixed, a Letter to Lord John Russell, Regarding the Slave Trade and the Improvement of Africa.*
18 Buxton, T. F., 7 August 1840 in BPP (Cd.472), *Papers Relative to the Expedition to the River Niger*, p. 17.
19 Allen, W., *Picturesque Views on the River Niger: sketched during Lander's last visit in 1832–33*, p. vi. On the 'picturesque' and its use in describing the depiction of overseas landscapes according to European aesthetic conventions, see Burnett, D. G., *Masters of All They Surveyed: Exploration, Geography, and a British El Dorado*, pp. 126–30, from whom I take the term 'strategies of imperial visibility'.
20 De Gramont, *Strong Brown God*, p. 211.
21 McWilliam, J. O., *Medical History of the Expedition to the Niger during the years 1841–2, comprising an account of the fever which led to its termination*, pp. 74 and 107, respectively.
22 De Gramont, *Strong Brown God*, p. 214.
23 Laird and Oldfield, *Narrative of an Expedition into the Interior of Africa*, p. 104.

24 Allen, W., and Thomson, T. R. H., *A Narrative of the Expedition sent by Her Majesty's Government to the River Niger in 1841, under the command of Captain H. D. Trotter*, p. 371.

25 BPP (Cd.472), *Papers Relative to the Expedition to the River Niger*, p. 139.

26 Richardson, J., *Travels in the Great Desert of Sahara, in the Years of 1845 and 1846*, I, pp. xii and xv, respectively.

27 Becroft, J., 'On Benin and the upper course of the River Quorra, or Niger'; Becroft, J., and King, J. B., 'Details of the exploration of the Old Calabar River, in 1841 and 1842', *Journal of the Royal Geographical Society of London* 14 (1844), pp. 260–83; MacQueen, J., 'Notes on African geography', *Journal of the Royal Geographical Society of London* 15 (1845), pp. 371–6; MacQueen, J., 'Notes on the present state of the geography of some parts of Africa', *Journal of the Royal Geographical Society of London* 20 (1850), pp. 235–52.

28 RGS–IBG, CB4/Petermann, Augustus Petermann to Norton Shaw, 5 December 1851.

29 Petermann, A., *African Discovery: A Letter Addressed to the President and Council of the Royal Geographical Society of London, by Augustus Petermann*, p. 8. For a fuller account of these relationships, see Withers, C. W. J., 'On trial – social relations of map production in mid-nineteenth-century Britain', *Imago Mundi* (2019), pp. 173–95.

30 See RGS–IBG, RGS Evening Minutes, November 1840–April 1856. Minutes of the Meeting of 9 January 1854, ff. 262–3. 'Account of the Steamer prepared to ascend the Niger and Chadda Rivers, by MacGregor Laird Esq FRGS'.

31 Baikie, W. B., *Narrative of an Exploring Voyage up the Rivers Kwo'ra and Binue (commonly known as the Niger and Tsadda) in 1854*, p. vii for Baikie's use of the term 'exploring expedition', and p. 385. Laird's instructions, issued on 8 May 1854, are given on p. 405.

32 RGS–IBG, JMS 1/40, W. B. Baikie to R. I. Murchison, 12 December 1857.

33 Baikie, W. B., 'On recent discovery in Central Africa and the reasons for renewed research', RGS–IBG, JMS, 6 August 1856, and 'Latest results of the Niger Expedition', Journal MS, 10–11 September 1861. See also Baikie, W. B., and May, D. T., 'Extract of Reports from the Niger Expedition', *Proceedings of the Royal Geographical Society of London* 2 (1857–8), pp. 83–101; Knowles, C., 'Ascent of the Niger in September and October 1864', *Proceedings of the Royal Geographical Society of London* 9 (1864–5), pp. 72–5.

34 Crowther, S., *Journal of an Expedition up the Niger and Tshadda Rivers, undertaken by MacGregor Laird Esq. in Connection with the British Government, in 1854*; Hutchinson, T., *Narrative of the Niger, Tshadda, & Binuë Exploration: including a Report on the Position and Prospect of Trade up those Rivers, with remarks on the malaria and fevers of Western Africa.* Hutchinson described MacGregor Laird as 'The Pioneer of the Niger Exploration' in the dedication of his book.

35 De Gramont, *Strong Brown God*, p. 239. De Gramont is sympathetic to Baikie's role as a 'visionary', the first resident British colonial civil servant in what

became Nigeria. Baikie's quote is from his *Narrative of an Exploring Voyage*, p. 388.

36 Kennedy, D., *The Last Blank Spaces: Exploring Africa and Australia*, p. 137.

Chapter 12

1 NLS, MS 10782, f. 181, 20 May 1794.

2 The phrase is from Hutton, P., *History as an Art of Memory*, p. 161. This and Hutton's more recent work are good guides to the burgeoning field of memory studies: Hutton, P., *The Memory Phenomenon in Contemporary Historical Writing: How the Interest in Memory has Influenced our Understanding of History*. I take the phrase 'commemorative trajectory' from Olick, J. K., 'Introduction: memory and the nation – continuities, conflicts, and transformations', *Social Science History* 22 (1998), p. 385. Elements of what follows on memory in relation to Park are taken from Withers, C. W. J., 'Memory and the history of geographical knowledge: the commemoration of Mungo Park, African explorer', *Journal of Historical Geography* 30 (2004), pp. 316–39.

3 [Park, M.], *Travels in the Interior of Africa. Abridged from the Original Work*. A second edition of this appeared in 1800.

4 I take these totals from work presented by Professor Katherine Marsters in her 2005 seminar 'Park in the nineteenth century' to the Institute for Advanced Studies in the Humanities, University of Edinburgh. I have been unable to trace Professor Marsters and hope that she will accept this acknowledgement of her work.

5 Quoted in the entry for 'John Leyden', in Chambers, R. (ed.), *A Biographical Dictionary of Eminent Scotsmen*, III, p. 427.

6 Murray, H., 'Observations on the information collected by the Ashantee Mission regarding the course of the Niger, and the interior of Africa', *Edinburgh Philosophical Journal* 1 (1819), pp. 163–70 [on Bowdich]; Murray, H., 'Recent discoveries in Africa', *Edinburgh Review* 44 (1826), pp. 173–219 [largely a review of the Denham, Clapperton, and Oudney Bornu Mission]; Murray, H., 'Interior of Africa', *Edinburgh Review* 49 (1827), pp. 127–49 [on Laing and Cailleé]; Murray, H., 'Landers' voyage and discoveries on the Niger', *Edinburgh Review* 55 (1832), pp. 397–421 [a discussion of the Landers' achievement and the benefits likely to follow to British trade]. Murray's expanded version of Leyden's text went into a second edition in 1818 and, like Park's 1799 book, was translated into French, in 1821. Jameson, R., White, J., and Murray, H., *Narrative of Discovery and Adventure*.

7 Wakefield, P., *The Traveller in Africa*; Taylor, I., *African Scenes for Tarry-at-Home Travellers*; Mossman, S., *Heroes of Discovery*; Mossman, S., *Gems of Womanhood, or, Sketches of Distinguished Women in Various Ages and Nations*, pp. 94–106.

8 This was printed in 1824 in the *Ladies' Garland* and in Loeve-Veimars, F. A., *Popular Ballads and Songs, from Traditional Manuscripts, and Scarce Editions*.

9 This paragraph on the poetic response to Park in the early nineteenth century

draws from Fulford, T., Lee, D., and Kitson, P. J., *Literature, Science and Exploration in the Romantic Era: Bodies of Knowledge*, pp. 95–104, and Thompson, C., *The Suffering Traveller and the Romantic Imagination*, pp. 172–81.

10 *Anniversary Dinner of the Royal Society, November 30, 1887: Sir Joseph Hooker's Reply to the Toast proposed by the Treasurer, R.S.* (Privately printed and distributed by John Bellows of Gloucester, n.d. [but 1887]), pp. 11–12. I am grateful to Henry Noltie for drawing this to my attention.

11 The Royal Botanic Garden Edinburgh has two 'isotypes' of the Park moss, both of which appear to have come via William Hooker. David Long of the RBGE thinks it likely that Hooker would have got these specimens either when he was in Glasgow, or more likely before he took up the chair in Glasgow [February 1820]. I am grateful to David Long of the RBGE for his insights.

12 Gibbs, J. M., *Performing the Temple of Liberty: Slavery, Theater, and Popular Culture in London and Philadelphia, 1760–1850*, pp. 98–103.

13 I take these theatrical performances and their descriptions and dates from the listings in Nicol, A., *A History of Early Nineteenth-Century Drama 1800–1850.*

14 BL, Add MS 42954, ff. 381–405, *Mungo Park, or the Arab of the Niger.*

15 I am indebted to Martin Lyngbo, the director, and to Matthew Zajac, who played Park, for their gracious response to my email queries following the 2016 production in Edinburgh: I draw upon their words in this paragraph.

16 NLS, MS Acc 8078, no. 493. Script by Kathleen Fidler.

17 Gray, J., *Park: A Fantastic Story*. The book was twice reissued in the 1960s, the 1966 edition with an introduction by Bernard Bergonzi.

18 Boyle, T. C., *Water Music*, Apologia, p. ix, and Coda, p. 436.

19 NLS, MS 20311, f. 31, Adam Park to Thomas Anderson, 27 November 1821.

20 NLS, Acc 9834/5, Charles Napier to Thomas Anderson, 3 May 1828.

21 *Kelso Chronicle*, 31 December 1841, p. 7.

22 'The Lander Monument', *West Briton*, 27 May 1836.

23 On the statue by Currie, see BCA, MSS SC/S/16/1/3, 3 June 1858; *Southern Reporter*, 3 April 1858.

24 The subscribers' details are given in BCA, MSS SC/S/12/18/2.

25 *Southern Reporter*, 10 February 1859, p. 2; 24 February 1859, p. 2. Locals take the name 'Foresters' from the surrounding Ettrick Forest.

26 *Southern Reporter*, 3 March 1859, p. 4 [Johnstone]; *Kelso Chronicle*, 18 March 1859 [np] [the local woman].

27 Johnston, H. H., 'The Niger basin and Mungo Park', *Scottish Geographical Magazine* 23 (1907), pp. 58–72. On Johnston's admission, see RSGS Archives, uncatalogued letter of Sir Harry Johnston to the RSGS, 14 December 1907.

28 [Anon.], 'The Park-Lander memorial', *Geographical Journal* 68 (1926), p. 456 and *Geographical Journal* 1929 (74), pp. 470–1.

29 RSGS Archives, Archive Series 31.48, Letters to Lord Stair, 29 January 1912 [Glasgow]; 25 January 1912 [Edinburgh]; and 13 February 1912 [Ayr]. Notices of the appeal on behalf of Baroness Conrad-Philips appear in the Society's Council Minutes for 23 June 1925 and 16 July 1925. A joint committee of

the RGS and the RSGS tasked with further funding for the memorial was in operation in 1933.

30 UoE, CRC, Gen. 2248/3, Minutes of the Mungo Park Memorial Committee. Shepperson's remarks were made in a letter of 17 February 1970. Public support is clear from *The Scotsman*, 15 August 1971.

31 Thomson, J., *Mungo Park and the Niger*, p. vi and endpapers.

32 Driver, F., 'Henry Morton Stanley and his critics: geography, exploration and empire', *Past and Present* 133 (1991), pp. 134–71; Jeal, T., *Stanley: The Impossible Life of Africa's Greatest Explorer*.

33 Stanley, H. M., 'Inaugural address delivered before the Scottish Geographical Society at Edinburgh, 3 December 1884', *Scottish Geographical Magazine* 1 (1885), pp. 1–17, quote from p. 5.

34 White, A. S., 'On the achievements of Scotsmen during the nineteenth century in the fields of exploration and research', *Scottish Geographical Magazine* 5 (1889), pp. 480–97; Darwin, L., 'On the personal characteristics of great explorers', *Scottish Geographical Magazine* 28 (1912), pp. 134–46, quote from p. 141.

35 Hewitt, W. H., *Mungo Park*; Scott, G., *Mungo Park*.

36 Miller, R., 'A Mungo Park anniversary', *Scottish Geographical Magazine* 71 (1955), pp. 147–56, quote from p. 156; Miller, R., 'Mungo Park 1771–1971', *Scottish Geographical Magazine* 87 (1971), pp. 159–65, quote from p. 164; Harold Shepperson in UoE, CRC, MS Gen 2248/3, Letter of 4 February 1971. See also Barczewski, S., *Heroic Failure and the British*: on Park, pp. 36–43. The quotes are from pp. 43 and 44.

Chapter 13

1 Syme, R., *I, Mungo Park*; [H.B.], *The Life of Mungo Park*; Maclachlan, T. B., *Mungo Park*; Gwynn, S., *Mungo Park and the Quest of the Niger*; Gibbon, L. G., *Niger: The Life of Mungo Park*; Lupton, K., *Mungo Park The African Traveler*; Duffill, M., *Mungo Park West African Explorer*. This list takes no account of Park's treatment as an explorer as discussed in chapter 12 – Thomson's *Mungo Park and the Niger* (1890) might equally be treated as a biography – nor of those works which summarise Park's life and his Niger journeys but do not reflect either on Park the man or upon the Niger's exploration: see, for example, Millar, R., *Travels of Mungo Park*; Brent, P., *Black Nile: Mungo Park and the Search for the Niger*. Even at the end of the nineteenth century, compilation volumes placed Park within the pantheon of Africa's explorers: see [Anon.], *Africa and its exploration as told by its explorers*. Park is included alongside Clapperton, Burton, Livingstone, Stanley, Speke, and others.

2 See, for example, Shortland, M., and Yeo, R. (eds), *Telling Lives in Science: Essays on Scientific Biography* and as an example of an explorer's metabiography, Livingstone, J., *Livingstone's 'Lives': A Metabiography of a Victorian Icon*.

3 Nwelme , O., and Okeh, I. K., *Saving Mungo Park*; Fagbule, F., and Fawehinmi, F., *Formation: The Making of Nigeria from Jihad to Amalgamation*, pp. 103–40.

4 Whishaw, J., 'Account of the Life of Park', in Park, M., *The Journal of a Mission to the Interior of Africa, in the Year 1805*, p. lxxxix; H.B., *The Life of Mungo Park*, pp. 281, 283; Maclachlan, T. B., *Mungo Park*, p. 21.
5 Gwynn, *Mungo Park and the Quest of the Niger*, p. 33; Gibbon, *Niger: The Life of Mungo Park*, p. 11.
6 Withers, C. W. J., *Geography, Science and National Identity: Scotland since 1520*, pp. 220–5.
7 De Gramont, S., *The Strong Brown God: The Story of the Niger River*, pp. 70 and 71.
8 Ruskin, J., *Fors Clavigera: Letters to the Workmen and Laborers of Great Britain*, VIII: Letter 92, pp. 197–8.
9 Duffill, *Mungo Park West African Explorer*, p. 17.
10 These words appear in the 'Advertisement' to Park, M., *The Journal of a Mission to the Interior of Africa, in the Year 1805*, no pagination but p. 3. The advertisement is dated 1 March 1815.
11 Mitchell, J. L., and Gibbon, L. G., *Nine Against the Unknown: A Record of Geographical Exploration*, p. 263. The 1934 original was published in London by Jarrolds. It is not clear why J. Leslie Mitchell chose to use his given name and his pseudonym, Lewis Grassic Gibbon, in the authorship of this work.
12 Mungo Park's letter to Lord Camden, from Sansanding on the Niger, 17 November 1805, in Whishaw's 'Account of the Life of Mungo Park', in Park, *Journal of a Mission*, p. lxxx.
13 AA, SFP/NLS/Box4/Z.AT.991, 'Letter of 7 March 1825 from Thomas Anderson detailing particulars of Park's Estate', and 'Letter, 1 April 1825 from Mr Thomas Anderson to Andrew Lang, Writer, Selkirk'. This is a bundle of correspondence and documents apparently once in the possession of Andrew Lang. SFP is shorthand for the Scott Family Papers, cared for by The Abbotsford Trust. The quotation and reference is published with their permission. The letter from Allison Park to Archibald Buchanan with her remark about the young Mungo Park is NLS, Acc 9834/3, Allison Park to Archibald Buchanan, 19 September 1815.
14 [H.B.], *The Life of Mungo Park*, p. 259; Maclachlan, *Mungo Park*, p. 129; Lupton, *Mungo Park The African Traveler*, Appendix III, 'Stories of Mungo Park at Yauri and of his death at Bussa', pp. 231–9.
15 Thomson, J., *Mungo Park and the Niger*, p. 255.
16 Maclachlan, *Mungo Park*, p. 133; Thomson, *Mungo Park and the Niger*, pp. 258, 259, 261.
17 Campbell, R., *A Pilgrimage to my Motherland: An Account of a Journey among the Egbas and Yorubas of Central Africa, in 1859–60*; Burton, R. F., and MacQueen, J., *The Nile Basin. Part I. Showing Tanganyika to be Ptolemy's Western Lake Reservoir: A Memoir read before the Royal Geographical Society, November 1864. With prefatory Remarks. Part II. Captain Speke's Discovery of the Source of the Nile. A Review by James MacQueen.*

Bibliography

Primary unpublished sources

Abbotsford Archives

SFP/NLS/Box4/Z.AT.991

British Library

Add MSS 34614
Add MSS 38233
Add MSS 42954
King's Topographical Collections, Maps K. Top. 117. 24.1.c.
King's Topographical Collections, Maps K. Top. 117. 24.1.g.

Chatsworth House Archives

DF 12/1/1/5

Borders Council Archives

MSS SC/5/56
MSS SC/S/12/18/2
MSS SC/S/16/1/3

British Museum of Natural History, London

Dawson Turner Collection
Banks MSS A: 3

Edinburgh City Archives

Bell and Bradfute Records, SL138/7/9–SL138/7/14

National Library of Scotland

Acc 6207
Acc 8078
Acc 9834
Acc 12055
Acc 13149
MS 581
MS 786
MS 3278
MS 4005
MS 4011
MS 4017
MS 4019
MS 4030
MS 5308
MS 10290
MS 10708
MS 10782
MS 19390
MS 20311
MS 40057
MS 40555
MS 40668
MS 40908
MS 40909
MS 41263
MS 41908
MS 42326
MS 42725

Royal Geographical Society (with the Institute of British Geographers) Archives

CB4
DD/14-DD/18
Evening Minutes, November 1840–April 1856
JMS 1/3
JMS 1/40
MS 8/8.37

Royal Scottish Geographical Society Archives

Archive Series 31.48
Uncatalogued letter of Sir Harry Johnston, 14 December 1907

The National Archives

ADM 1/1267
ADM 1/2617
ADM 1/4234
CO 2/5
CO 2/8
CO 2/12
CO 2/13
CO 2/14
CO 2/16
CO 2/17
CO 2/20
CO 2/115
CO 267/10
CO 267/41
CO 323/14
CO 323/144
CO 324/3
CO 324/83
CO 392/3
FO 8/7
FO 8/8

University of Edinburgh Centre for Research Collections

Coll-1587
Coll-1929 (Gen.786)
Gen. 2248/3
MS Dc.1.18

Newspapers

Charleston Courier
Edinburgh Evening Courant
Kelso Chronicle

Southern Reporter
The Scotsman
The Times
West Briton

British Parliamentary Papers

(Cd.472), *Papers Relative to the Expedition to the River Niger*, London: W. Clowes for Her Majesty's Stationery Office, 1843

Primary published sources

Adams, R., *The Narrative of Robert Adams*, London: John Murray, 1816

[African Association], *Proceedings of the Association for Promoting the Discovery of the Interior Parts of Africa*, London: C. Macrae, 1790

[African Association], *Proceedings of the Association for Promoting the Discovery of the Interior Parts of Africa*, London: W. Bulmer and Son, 1798

[African Association], *Proceedings of the Association for Promoting the Discovery of the Interior Parts of Africa*, two vols, London: W. Bulmer and Co., 1810

[African Institution], *Rules and Regulations of the African Institution: Formed on the 14th April 1807*, London: William Phillips, 1807

Allen, W., 'Is the Old Calabar a branch of the River Quorra', *Journal of the Royal Geographical Society of London* 7 (1837), pp. 198–203

Allen, W., 'On a new construction of a map of a portion of West Africa, showing the possibility of the Rivers Yéu and Chadda being the outlet of the Lake Chad', *Journal of the Royal Geographical Society of London* 8 (1838), pp. 289–307

Allen, W., *Picturesque Views on the River Niger: sketched during Lander's last visit in 1832–33*, London: John Murray, 1840

Allen, W., and Thomson, T. R. H., *A Narrative of the Expedition sent by Her Majesty's Government to the River Niger in 1841, under the command of Captain H. D. Trotter*, London: R. Bentley, 1848

Anderson, J., 'Hints respecting the study of Geography', *The Bee* 6 (December 1791), pp. 208–14

Anderson, J., 'Philosophical Geography', *The Bee* 6 (December 1791), pp. 335–41

[Anon.], *Travels in the Interior of Africa, by Mungo Park. And in Southern Africa, by John Barrow. Interspersed with notes and observations, geographical, commercial, and philosophical. By the author of the* New System of Geography, Glasgow: A. Napier, 1812

[Anon.], 'Review of *The Journal of a Mission to the Interior of Africa in the Year 1805, by Mungo Park: Together with other Documents, Official and Private, relative to the same Expedition: To which is prefixed, an Account of the Life of Mr. Park*', London: John Murray, 1815, *Edinburgh Review* 24 (1815), pp. 471–90.

[Anon.], 'Remarks on the two accounts of the death of Mungo Park', *Asiatic Journal and Monthly Miscellany* 11 (1821), pp. 453–6

Baikie, W. B., *Narrative of an Exploring Voyage up the Rivers Kwo'ra and Binue (commonly known as the Niger and Tsadda) in 1854*, London: John Murray, 1856

Baikie, W. B., and May, D. T., 'Extract of Reports from the Niger Expedition', *Proceedings of the Royal Geographical Society of London* 2 (1857–8), pp. 83–101

[Barrow, J.], 'Review of *The Journal of a Mission to the Interior of Africa, in the Year 1805, by Mungo Park*', *Quarterly Review* 13 (1815), pp. 120–51

[Barrow, J.], 'Review of *Voyage dans l'Intérieur de l'Afrique aux Sources du Sénégal et de la Gambie, fait en 1818, par ordre du Gouvernement Francais.* Par. G. Mollien', *Quarterly Review* XXIII (1820), pp. 225–44

[Barrow, J.], 'Review of 1. *Mission from Cape Coast Castle to Ashantee, with a Statistical Account of that Kingdom, and Geographical Notices of other Parts of the Interior of Africa.* By T. Edward Bowdich, Esq. Conductor. 1819. 2. *The African Committee.* By T. Edward Bowdich, Esq. Conductor of the Mission to Ashantee. 1819', *Quarterly Review* 22 (1820–1), pp. 273–302

[Barrow, J.], 'Review of *A Narrative of Travels in Northern Africa, in the Years, 1818, 19, and 20; accompanied by Geographical Notices of Soudan, and the Course of the Niger. With a chart of the routes, and a variety of coloured plates, illustrative of the costumes of the several natives of Northern Africa.* By G. F. Lyon', *Quarterly Review* 25 (1821), pp. 25–50

[Barrow, J.], 'Review of *A Geographical and Commercial View of Northern Central Africa; containing a particular Account of the Course and Termination of the Great River Niger in the Atlantic Ocean.* By James MacQueen, Edinburgh, 1821', *Quarterly Review* 26 (1822), pp. 51–81

[Barrow, J.], 'Review of Adams, R., *Sketches taken during Ten Voyages to Africa*', *Quarterly Review* 29 (1823), pp. 508–24

[Barrow, J.], 'Recent discoveries in Africa, made in the years 1823 and 1824, by Major Denham, Captain Clapperton, R.N. and the late Dr Oudney, extending across the Great Desert to the tenth degree of northern latitude, and from Kouka in Bornou to Sackatoo, the capital of the Soudan Empire', *Quarterly Review* 33 (1826), pp. 519–49

[Barrow, J.], 'Review of *Dissertation on the Course and Probable Termination of the Niger*. By Lt.-Gen. Rufane Shaw Donkin, London, 1829', *Quarterly Review* 41 (1829), pp. 226–39

[Barrow, J.], 'Review of *Journal of a Second Expedition into the Interior of Africa, from the Bight of Benin to Soccatoo*', *Quarterly Review* 39 (1829), pp. 143–83

[Barrow, J.], 'Art VI. Journal d'un Voyage à Temboctoo et à Jenné, dans l'Afrique Centrale, &c.', *Quarterly Review* 42 (1830), pp. 450–75

[Barrow, J.], 'Review of *The Journal of the Royal Geographical Society of London for the Year 1830–1831*', *Quarterly Review* 46 (1832), pp. 55–80

Becroft, J., 'On Benin and the upper course of the River Quorra, or Niger', *Journal of the Royal Geographical Society of London* 11 (1841), pp. 184–92

Becroft, J., and King, J. B., 'Details of the exploration of the Old Calabar River, in 1841 and 1842', *Journal of the Royal Geographical Society of London* 14 (1844), pp. 260–83

Beechey, F. W., and Beechey, H. W., *Proceedings of the Expedition to Explore the Northern Coast of Africa, from Tripoly Eastward; in MDCCCXXI. and MDCCCXXII. comprehending an account of the Greater Srytis and Cyrenaica; and of the Ancient Cities comprising the Pentapolis*, London: John Murray, 1829

Bovill, E. W. (ed.), *Missions to the Niger I: The Journal of Friedrich Hornemann's Travels and the Letters of Alexander Gordon Laing*, Cambridge: Cambridge University Press, 1964

Bovill, E. W. (ed.), *Missions to the Niger II: The Bornu Mission, 1822–25 Part 1*, Cambridge: Cambridge University Press, 1966

Bovill, E. W. (ed.), *Missions to the Niger III: The Bornu Mission, 1822–25 Part 2*, Cambridge: Cambridge University Press, 1966

Bovill, E. W. (ed.), *Missions to the Niger IV: the Bornu Mission, 1822–25 Part 3*, Cambridge: Cambridge University Press, 1966

Bowdich, T. E., *Mission from Cape Coast Castle to Ashantee, with a Statistical Account of that Kingdom, and Geographical Notices of other Parts of the Interior of Africa*, London: John Murray, 1819

[Brougham, H.], [Review of *The Journal of a Mission to the Interior of Africa, in the Year 1805, by Mungo Park*], *Edinburgh Review* 24 (1814–15), pp. 471–90

Brown, W., 'Account of the Correspondence between Mr Park and Mr Maxwell, respecting the identity of the Congo and the Niger. By the Rev. William Brown, D. D. Minister at Eskdalemuir', *Edinburgh Philosophical Journal* 3 (1820), pp. 102–14 and pp. 205–18

Browne, W. G., *Travels in Africa, Egypt and Syria from the Year 1792 to 1798*, London: Cadell and Davies, Longman and Rees, 1799

Bruce, J., *Travels to Discover the Source of the Nile; in the years 1768, 1769, 1770, 1771, 1772, and 1773*, Edinburgh: Printed by J. Ruthven for G. G. J. and J. Robinson, London, 1790

Burckhardt, J. J., *Travels in Nubia*, London: John Murray, 1819

Burckhardt, J. L., *Travels in Syria and the Holy Land, by the late John Lewis Burckhardt published by the Association for Promoting the Discovery of the Interior Parts of Africa*, London: John Murray, 1822

Buxton, T. F., *The African Slave Trade and its Remedy*, London: John Murray, 1840

Chambers, R. (ed.), *A Biographical Dictionary of Eminent Scotsmen*, four vols, Glasgow: Blackie and Son, 1835

Chirurgo-Physical Society, *Laws and Regulations of the Chirurgo-Physical Society*, Edinburgh: Printed by Campbell Denovan, 1791

Clapperton, H., *Journal of a Second Expedition into the Interior of Africa, from the Bight of Benin to Socatoo, to which is added The Journal of Richard Lander from Kano to the Coast, partly by a more eastern route*, London: John Murray, 1829

Cory, J., *A Geographical Historie of Africa, written in Arabicke and Italian, by John Leo a More, borne in Granada and brought up in Berberie*, London: George Bishop, 1600

Crowther, S., *Journal of an Expedition up the Niger and Tshadda Rivers, undertaken by MacGregor Laird Esq. in Connection with the British Government, in 1854*, London: Church Missionary Society and Seeley, Jackson, and Halliday, 1855

D'Anville, J.-B., 'Mémoire concernant les rivières de l'interieur de l'Afrique, sur les notions tirées des Anciens & des Modernes', *Mémoires de Litterature de l'Académie Royale des Inscriptions et Belles-Lettres* 26 (1759), pp. 64–81

Denham, D., Clapperton, H., and Oudney, W., *Narrative of Travels in Northern and Central Africa, in the years 1822, 1823, and 1824, by Major Denham, F.R.S., Captain Clapperton, and the late Dr Oudney*, two vols, London: John Murray, 1826

Dickson, J., *Fasciculus Plantarum Cryptogamicarum Britanniae*, four vols, London: G. Nicol, 1785–1801

Dickson, J., 'An account of some plants newly discovered in Scotland', *Transactions of the Linnean Society* II (1794), pp. 286–91

Donkin, R. S., *Dissertation on the Course and Probable Termination of the Niger*, London: John Murray, 1829

[Donkin, R. S.], *A Letter to the Publisher of the Quarterly Review*, London: Saunders and Otley, 1829

Dudley, J., *A Dissertation showing the Identity of the Rivers Niger and Nile;*

chiefly from the Authority of the Ancients, London: Longman, Hurst, Rees, Orme, and Brown, 1821

Dupuis, J., *Journal of a Residence in Ashantee*, London: Henry Colburn, 1824

Edwards, B., *History, Civil and Commercial, of the British Colonies in the West Indies*, two vols, London: Printed for John Stockdale, 1793

Gibbon, L. G., *Niger: The Life of Mungo Park*, Edinburgh: The Porpoise Press, 1934

Gray, J., *Park: A Fantastic Story*, London: Sheed and Ward, 1932

Gray, W., and Dochard, D., *Travels in Western Africa, in the Years 1818, 19, 20, and 21, from the River Gambia, through Woolli, Bondoo, Galam, Kasson, and Foolidoo, to the River Niger*, London: John Murray, 1825

Guthrie, W., *A New Geographical, Historical, and Commercial Grammar*, London: Printed for J. Knox, 1770

Gwynn, S., *Mungo Park and the Quest of the Niger*, London: John Lane and the Bodley Head Limited, 1934

Hallett, R., *Records of the African Association 1788–1831*, London: Thomas Nelson and Sons, 1964

[H.B.], *The Life of Mungo Park*, Edinburgh: Fraser and Co., 1835

Hewitt, W. H., *Mungo Park*, London: Sheldon Press, 1923

Hornemann, F., *The Journal of Frederick Hornemann's Travels: From Cairo to Mourzouk, the Capital of the Kingdom of the Fezzan, in Africa, in the Years 1797–8*, London: G. and W. Nicol, 1802

Hutchinson, T., *Narrative of the Niger, Tshadda, & Binuë Exploration: including a Report on the Position and Prospect of Trade up those Rivers, with remarks on the malaria and fevers of Western Africa*, London: Longman, Brown, Green, and Longmans, 1855

Hutton, W., *A Voyage to Africa: including A Narrative of an Embassy to one of the Interior Kingdoms in the Year 1820; with remarks on the course and termination of the Niger, and other principal rivers in that country*, London: Longman, Hurst, Rees, Orme, and Brown, 1821

Jackson, J. G., *An Account of the Empire of Marocco, and the District of Suse*, London: W. Bulmer, 1809

Jackson, J. G., *An Account of Timbuctoo and Housa, Territories in the Interior of Africa*, London: Longman, Hurst, Rees, Orme, and Brown, 1820

Jameson, R., White, J., and Murray, H., *Narrative of Discovery and Adventure*, Edinburgh: Oliver and Boyd, 1830

Knowles, C. 'Ascent of the Niger in September and October 1864', *Proceedings of the Royal Geographical Society of London* 9 (1864–5), pp. 72–5

Labat, J.-B., *Nouvelle relation de l'Afrique Occidentale*, Paris: Guillaume Cavalier, 1728

Laing, A. G., *Travels in the Timannee, Kooranko, and Soolima countries in western Africa*, London: John Murray, 1825

Laird, M., and Oldfield, R. A. K., *Narrative of an Expedition into the Interior of Africa, by the River Niger, in the steam-vessels Quorra and Alburkah in 1832, 1833 and 1834*, two vols, London: Richard Bentley, 1837

Lander, R., *Records of Captain Clapperton's Last Expedition to Africa*, two vols, London: Henry Colburn and Richard Bentley, 1830

Lander, R. and J., *Journal of an Expedition to Explore the Course and Termination of the Niger; with a Narrative of a Voyage down that River to its Termination*, three vols, London: John Murray, 1832

Leake, W. M. 'Is the Quorra, which has lately been traced to its Discharge into the Sea, the same River as the Nigir of the Ancients?', *Journal of the Royal Geographical Society of London* 2 (1832), pp. 1–28

Ledyard, J., *A Journal of Captain Cook's Last Voyage to the Pacific Ocean*, Hartford Conn: Printed and sold by Nathaniel Patten, 1783

Lupton, K., *Mungo Park The African Traveler*, Oxford: Oxford University Press, 1979

Lyon, G. F., *A Narrative of Travels in Northern Africa, in the Years, 1818, 19, and 20; accompanied by Geographical Notices of Soudan, and the Course of the Niger. With a chart of the routes, and a variety of coloured plates, illustrative of the costumes of the several natives of Northern Africa*, London: John Murray, 1821

Maclachlan, T. B., *Mungo Park*, Edinburgh and London: Oliphant Anderson & Ferrier, 1898

MacQueen, J., *A Geographical and Commercial View of Northern Central Africa: containing a Particular Account of the Course and Termination of the great River Niger in the Atlantic Ocean*, Edinburgh: Blackwood and Sons, 1821

MacQueen, J., 'M'Queen on the course and termination of the Niger', *Blackwood's Edinburgh Magazine* 13 (1823), pp. 417–32

MacQueen, J., 'Geography of Central Africa: Denham and Clapperton's Journal', *Blackwood's Edinburgh Magazine* 19 (1826), pp. 687–709

MacQueen, J., 'The River Niger – Termination in the Sea', *Blackwood's Edinburgh Magazine* 30 (1831), pp. 130–6

MacQueen, J., 'Geography of Africa – Quarterly Review. Letter from James MacQueen', *Blackwood's Edinburgh Magazine* 31 (1832), pp. 201–16

MacQueen, J., *A Geographical Survey of Africa: Its Rivers, Mountains, Production, States, Population, &c, with a Map on an Entirely New Construction to Which is Prefixed, a Letter to Lord John Russell, Regarding the Slave Trade and the Improvement of Africa*, London: B. Fellowes, 1840

MacQueen, J., 'Notes on African geography', *Journal of the Royal Geographical Society of London* 15 (1845), pp. 371–6

MacQueen, J., 'Notes on the present state of the geography of some parts of Africa', *Journal of the Royal Geographical Society of London* 20 (1850), pp. 235–52

McWilliam, J. O., *Medical History of the Expedition to the Niger during the years 1841–2, comprising an account of the fever which led to its termination*, London: John Churchill, 1843

Marsden, W., *The History of Sumatra, containing an account of the government, laws, customs, and manners of the native inhabitants, with a description of the natural productions, and a relation of the ancient political state of that island*, London: Printed for the author, 1783

Monrad, H. C., *Efterretninger om det indre Afrika, uddragne af de af det i London oprettede afrikanske Selskab udgivne Oplysninger om Houghtons og Parks Reiser, som Foreløber for denne sidstes større Reisebeskrivelse*, Copenhagen: Seidelin, 1799

Moore, F., *Travels into the Inland Parts of Africa*, London: Printed by Edward Cave, at St John's Gate, 1738

Mossman, S., *Heroes of Discovery*, Edinburgh and London: Oliphant Anderson & Ferrier, 1868

Mossman, S., *Gems of Womanhood, or, Sketches of Distinguished Women in Various Ages and Nations*, Edinburgh: Gall and Inglis, 1870

Murray, H., 'Observations on the information collected by the Ashantee Mission, respecting the course of the Niger, and the interior of Africa', *Edinburgh Philosophical Journal* 1 (1819), pp. 163–70

Murray, H., 'Recent discoveries in Africa', *Edinburgh Review* 44 (1826), pp. 173–219

Murray, H., 'Interior of Africa', *Edinburgh Review* 49 (1827), pp. 127–49

Murray, H., 'Landers' Voyage and discoveries on the Niger', *Edinburgh Review* 55 (1832), pp. 397–421

Nicholls, E., Calabar, D., and Coulthurst, C. H., 'Failure of another expedition to explore the interior of Africa', *Journal of the Royal Geographical Society* 2 (1832), pp. 305–12

Oldfield, R. A. K., 'A brief account of the ascent of the Old Calabar river in 1836,' *Journal of the Royal Geographical Society of London* 7 (1837), pp. 195–8

Park, M., 'Descriptions of eight new fishes from Sumatra', *Transactions of the Linnean Society* 3 (1797), pp. 33–8

Park, M., *Travels in the Interior Districts of Africa: Performed under the Direction and Patronage of the African Association, in the Years 1795, 1796, and 1797*, London: G. and W. Nicol, 1799

[Park, M.], *Travels in the Interior of Africa. Abridged from the Original Work*, London: Printed by J. W. Myers for Crosby and Letterman, 1799

Park, M., *The Journal of a Mission to the Interior of Africa, in the Year 1805*, London: John Murray, 1815

Petermann, A., *African Discovery: A Letter Addressed to the President and Council of the Royal Geographical Society of London, by Augustus Petermann*, London: Edward Stanford, 1854

Playfair, J., *A System of Geography, Ancient and Modern*, six vols, Edinburgh: Peter Hall, 1810–14

Reichard, C. G., 'Ueber die vermuthung des Dr. Seetzen zu Jever, dass sich *Niger* in Afrika vielleicht mit dem Zaire vereingen konne', *Zach's Monatliche Correspondenz* (May 1802), pp. 409–15

Reichard, C. G., 'Ueber den angekündigten, nun bald erscheinenden, *Atlas des ganzen Erdkreises*', *Allgemeine Geographische Ephermeriden* 12 (1803), pp. 129–70

Rennell, J., *Memoir of the Map of Hindoostan*, London: M. Brown, for the author, 1788

Rennell, J., 'On the rate of travelling, as performed by camels; and its application, as a scale, to the purpose of geography', *Philosophical Transactions of the Royal Society of London* 81 (1791), pp. 129–45

[Rennell, J.], *Elucidations of the African Geography; from the Communications of Major Houghton, and Mr Magra; 1791*, London: W. Bulmer and Co., 1793

Rennell, J., *The Geographical System of Herodotus, Examined, and Explained, by a Comparison with those of other Ancient Authors, and with Modern Geography*, London: W. Bulmer, 1800

Richardson, J., *A dictionary, Persian, Arabic, and English*, two vols, Oxford: Printed at the Clarendon Press, 1777–80

Richardson, J., *Travels in the Great Desert of Sahara, in the Years of 1845 and 1846*, two vols, London: Richard Bentley, 1848

Riley, J., *Loss of the American Brig* Commerce, London: John Murray, 1817

Robertson, G. A., *Notes on Africa: Particularly Those Parts which are Situated between Cape Verd and the River Congo*, London: Sherwood, Neely, & Jones, 1819

Salamé, A., *A Narrative of the Expedition to Algiers in the Year 1816, under the Command of the Right Hon. Admiral Lord Viscount Exmouth*, London: John Murray, 1819

Salmon, T., *A New Geographical and Historical Grammar*, twelfth edition, London: Printed for C. Bathurst; W. Strahan; J. and F. Rivington; W. Johnston; J. Hinton; K, Hawes, W. Clarke, and R. Collins; T. Davies; S. Crowder; T. Longman; T. Caslon; B. Law; T. Lowndes; J. and T. Pote; L. Stuart; W. Nicoll; G. Robinson; T. Cadell; and R. Baldwin, 1772

Scott, G., *Mungo Park*, A. & C. Black: London, 1930

Sparrman, A., *A Voyage to the Cape of Good Hope, towards the Antarctic Polar Circle and round the World: But chiefly in the Country of the Hottentots and Caffres, from the year 1772 to 1776*, two vols, London: Printed for G. G. J. and J. Robinson, 1785

Taylor, I., *African Scenes for Tarry-at-Home Travellers*, London: Samuel and Richard Bentley, 1821

Thomson, J., *Mungo Park and the Niger*, London: George Philip & Son, 1890

[Thomson, T.], 'Isaaco's Journal of a Voyage after Mr. Mungo Park, to ascertain his Life or Death', *Annals of Philosophy* (November 1814), Article XI, pp. 369–86

[Traill, T. S.], 'Account of the captivity of Alexander Scott, among the wandering Arabs of the Great African Desert, for a period of nearly six years. Drawn up by T. S. Traill, M. D., F.R.S.E. With geographical observations on his routes, and remarks on the currents of the ocean on the north-western coast of Africa, by Major Rennell, F.R.S.', *Edinburgh Philosophical Journal* 4 (1821), pp. 38–54, 225–34, and 235–41

Tuckey, J. H., *Narrative of an Expedition to Explore the River Zaire*, London: John Murray, 1818

Wakefield, P., *The Traveller in Africa*, two vols, London: James Swann, 1814

Washington, J., 'A brief account of a Mandingo, native of Nyáni-Marú, on the River Gambia, in western Africa', *Report of the Eighth Meeting of the British Association for the Advancement of Science*, London: John Murray, 1839

Wilson, S. A., *Real Stories: Taken from the Narratives of Various Travellers*, London: Harvey and Darton, 1827

Secondary sources

Anniversary Dinner of the Royal Society, November 30, 1887: Sir Joseph Hooker's Reply to the Toast proposed by the Treasurer, R.S. (Privately printed and distributed by John Bellows of Gloucester, n.d. [but 1887])

[Anon.], *Africa and its Exploration as told by its Explorers*, London: Sampson Low, n.d., *c.*1891

[Anon.], 'The Park-Lander Memorial', *Geographical Journal* 68 (1926), p. 456

[Anon.], 'The Park-Lander Memorial', *Geographical Journal* 74 (1929), pp. 470–1

Arnold, D., *The Problem of Nature: Environment, Culture and European Expansion*, Oxford: Blackwell, 1995

Baird, D., *Thing Knowledge: A Philosophy of Scientific Instruments*, Berkeley, CA: University of California Press, 2004

Baker, A., '"Precision", "Perfection" and the reality of British scientific instruments on the move during the 18th century', *Material Culture Review* 74–5 (2012), pp. 14–29

Baker, J. N. L., 'Major James Rennell, 1742–1830, and his place in the history of geography', in J. N. L. Baker, *The History of Geography*, Oxford: Oxford University Press, 1963, pp. 130–57

Barczewski, S., *Heroic Failure and the British*, New Haven and London: Yale University Press, 2016

Barnett, C., 'Impure and worldly geography: the Africanist discourse of the Royal Geographical Society, 1831–1873', *Transactions of the Institute of British Geographers* 23 (1998), pp. 239–51

Bassett, T. J., and Porter, P. W., '"From the Best Authorities": the Mountains of Kong in the cartography of West Africa', *Journal of African History* 32 (1991), pp. 367–413

Bickers, R. A. (ed.), *Ritual and Diplomacy: The Macartney Mission to China, 1792–1794*, London: British Association for Chinese Studies in association with Wellsweep Press, 1993

Blouet, O. M., 'Bryan Edwards, F.R.S., 1743–1800', *Notes and Records of the Royal Society of London* 54 (2000), pp. 215–22

Boahen, A., *Britain, The Sahara, and the Western Sudan 1788–1861*, Oxford: Clarendon Press, 1964

Bourguet, M.-N., 'The explorer', in M. Vovelle (ed.), *Enlightenment Portraits*, translated by Lydia Cochrane, Chicago: University of Chicago Press, 1997, pp. 257–315

Bourguet, M.-N., 'A portable world: the notebooks of European travellers (eighteenth to nineteenth centuries)', *Intellectual History Review* 20 (2010), pp. 377–400

Bourguet, M.-N., Licoppe, C., and Sibum, H. O. (eds), *Instruments, Travel and Science: Itineraries of Precision from the Seventeenth to the Twentieth Century*, London: Routledge, 2002

Bovill, E. W., *The Niger Explored*, Oxford: Oxford University Press, 1968

Boyle, T. C., *Water Music*, London: Panther Books, 1983 edition

Bravo, M., 'Precision and curiosity in scientific travel: James Rennell and the Orientalist geography of the new imperial age (1760–1830)', in J. Elsner and J.-P. Rubieś (eds), *Voyages and Visions: Towards a Cultural History of Travel*, London: Reaktion Books, 1999, pp. 162–83

Bravo, M., 'Ethnographic navigation and the geographical gift', in D. N. Livingstone and C. W. J. Withers (eds), *Geography and Enlightenment*, Chicago: University of Chicago Press, 1999, pp. 199–235

Brent, P., *Black Nile: Mungo Park and the Search for the Niger*, London: Gordon and Cremonesi, 1977

Bridges, R. C., 'W. D. Cooley, the Royal Geographical Society and African geography in the nineteenth century', *Geographical Journal* 142 (1976), pp. 27–47 and 274–86

Broadie, A. (ed.), *The Cambridge Companion to the Scottish Enlightenment*, Cambridge: Cambridge University Press, 2003

Burnett, D. G., *Masters of All They Surveyed: Exploration, Geography, and a British El Dorado*, Chicago: University of Chicago Press, 2000

Burton, R., and MacQueen, J., *The Nile Basin. Part I. Showing Tanganyika to be Ptolemy's Western Lake Reservoir: A Memoir read before the Royal Geographical Society, November 1864. With prefatory Remarks. Part II. Captain Speke's Discovery of the Source of the Nile. A Review by James MacQueen*, London: Tinsley Brothers, 1864.

Cameron, J. M. R., 'Sir John Barrow as a *Quarterly* reviewer, 1809–1843', *Notes and Queries* 43 (1998), pp. 34–7

Cameron, J. M. R., 'John Barrow, the *Quarterly*'s imperial reviewer', in J. Cutmore (ed.), *Conservatism and the Quarterly Review*, London: Pickering and Chatto, 2007, pp. 133–49

Campbell, R., *A Pilgrimage to my Motherland: An Account of a Journey among the Egbas and Yorubas of Central Africa, in 1859–60*, Philadelphia: Thomas Hamilton, 1861

Cavell, J., *Tracing the Connected Narrative: Arctic Exploration in British Print Culture, 1818–1860*, Toronto: University of Toronto Press, 2008

Cook, A. S., 'Major James Rennell and the Bengal Atlas (1780 and 1781)', in *India Office Library and Records Report for the Year 1976*, London: Foreign and Commonwealth Office, 1978, pp. 5–42

Craciun, A., 'What is an explorer?' *Eighteenth-Century Studies* 41 (2011), pp. 29–51

Craciun, A., *Writing Arctic Disaster: Authorship and Exploration*, Cambridge: Cambridge University Press, 2014

Curtin, P. D., *The Image of Africa: British Ideas and Action, 1780–1850*, Madison: University of Wisconsin Press, 1964

Cutmore, J. B., 'Sir John Barrow's contributions to the *Quarterly Review* 1809–24', *Notes and Queries* 41 (1996), pp. 326–8

Darwin, L., 'On the personal characteristics of great explorers', *Scottish Geographical Magazine* 28 (1912), pp. 134–46

Daston, L. (ed.), *Biographies of Scientific Objects*, Chicago: University of Chicago Press, 1999

De Almeida Gebara, A. L., 'The British search for the course of the Niger River: from a geographical problem to potential possession', *História* (São Paulo) 31 (2012), pp. 146–70

De Gramont, S., *The Strong Brown God: The Story of the Niger River*, Boston: Houghton Mifflin, 1975

Dolnick, E., *The Great Unknown: John Wesley Powell's 1869 Journey of Discovery and Tragedy through the Grand Canyon*, London: Harper Perennial, 2002

Dritsas, L., 'Expeditionary science: conflicts of method in mid-nineteenth-century geographical discovery', in D. N. Livingstone and C. W. J. Withers (eds), *Geographies of Nineteenth-Century Science*, Chicago: University of Chicago Press, 2011, pp. 255–77

Driver, F., 'Henry Morton Stanley and his critics: geography, exploration and empire', *Past and Present* 133 (1991), pp. 134–71

Driver, F., *Geography Militant: Cultures of Exploration and Empire*, Oxford: Blackwell, 2001

Driver, F., '*Missionary Travels*: Livingstone, Africa and the book', *Scottish Geographical Journal* 129 (2013), pp. 164–78

Driver, F., and Jones, L., *Hidden Histories of Exploration: Researching the RGS-IBG Collections*, London: Royal Holloway, University of London, 2009

Duffill, M., *Mungo Park West African Explorer*, Edinburgh: National Museums of Scotland Publishing, 1999

Duffill, M., 'Notes on a collection of letters written by Mungo Park between 1790 and 1794, with some remarks concerning his later published observations on the African slave trade', *Hawick Archaeological Society Transactions* [no volume] (2001), pp. 35–55

Duncan, A., *Heads of Lectures on Medical Jurisprudence, or the Institutiones Medicinae Legalis*, Edinburgh: Printed by Neil & Co., 1792

Eddy, M. D., *The Language of Mineralogy: John Walker, Chemistry and the Edinburgh Medical School, 1750–1800*, Farnham and Burlington VT: Ashgate, 2008

Edney, M. H., *Mapping an Empire: The Geographical Construction of British India, 1765–1843*, Chicago: University of Chicago Press, 1997

Edney, M. H., 'Reconsidering Enlightenment geography and map making: reconnaissance, mapping, archive', in D. N. Livingstone and C. W. J. Withers (eds), *Geography and Enlightenment*, Chicago: University of Chicago Press, 1999, pp. 165–98

Eze, E. C. (ed.), *Race and the Enlightenment: A Reader*, Oxford: Basil Blackwell, 1997

Fabian, F., *Out of Our Mind: Reason and Madness in the Exploration of Central Africa*, Berkeley and Los Angeles: University of California Press, 2000

Fagbule, F., and Feyi Fawehinmi, F., *Formation: The Making of Nigeria from Jihad to Amalgamation*, Abuja and London: Cassava Republic Press, 2021

Finkelstein, D., *The House of Blackwood: Author-Publisher Relations in the Victorian Era*, University Park PA: Pennsylvania State University Press, 2002

Finkelstein, D., 'Unraveling Speke: the unknown revision of an African exploration classic', *History in Africa* 30 (2003), pp. 117–32

Fleetwood, L., 'No former travellers having attained such a height on the Earth's surface: instruments and inscriptions in the Himalaya, 1800–1830', *History of Science* 56 (2018), pp. 3–34

Fleming, F. *Barrow's Boys: A Stirring Story of Daring, Fortitude and Outright Lunacy*, London: Granta Books, 1998

Flinn, M. (ed.), *Scottish Population History from the 17th century to the 1930s*, Cambridge: Cambridge University Press, 1977

Fox, A., 'Printed questionnaires, research networks and the discovery of the British Isles, 1650–1800', *Historical Journal* 53 (2010), pp. 593–621

Fulford, T., Lee, D., and Kitson, P. J., *Literature, Science and Exploration in the Romantic Era: Bodies of Knowledge*, Cambridge: Cambridge University Press, 2004

Ganeri, A., *Horrible Geography: Stunning Scotland*, London: Scholastic, in association with the Royal Scottish Geographical Society, 2021

Gascoigne, J., *The Enlightenment and the Origins of European Australia*, Cambridge: Cambridge University Press, 2002

Geertz, C., *Local Knowledge: Further Essays in Interpretive Anthropology*, New York: Basic Books, 1983

Gibbs, J. M., *Performing the Temple of Liberty: Slavery, Theater, and Popular Culture in London and Philadelphia, 1760–1850*, Baltimore MD: Johns Hopkins University Press, 2014

Godlewska, A. M. C., *The Napoleonic Survey of Egypt: A Masterpiece of Cartographic Compilation and Early Nineteenth-Century Fieldwork*, Toronto: University of Toronto Press, 1988

Goodman, M., 'Proving instruments credible in the early nineteenth century: the British Magnetic Survey and site-specific experimentation', *Notes and Records of the Royal Society of London* 70 (2016), pp. 1–18

Gray, E., *The Making of John Ledyard: Empire and Ambition in the Life of an Early American Traveler*, New Haven: Yale University Press, 2007

Grieve, H. J. C. (ed.), *The Letters of Sir Walter Scott*, twelve vols, London: Constable, 1932–7

Haddad, E. A., 'Body and belonging(s): property in the captivity of Mungo Park', in G. Harper (ed.), *Colonial and Post-Colonial Incarceration*, New York: Continuum, pp. 124–43

Hallett, R., *The Penetration of Africa: European Enterprise and Exploration, Principally in Northern and Western Africa up to 1830*, London: Routledge and Kegan Paul, 1965

Hallett, R. (ed.), *The Niger Journal of Richard and John Lander*, London: Routledge and Kegan Paul, 1965

Hancock, E. G., and Wheeler, A., 'An addition to the archival record of Mungo Park (1771–1806), African explorer', *Archives of Natural History* 15 (1988), pp. 311–15

Heffernan, M. '"A Dream as Frail as Those of Ancient Time": The in-credible geographies of Timbuctoo', *Environment and Planning D: Society and Space* 19 (2001), pp. 203–25

Higgitt, R., *Recreating Newton: Newtonian Biography and the Making of Nineteenth-Century History of Science*, London: Pickering and Chatto, 2007

Hopkins, D., 'Books, geography and Denmark's colonial undertaking in west Africa, 1790–1850', in M. Ogborn and C. W. J. Withers (eds), *Geographies of the Book*, Farnham: Ashgate, 2010, pp. 221–46

Hudson, P., *Two Rivers: Travels in West Africa on the Trail of Mungo Park*, London: Chapman, 1991

Hutton, P., *History as an Art of Memory*, London and Hanover: The University Press of New England, 1993

Hutton, P., *The Memory Phenomenon in Contemporary Historical Writing: How the Interest in Memory has Influenced our Understanding of History*, London: Palgrave MacMillan, 2016

Isaac, P., *William Bulmer: The Fine Printer in Context*, London: Bain and Williams, 1993

Jeal, T., *Stanley: The Impossible Life of Africa's Greatest Explorer*, London: Faber and Faber, 2007

Johnston, H. H., 'The Niger basin and Mungo Park', *Scottish Geographical Magazine* 23 (1907), pp. 58–72

Johnston, H. H., *Pioneers in West Africa*, Glasgow: Blackie & Son Limited, 1912

Juengel, S. J., 'Mungo Park's artificial skin: or, the year the white man passed', *The Eighteenth Century* 47 (2006), pp. 19–38

Keighren, I., Withers, C. W. J., and Bell, B., *Travels into Print: Exploration, Writing, and Publishing with John Murray 1773–1859*, Chicago: University of Chicago Press, 2015

Kennedy, D., *The Last Blank Spaces: Exploring Africa and Australia*, Cambridge MA: Harvard University Press, 2013

Kennedy, D., 'Introduction: Reinterpreting exploration', in D. Kennedy (ed.), *Reinterpreting Exploration: The West in the World*, Oxford and New York: Oxford University Press, 2014, pp. 1–18

Kennedy, D., 'Lost in place: two expeditions gone awry in Africa', in R. J. Mayhew and C. W. J. Withers (eds), *Geographies of Knowledge: Science, Scale, and Spatiality in the Nineteenth Century*, Baltimore: Johns Hopkins University Press, 2020, pp. 235–54

Koelsch, W., *Geography and the Classical World: Unearthing Historical Geography's Hidden Past*, London: I. B. Tauris, 2013

Konishi, S., Nugent, M., and Shellam, T. (eds), *Indigenous Intermediaries: New Perspectives on Exploration Archives*, Acton: Australian National University Press and Aboriginal History Inc., 2015

Laithwaite, G., 'Obituary: E. W. Bovill', *Geographical Journal* 133 (1967), pp. 280–1

Lambert, D., *Mastering the Niger: James MacQueen's African Geography & the Struggle over Atlantic Slavery*, Chicago: University of Chicago Press, 2013

Leask, N., *Curiosity and the Aesthetics of Travel Writing 1770–1840*, Oxford: Oxford University Press, 2002

Livingstone, J., *Livingstone's 'Lives': A Metabiography of a Victorian Icon*, Manchester: Manchester University Press, 2014

Lockhart, J. B., 'In the raw: some reflections on transcribing and editing Lieutenant Hugh Clapperton's writings on the Borno Mission of 1822–25', *History in Africa* 26 (1999), pp. 157–95

Lockhart, J. B., *A Sailor in the Sahara: The Life and Travels in Africa of Hugh Clapperton, Commander R. N.*, London: I. B. Tauris, 2008

Loeve-Veimars, F. A., *Popular Ballads and Songs, from Traditional Manuscripts, and Scarce Editions*, Paris: L'Harmattan, 1825

MacLaren, I. S., 'Exploration/travel literature and the evolution of the author', *International Journal of Canadian Studies* 5 (1992), pp. 39–68

MacLaren, I. S., 'In consideration of the evolution of explorers and travellers into authors: a model', *Studies in Travel Writing* 15 (2011), pp 221–41

Markham, C. R., *Major James Rennell and the Rise of Modern English Geography*, London: Cassell, 1895

Marsters, K. F., 'Introduction,' to K. F. Marsters, *Travels in the Interior Districts of Africa, by Mungo Park*, Durham NC: Duke University Press, 2000, pp. 1–28

Martin, A. E., *Nature Translated: Alexander von Humboldt's Works in Nineteenth-Century Britain*, Edinburgh: Edinburgh University Press, 2018

Martin, P. R., and Armston-Sheret, F., 'Off the beaten track?: critical approaches to exploration studies', *Geography Compass* 14 (2020), pp. 1–14

Mayhew, R. 'The character of English geography *c.*1660–1800: a textual approach', *Journal of Historical Geography* 24 (1998), pp. 385–412

Mayhew, R., 'Mapping science's imagined community: geography as a Republic of Letters, 1600–1800', *British Journal for the History of Science* 38 (2005), pp. 73–92

Millar, R., *Travels of Mungo Park*, London: J. M. Dent & Sons Ltd., 1954
Miller, R., 'A Mungo Park anniversary', *Scottish Geographical Magazine* 71 (1955), pp. 147–56
Miller, R., 'Mungo Park 1771–1971', *Scottish Geographical Magazine* 87 (1971), pp. 159–65
Mitchell, J. L., and Gibbon, L. G., *Nine Against the Unknown: A Record of Geographical Exploration*, Edinburgh: Polygon, 2000
Morgan, K., *Matthew Flinders, Maritime Explorer of Australia*, London: Bloomsbury, 2016
Morrell, J. B., 'Professionalisation', in R. C. Olby, G. N. Cantor, J. R. R. Christie, and M. J. S. Hodge (eds), *Companion to the History of Modern Science*, London: Routledge, 1990, pp. 980–9
Mouser, B. L. (ed.), *The Forgotten Peddie/Campbell Expedition into Fuuta Jaloo, West Africa, 1815–17*, Madison: University of Wisconsin Press, 2007
Mouser, B. L., 'Forgotten expedition into Guinea, West Africa, 1815–1817: an editor's comments', *History in Africa* 35 (2008), pp. 481–9
Mouser, B. L., 'The trial of Samuel Samo and the trading syndicates of the Rio Pongo, 1797–1812', *International Journal of African Historical Studies* 46 (2013), pp. 423–41
Mullett, C. F., 'A village Aristotle and the harmony of interests: James Anderson (1739–1808) of Monks Hill', *Journal of British Studies* 8 (1968), pp. 94–118
Nicol, A., *A History of Early Nineteenth-Century Drama 1800–1850*, two vols, Cambridge: Cambridge University Press, 1930
Nielsen, K. H., Harbsmeier, M., and Ries, C. J. (eds), *Scientists and Scholars in the Field: Studies in the History of Fieldwork and Expeditions*, Aarhus: Aarhus University Press, 2012
Nwelme, O., and Chinedu Okeh, I., *Saving Mungo Park*, New Delhi: Hattus Books, 2021
Oguejiofor, J. O., 'The Enlightenment gaze: Africans in the minds of western philosophy', *Philosophia Africana* 10 (2007), pp. 31–6
Olick, J. K., 'Introduction: memory and the nation – continuities, conflicts, and transformations', *Social Science History* 22 (1998), pp. 377–87
Outram, D., 'On being Perseus: new knowledge, dislocation, and Enlightenment exploration', in D. N. Livingstone and C. W. J. Withers (eds), *Geography and Enlightenment*, Chicago: University of Chicago Press, 1999, pp. 281–94
Pratt, M. L., *Imperial Eyes: Travel Writing and Transculturation*, London and New York: Routledge, 1992
Rae, E., Souch, C., and Withers, C. W. J. "'Instruments in the Hands of

Others": the life and liveliness of instruments of British geographical exploration, *c.*1860-*c.*1930', in F. MacDonald and C. W. J. Withers (eds), *Geography, Technology and Instruments of Exploration*, Farnham: Ashgate, 2015, pp. 139–61

Raj, K., 'When human travellers became instruments: the Indo-British exploration of Asia in the nineteenth century', in M.-N. Bourguet, C. Licoppe and H. O. Sibum (eds), *Instruments, Travel and Science: Itineraries of Precision from the Seventeenth to the Twentieth Century*, London: Routledge, 2002, pp. 156–88

Regard, F., 'Introduction: articulating empire's unstable zones', in F. Regard (ed.), *British Narratives of Exploration*, London: Pickering and Chatto, 2009, pp. 1–17

Relano, F., *The Shaping of Africa: Cosmographic Discourse and Cartographic Science in Late Medieval and Early Modern Europe*, Aldershot: Ashgate, 2002

Richardson, R., *Story of the Niger: A Record of Travel and Adventure from the Dogs of Mungo Park to the Present Time*, London: Thomas Nelson, 1888

Robertson, T., 'Parish of Selkirk', in Sir John Sinclair (ed.), *The Statistical Account of Scotland: Drawn up by the Communications of the Ministers of the Different Parishes*, Edinburgh: William Creech, 1791–9, II (1792), pp. 434–48

Rodd, F. R., 'Rennell's comments upon the journeys of Park and Laing to the Niger', *Geographical Journal* 86 (1935), pp. 28–31

Roy, G. R. (ed.), *The Letters of Robert Burns*, Oxford: Oxford University Press, 1985

Rupke, N., *Alexander von Humboldt: A Metabiography*, Chicago: University of Chicago Press, 2008

Ruskin, J., *Fors Clavigera: Letters to the Workmen and Laborers of Great Britain*, eight vols, London: G. Allen, 1883

Russell, P., *Prince Henry 'the Navigator': A Life*, New Haven, PA: Yale University Press, 2001

Sattin, A., *The Gates of Africa: Death, Discovery and the Search for Timbuktu*, London: Harper Perennial, 2003

Schaffer, S., Roberts, L., Raj, K., and Delbourgo, J. (eds), *The Brokered World: Go-Betweens and Global Intelligence, 1770–1820*, Sagamore Beach MS: Science History Publications, 2009

Schaffer, S., 'Easily cracked: scientific instruments in states of disrepair', *Isis* 102 (2011), pp. 706–17

Schwartz, J., *Robert Brown and Mungo Park: Travels and Explorations in Natural History for the Royal Society*, New York: Springer, 2021

Scott, H. W. (ed.), *Lectures on Geology by John Walker*, Chicago: University of Chicago Press, 1966

Secord, J. A., *Victorian Sensation: The Extraordinary Publication, Reception, and Secret Authorship of* Vestiges of the Natural History of Creation, Chicago: University of Chicago Press, 2000

Secord, J. A., *Visions of Science: Books and Readers at the Dawn of the Victorian Age*, Chicago: University of Chicago Press, 2014

Shapin, S., *A Social History of Truth: Civility and Science in Seventeenth-Century England*, Chicago: University of Chicago Press, 1994

Shea, R. W., and Bell, D., 'Charting the unknown: Islamic cartography and visions of Africa in the Abbasid era', *History in Africa* 46 (2019), pp. 37–56

Sher, R. B., *The Enlightenment & the Book: Scottish Authors & Their Publishers in Eighteenth-Century Britain, Ireland & America*, Chicago: University of Chicago Press, 2006

Shortland, M., and Yeo, R. (eds), *Telling Lives in Science: Essays on Scientific Biography*, Cambridge: Cambridge University Press, 1996

Sloan, P., 'The gaze of natural history', in C. Fox, R. Porter and R. Wokler (eds), *Inventing Human Science: Eighteenth-Century Domains*, Berkeley CA: University of California Press, 1995, pp. 112–51

Stanley, H. M., 'Inaugural address delivered before the Scottish Geographical Society at Edinburgh, 3 December 1884', *Scottish Geographical Magazine* 1 (1885), pp. 1–17

Stone, J. C., *A Short History of the Cartography of Africa*, Lewiston, NY: Edward Mellen Press, 1995

Syme, R., *I, Mungo Park*, London: Burke, 1951

Thompson, C., *The Suffering Traveller and the Romantic Imagination*, Oxford: Oxford University Press, 2007

Thompson, C., 'Nineteenth-century travel writing', in N. Das and T. Youngs (eds), *The Cambridge History of Travel Writing*, Cambridge: Cambridge University Press, 2019, pp. 108–24

Tooley, R. V., *Collectors' Guide to Maps of the African Continent and Southern Africa*, London: Carta Press, 1969

Topham, J. M., 'Science, print, and crossing borders: importing French science books into Britain, 1790–1815', in D. N. Livingstone and C. W. J. Withers (eds), *Geographies of Nineteenth-Century Science*, Chicago: University of Chicago Press, 2011, pp. 311–44

Wess, J., 'The role of instruments in exploration: A study of the Royal Geographical Society, 1830–1930', Unpublished PhD, University of Edinburgh, 2017

Wess, J., and Withers, C. W. J., 'Instrument provision and geographical science: the work of the Royal Geographical Society, 1830–*ca*1930', *Notes and Records of the Royal Society of London* 73 (2019), pp. 223–42

White, A. S., 'On the achievements of Scotsmen during the nineteenth century in the fields of exploration and research', *Scottish Geographical Magazine* 5 (1889), pp. 480–97

Williamson, K., and Duffill, M., 'Mungo Park, man of letters', *Scottish Literary Review* 3 (2011), pp. 55–79

Wise, C., 'Writing Timbuktu: Park's hat, Laing's hand', in C. Wise (ed.), *The Desert Shore: Literatures of the Sahel*, London: Lynne Reiner, pp. 175–200

Withers, C. W. J., *Geography, Science and National Identity: Scotland since 1520*, Cambridge: Cambridge University Press, 2001

Withers, C. W. J., 'Memory and the history of geographical knowledge: the commemoration of Mungo Park, African explorer', *Journal of Historical Geography* 30 (2004), pp. 316–39

Withers, C. W. J., 'James Playfair (1738–1819)', *Geographers Biobibliographical Studies* 24 (2005), pp. 79–85

Withers, C. W. J., *Placing the Enlightenment: Thinking Geographically about the Age of Reason*, Chicago; University of Chicago Press, 2007

Withers, C. W. J., 'Geography, enlightenment and the book: authorship and audience in Mungo Park's African texts', in M. Ogborn and C. W. J. Withers (eds), *Geographies of the Book*, Farnham: Ashgate, 2010, pp. 191–220

Withers, C. W. J., *Geography and Science in Britain, 1831–1939: A Study of the British Association for the Advancement of Science*, Manchester: Manchester University Press, 2010

Withers, C. W. J., 'Science, scientific instruments and questions of method in nineteenth-century British geography', *Transactions of the Institute of British Geographers* 38 (2013), pp. 167–79

Withers, C. W. J., *Zero Degrees: Geographies of the Prime Meridian*, Cambridge MA: Harvard University Press, 2017

Withers, C. W. J., 'Geography and "thing knowledge": instrument epistemology, failure and narratives of 19th-century exploration', *Transactions of the Institute of British Geographers* 44 (2019), pp. 676–91

Withers, C. W. J., 'On trial – social relations of map production in mid-nineteenth-century Britain', *Imago Mundi* 74 (2019), pp. 173–95

Withers, C. W. J., 'Map making, defamation, and credibility: the case of the *Athenaeum*, Charles Tilstone Beke, and W. & A. K. Johnston's *Edinburgh Educational Atlas* (1874)', *Imago Mundi* 73 (2021), pp. 46–63

Withers, C. W. J., 'Disguise – trust and truth in travel writing', *Terrae Incognitae* 52 (2021), pp. 48–64

Withers, C. W. J., and Wood, P. (eds), *Science and Medicine in the Scottish Enlightenment*, Edinburgh, Tuckwell Press, 2002

Wood, P. (ed.), *The Scottish Enlightenment: Essays in Reinterpretation*, Rochester NY: University of Rochester Press, 2000

Worster, D., *A River Running West: The Life of John Wesley Powell*, Oxford: Oxford University Press, 2001

Index

Page numbers in **bold** refer to figures